Dynamics of Urbanization and Ecosystem Services Provision II

Dynamics of Urbanization and Ecosystem Services Provision II

Guest Editors

Luca Congedo
Francesca Assennato
Michele Munafò

MDPI

Basel • Beijing • Wuhan • Barcelona • Belgrade • Novi Sad • Cluj • Manchester

Guest Editors
Luca Congedo
ISPRA
Italian Institute for
Environmental Protection
and Research
Rome
Italy

Francesca Assennato
ISPRA
Italian Institute for
Environmental Protection
and Research
Rome
Italy

Michele Munafò
ISPRA
Italian Institute for
Environmental Protection
and Research
Rome
Italy

Editorial Office
MDPI AG
Grosspeteranlage 5
4052 Basel, Switzerland

This is a reprint of the Special Issue, published open access by the journal *Land* (ISSN 2073-445X), freely accessible at: https://www.mdpi.com/journal/land/special_issues/urban_ecosystem_service2.

For citation purposes, cite each article independently as indicated on the article page online and as indicated below:

Lastname, A.A.; Lastname, B.B. Article Title. *Journal Name* **Year**, *Volume Number*, Page Range.

ISBN 978-3-7258-4127-1 (Hbk)
ISBN 978-3-7258-4128-8 (PDF)
https://doi.org/10.3390/books978-3-7258-4128-8

Contents

land

MDPI

Article

Assessing the Cooling Potential of Vegetation in a Central European Rural Landscape: A Local Study

Tereza Pohanková and Vilém Pechanec *

Department of Geoinformatics, Faculty of Science, Palacký University Olomouc, 17. listopadu 50, CZ-771 46 Olomouc, Czech Republic; tereza.pohankova@upol.cz
* Correspondence: vilem.pechanec@upol.cz

Abstract: This study investigates the cooling potential of vegetation in rural landscapes of the Czech Republic to mitigate heat-related issues. Using remote sensing, the Cooling Capacity Index (*CCI*) is assessed to measure green spaces' ability to lower air temperatures using evapotranspiration and shading. Landsat 8/9 and meteorological data are utilised, with *CCI* calculated based on vegetation cover, albedo, and evapotranspiration. Our results demonstrate significant variations in cooling capacity across different land use types. Forests exhibited the highest cooling potential, while urban areas, characterised by heat-absorbing materials, displayed the least. We analysed temporal and spatial variations in cooling capacity using various visualisation tools and validated the results against the InVEST software (v3.14.0). This study highlights the effectiveness of remote sensing in quantifying ecosystem functions, particularly the cooling services provided by vegetation. Our findings emphasise the crucial role of vegetation in mitigating urban heat islands and addressing climate change. This research provides valuable insights for developing climate change adaptation strategies in rural landscapes.

Keywords: cooling function; remote sensing; spatiotemporal analysis; vegetation

Citation: Pohanková, T.; Pechanec, V. Assessing the Cooling Potential of Vegetation in a Central European Rural Landscape: A Local Study. *Land* **2024**, *13*, 1685. https://doi.org/10.3390/land13101685

Academic Editors: Luca Congedo, Francesca Assennato and Michele Munafò

Received: 10 September 2024
Revised: 7 October 2024
Accepted: 12 October 2024
Published: 16 October 2024

1. Introduction

Heat plays a critical role in regulating various natural processes, and its effects are one of the most important aspects of life on Earth [1]. Extreme climatic events, not only heat-related, are the results of long-term global phenomena, often described as climate change, caused by changes in atmospheric conditions.

As global temperatures rise, the frequency and intensity of extreme weather events such as heatwaves, droughts, and floods increase, further amplifying the significance of heat regulation for both natural and human systems [2]. Surface cooling is an important regulating ecosystem function where vegetation plays a vital role [3]. Vegetation mitigates the effects of heat by providing surface shading and through the process of evapotranspiration, which cools the surrounding air and reduces surface temperatures [4]. Through these mechanisms, vegetation contributes to thermal regulation, depending on the structure, current physiological conditions, and overall health status [5].

Evapotranspiration (ET; transferring water from the surface of the Earth to the atmosphere) is a critical factor affecting the climate from several points of view. It not only influences local weather patterns but also plays a key role in determining water availability.In terms of water availability, in places with higher ET than precipitation, water resources may be scarce, which then affects the water cycles [6]. This scarcity can have wide-ranging consequences, especially for agriculture, as it can limit water supply for irrigation, significantly reducing crop yields and the ability to sustain plant growth [7], or climate (balancing water vapours in the atmosphere has a great impact on life comfort) [8]. ET can be divided into three parts: potential (PET, the maximum amount of water that the surface can produce), actual (AET; the real amount of evapotranspiration generated by a cover over a specified time), and reference (ET0; the hypothetical amount of water that could be evaporated from a reference surface under optimal conditions) [6].

Temperature is deeply affected by the amount of evapotranspiration. Land surface temperature (LST; thermodynamical state of an object and a direct consequence of energy changes) can indicate the amount of energy left for evapotranspiration. Air temperature affects the evapotranspirational rate in soil moisture and plant water, which depends on air temperature [6].

Many different types of research regarding heat crises have been presented to mitigate the effects of climate change [9–11]. Most of the issue is caused by built-up areas filled with impermeable and unsustainable materials. Using these materials causes the UHI (Urban Heat Island) in densely populated areas with less arable land. In these urban parts, the temperature is higher than in the surrounding regions due to using heat-absorbing materials [12]. In 2022, an article by Neto et al. [13] concluded that land use and land cover changes significantly decrease the ability of surfaces to evapotranspiration. Determining ET remains a complex task. There are several ways to assess ET, mostly based on analyses of meteorological conditions [14]. A number of these require field measuring over a standardised grass-covered surface. One of the measures is a Crop Coefficient (Kc). The Kc is used to express plant transpiration and soil evaporation combined. However, even with this measure, we cannot analyse the air-conditioning effect thoroughly because it does not offer to determine the spatiotemporal variability of vegetation.

Field measurements are usually performed using eddy covariance stations or with lysimeters [15]; advantages to these solutions are high precision and the possibility of calibration specific to each surface. On the other hand, they provide point data and can be financially expensive.

Keeping in mind the landscape dynamics, the analysis of the cooling effect of vegetation needs to work with spatially oriented tools and data to capture and locate the different vegetation types and their different contributions to surface cooling. A promising approach is to use remote sensing data, which are spatial, non-destructive to vegetation, repeatable and allow the analysis of a range of biophysical properties of vegetation [16].

Blended methods use multiple means of determination, including a mix of remote sensing or surface energy balance modelling. In remote sensing, it is possible to determine various characteristics, e.g., Bowen Index [17], Cooling Capacity Index [18] or Latent Heat Flux [19]. Stisen [20], in 2021, developed a mixed model of eddy covariance and remote sensing data measurements to examine spatial patterns of ET in Spain and concluded that ET is significantly affected not only by vegetation or soil type but also by topography. In 2019, Scanlon developed a mixed model of eddy covariance and remote sensing data measurements to examine spatial patterns of ET in Spain. It concluded that ET is significantly affected not only by vegetation or soil type but also by topography. In 2019, Scanlon [21] published evapotranspiration research using the measurement of eddy covariance to analyse individual components of ET.

To determine the cooling function, an existing measure was used, namely, the Cooling Capacity Index, which was introduced by the Natural Capital Project developed by Stanford University as part of the Urban Cooling Model [18]. The *CCI* was chosen over other methods due to its integration of multiple environmental factors, allowing for a more holistic assessment of urban cooling efficiency. Unlike other metrics that can primarily focus on surface temperatures [22] or green/blue infrastructure [23], the *CCI* accounts for both evapotranspiration and shading cooling functions of vegetation, which makes it more versatile and accurate. This metric has been used in several other studies. Bosch et al. used the Urban Cooling Model to propose calibration of model parameters according to the best-fit observed for air temperature. J. E. Zawadska et al. concluded that the index is capable of depicting a portion of the thermal response of the surface.

Another advantage of the *CCI* is that it can be integrated with other environmental assessment indexes, such as the Building Intensity of Heat Mitigation Index, to provide a more detailed and layered analysis of urban heat mitigation strategies. This ability to combine datasets and methods enhances its applicability in complex urban environments [24]. This work aims to apply this index, emphasising filling the primary data with remote sensing approaches and with a minimum need for field measurements. This option would allow easy transferability of the calculation to other locations. Subsequently, it will enable the analysis of the cooling function variability for different land use types in the area of interest, which differ.

The choice of the *CCI* for this study was motivated by several factors. Firstly, the *CCI* provides a quantitative measure of the cooling capacity of different land cover types, which is essential for understanding the relationship between land use and urban heat island effects. Secondly, the *CCI* is based on well-established methodologies and has been successfully applied in various studies around the world. Thirdly, the availability of Landsat satellite imagery and ground-based meteorological data in Černovice made it possible to calculate the *CCI* for this specific location.

This study also extends the application of the Cooling Capacity Index (CCI) to a rural landscape. While the *CCI* has primarily been used in urban areas, this study demonstrates its applicability in rural settings. These innovations collectively enhance the understanding of vegetation cooling in rural landscapes and contribute to the development of effective climate change adaptation strategies.

2. Materials and Methods

This study presents an analysis of the *CCI* in the city of Černovice, Czech Republic. The *CCI*, a metric quantifying the cooling potential of land cover types, has not been previously applied to this specific area. Černovice, despite its small size, exhibits a diverse range of land cover types, making it an ideal location to investigate the impact of different land uses on local temperatures.

2.1. Study Area

The availability of a professional meteorological station influenced the choice of the study area. For this study, a relatively small city was chosen, but one that, based on its land use structure and surrounding environment, represents the rural landscape of the Czech Republic very well. A professional Czech Hydrometeorological Institute (CHMI) station is located in the cadastral area of the city, and we were given access to its data and allowed to place our own sensors to calibrate it. In this way, possible measurement errors due to non-representative conditions at the station site are significantly minimised.

The city of Černovice is located on the western edge of the Vysočina (Figure 1) region with an elevation between 470 and 710 m with various terrain conditions, predominantly highlands. Quitt [25] rated the climate conditions relatively homogeneous, mostly warm and moderately humid, with a climate largely moderately warm with the low and middle elevations falling within the climatic zone of moderately warm and the highest peaks within the cool zone.

The agricultural lands are used for vegetable growing and are influenced by rural forest management. Mostly fir and beech forests are present, with small portions of the herbaceous layer. Of the forest vegetation stages, the most abundant are flank and fir trees. The current composition of the vegetation and forest stands is influenced by intensive agricultural and forestry management, with a predominance of monocultural spruce, to a lesser extent pine, with a poor herbaceous cover.

There are two streams, Černovický stream and Včelnička, and several small artificial and natural water reservoirs scattered around the area of interest.

The current land cover/land use (LU/LC) of the area is described in Table 1. The Corine land cover layer in vector form has been used to describe land use. These data, valid for 2018, were revised by us during a repeat field survey in spring (April–May) 2022 [26]. A detailed habitat map at a scale of 1:10,000 was also produced during the field survey, which we also used for more detailed vegetation studies. However, it was not used in this case due to the resolution of the thermal channel of the Landsat satellite used (100 m/px). As a result, the updated Corine LC layer is more suitable for this study.

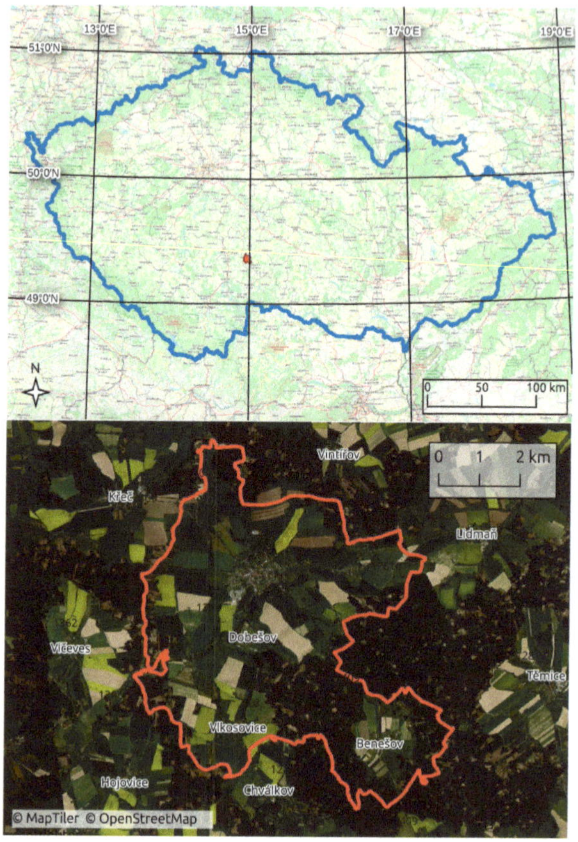

Figure 1. Study area—town Černovice.

Table 1. Land use/land cover in the study area.

TAG *	Category of LULC	Area [km^2]	Area [%]
112	Discontinuous urban fabric	1.27	0.55
211	Non-irrigated arable land	139.46	60.66
231	Pastures	3.86	1.67
242	Complex cultivation patterns	0.25	0.1
	Land principally occupied by agriculture, with significant areas of natural vegetation	2.75	1.19
312	Coniferous forest	79.68	34.64
313	Mixed forest	1.86	0.88
324	Transitional woodland/shrub	0.73	0.31
SUM		229.86	100

* Surface type code based on Corine Land Cover.

2.2. Data

2.2.1. Landsat Image Data

Landsat 8/9 (cooperated by the American organizations NASA and USGS) satellites were chosen for this study due to their free availability, long-term data record, and suitable spatial and temporal resolution. They are carrying an OLI (Operational Land Manager) sensor for acquiring multispectral images and TIRS (Thermal Infrared Sensor) for thermal images (Table 2) with spatial resolution varying from 15 to 100 m. It must be noted that Sentinel-2 and Sentinel-3 satellites would have been alternative options, but their thermal data are too coarse for this type of study. Sentinel-3′s thermal band has a resolution of 1 km/pixel, which is insufficient for detailed analysis of vegetation cooling at a small-scale level.

Table 2. Spatial resolution and band lengths of Landsat 8.

Band	Spectrum	Band Length (nm)	Spatial Resolution (m)
B1	Visible	430–450	30
B2	Visible	450–510	30
B3	Visible	530–590	30
B4	Red	640–670	30
B5	Near-Infrared	850–880	30
B6	SWIR 1	1570–1650	30
B7	SWIR 2	2110–2290	30
B8	Panchromatic	500–680	15
B9	Cirrus	1360–1380	30
B10	Thermal	10,600–11,900	100
B11	Thermal	11,500–12,510	100

Landsat 8/9 data were downloaded in Level 1TP and Level 2SP formats. Level 1TP is orthorectified using Ground Control Points and digital elevation models. Level 2SP is fully corrected by the provider. The products contain digital numbers (DN). Image path 191 and row 026 were used with a covering area of 180×185 km^2 [27,28].

However, thermal correction was necessary to obtain accurate land surface temperature (*LST*) values. Algorithms developed by Sobrino and Jimenez were used for this purpose.

Table 3 shows specific days of satellite overflight above the site. Since Landsat 8/9 is an optical satellite, the main limitation is the potential presence of clouds. Cloud assessment was initially performed through visual analysis due to the small number of usable satellite images. This process was then verified using the Landsat Quality Assessment Bands to ensure accurate identification and removal of cloud-contaminated pixels. These were filtered using the Landsat Quality Assessment Bands. Eight overpass days were clear enough to be used in this study. The first observation on 10th February was eliminated due to later discovered clouds.

Table 3. Dates of Landsat 8/9 overflights over the site.

Landsat 8	Landsat 9
14 March 2022	22 March 2022
18 June 2022	
20 July 2022	
5 August 2022	
6 September 2022	
27 December 2022	

2.2.2. Meteorological Data

The required meteorological data (Table 4) were acquired through the Czech Hydrometeorological Institute (CHMI), which publishes daily meteorological data from their meteorological stations in continuous series since 1961 [29]. The data are published as individual CSV tables, one file for one meteorological station. Data from the station Černovice-Dobešov, located in the cadastre of the town of Černovice, were used.

Table 4. Meteorological values used in the study.

Name	Symbol	Unit
Average daily air temperature	T_{avg}	°C
Maximum daily air temperature	T_{max}	°C
Minimum daily air temperature	T_{min}	°C
Average daily wind speed	RH	%
Average daily relative air humidity	μ	m/s

2.3. *Cooling Capacity Index (CCI)*

The Cooling Capacity Index (CCI) measures the ability of green spaces, such as parks and forests, to cool the surrounding area by reducing air temperatures through evapotranspirative processes and shading and is usually based on the from 0 to 1 range [24].

The ability of different objects to lower the air temperature can be described as a cooling function. The *CCI* depends on several factors, such as vegetation cover, surface roughness or imperviousness of surfaces, water stress or albedo. *CCI* compares the cooling capacity of a surface with the same-sized reference surface with no vegetation cover. The equation takes three inputs: vegetation shading (S; the percentage of a pixel covered by vegetation on a scale from 0 to 1), albedo (α; the proportion of solar radiation reflected by a surface, scaled from 0 to 1) and evapotranspiration index (*ETI*; a measure of the amount of water lost from a surface through evaporation and transpiration scaled as 0 to 1). *ETI* is calculated as the ratio of crop evapotranspiration to maximal reference evapotranspiration [30].

$$CC_i = 0.6 \times S + 0.2 \times \alpha + 0.2 \times ETI \quad (1)$$

The influence of shade and evapotranspiration is quantified by cooling capacity. This index is based on work by Zardo [31] (focused on the cooling capacity of green infrastructures) and Kunapo [32]. Albedo was added later, creating the Cooling Capacity Index. Each of its factors corresponds to one important heat mitigation strategy: the value of tree canopy (shade), potential evapotranspiration (evapotranspiration index), albedo (proportion of radiation) and crop coefficient (associated with land cover) [24].

Our proposed and validated approach to calculating the *CCI*, emphasising the use of remote sensing data, is shown in Figure 2. Our approach uses land surface temperature, Net Radiation (consisting of incoming and outgoing longwave and shortwave radiation) and albedo calculation, which is based on Liang [33]. Reference Evapotranspiration is based on Soil Heat Flux calculation. This variable's magnitude depends on various elements: surface cover, soil type, and solar radiance. Usually, it is used along with another similar value, such as net radiation or latent or sensible heat flux [34].

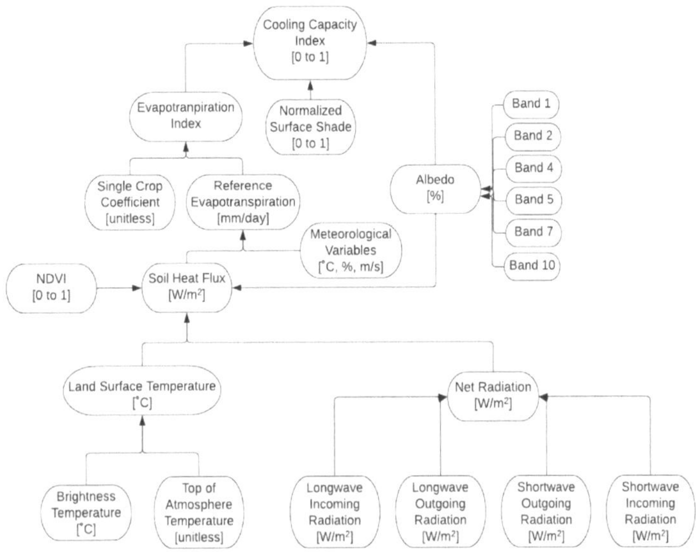

Figure 2. Workflow of our calculation *CCI*.

2.4. *Land Surface Temperature*

Land surface temperature (*LST*) is a crucial input parameter in the Cooling Capacity Index (CCI) calculation due to its direct relationship with several key components of the energy balance equation: surface emissivity (the material's ability to emit longwave radiation), albedo (higher albedo surfaces reflect more sunlight and absorb less heat, resulting in lower *LST*) and soil heat flux (*LST* is influenced by the exchange of heat between the surface and the subsurface, which is represented by soil heat flux). Those variables are later used in the calculation of reference evapotranspiration.

Land surface temperature (*LST*) describes the thermodynamical state of an object and is a direct consequence of energy changes due to detecting the amount of radiation emitted by the surface. Several factors, including solar radiation, atmospheric conditions, land cover type or soil moisture, do influence *LST*. It is one of the most important factors affecting the energy net budget of Earth [35]. For acquiring the surface temperature can be used either pyrometer (point-based), thermal camera (smaller areas) or airborne techniques such as UAVs or satellites [36]. Higher *LST* is usually found in places with impervious surfaces which absorb and store heat with no vegetation around, such as urban areas or wide roads [37]. The *LST* was calculated using Equation (2) [38].

$$LST = \frac{BT}{\left(1 + \frac{0.0015 \times BT}{1.4488}\right) \times lnE} \tag{2}$$

BT… Brightness Temperature [°C]. E… Surface Emissivity.

2.5. *Leaf Area Index (LAI)*

In this study, the crop coefficient (Kc) used in the *CCI* calculation was not derived from laboratory-measured values due to their unavailability in the study area. Instead, Kc values used in this *CCI* calculation were derived from the leaf area index (*LAI*; a measure of the total leaf area per unit of ground area), as recommended by InVEST (Kc is derived as a third of *LAI*, for *LAI* values over 3. Otherwise, Kc equals 1) [6]. By using *LAI* to estimate Kc, we were able to incorporate the influence of vegetation density and structure on evapotranspiration.

LAI has been defined as a dimensionless one-sided area of a leaf per ground unit [39]. It is a common attribute of ecological models and photosynthesis evaluation. Measuring can be performed directly (e.g., using an *LAI*-2000 analyser by LI-COR) and indirectly. There have been many different indirect measuring methods; for this study, we used the method based on SEBAL proposed by Bastiaansen [40] and implemented by Brom [41], which utilises the *SAVI* (Soil Adjusted Vegetation Index) threshold method instead of the NDVI (Equation (3)).

$$LAI = log\frac{\frac{(0.61-SAVI)}{0.51}}{-0.91} \quad (3)$$

The higher the *SAVI* value, the denser the vegetation with a rising significance of ground cover. The literature suggests that *SAVI* is more suitable for studies performed on smaller areas since it includes the L-factor for vegetation density [42,43].

Many models have been developed over the years to estimate *LAI* values [43]. They usually depend on a wide number of precise measurements. To eliminate this, machine learning techniques [44,45] and neural networks [46,47] have been implemented into several methods. InVEST recommends several ways of calculating *LAI* with various approaches based on Kristensen [48], Allen [6] and Wahid [49].

3. Results

The *CCI* maps can be used to identify specific areas that require targeted interventions to enhance cooling capacity, such as planting additional trees or improving green infrastructure. The results can further inform the authorities' planning and implementation of green and blue infrastructure initiatives to increase the cooling potential of urban and rural areas.

3.1. Cooling Capacity Index

The individual input parameters and the resulting *CCI* index were calculated for each imaging date during the growing season of 2022. Emphasis was placed on selecting the best image of the month (due to high cloud cover) and regular spacing between the imaging dates. The primary output of the analysis for each day analysed is a raster image.

In our view, the key output is creating a series of maps that capture the spatio–temporal evolution of *CCI* values in the area of interest (Figure 3). The maps show not only the evolution over time at a given location within a given category but also the actual variability of values within a category. Such a result allows the spatial targeting of adaptation measures, the addition and development of green and blue infrastructure, the modification of agro–technical practices, and the delineation of green centres in the further development of the area.

The next step was calculating summary statistics for each LU/LC category in the area. This was performed for quick reference and generalisation of the results. The minimum and maximum values for each region are shown in Table 5.

The average Cooling Capacity Index (Table 6) ranges from 0.545 to 0.584. Lower values indicate a lower cooling potential associated with built-up areas, while higher values indicate a higher one.

A similar progression of values throughout the year is observed for most of the selected land cover categories (Figure 4). While the forests (312, 313, 324) tend to maintain a higher cooling capacity (from a peak in June to a drop in September), the non-irrigated arable land (211) changes its cooling capacity during the year. An increase in values is observed in spring and at the end of summer.

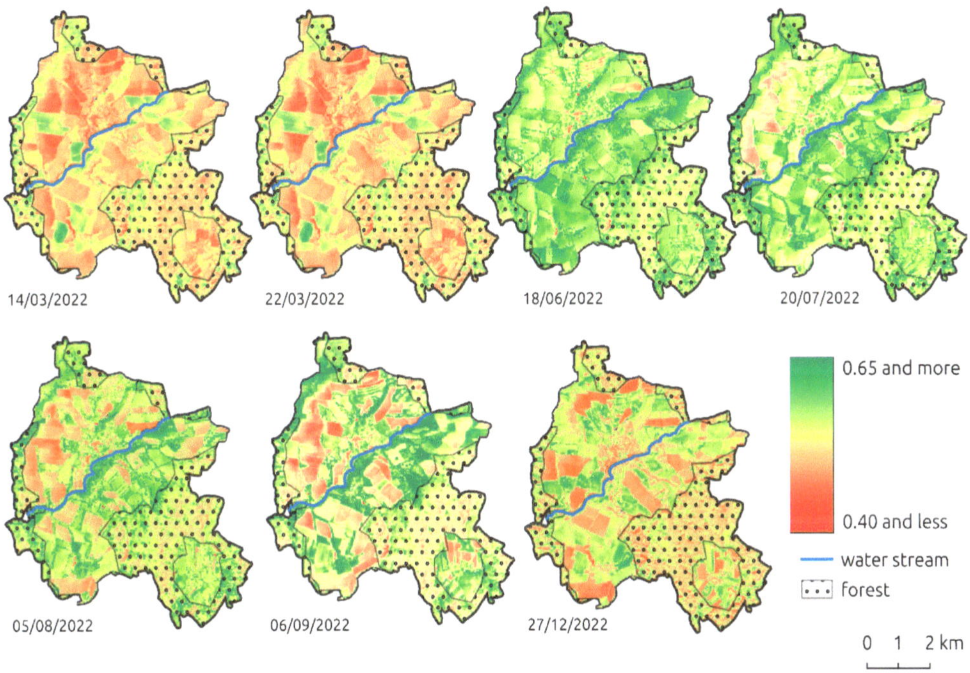

Figure 3. Visualisation development of Cooling Capacity Index during the analysed period.

Table 5. Cooling Capacity Index—maximum and minimum values per LU/LC category.

TAG *	14th March		22nd March		18th March		20th March		5th March		6th March		27th March	
	Max	Min	Max	Min	Max	Min	Max	Min	Max	Min	Max	Min	Max	Min
112	0.608	0.428	0.599	0.433	0.688	0.48	0.684	0.465	0.685	0.455	0.676	0.481	0.681	0.395
211	0.635	0.413	0.64	0.426	0.697	0.43	0.695	0.436	0.69	0.431	0.691	0.472	0.663	0.399
231	0.594	0.441	0.596	0.438	0.691	0.444	0.685	0.431	0.692	0.43	0.699	0.488	0.636	0.41
242	0.583	0.483	0.591	0.487	0.68	0.541	0.687	0.541	0.677	0.536	0.676	0.549	0.627	0.439
243	0.606	0.416	0.614	0.418	0.736	0.432	0.745	0.428	0.732	0.429	0.738	0.461	0.64	0.362
312	0.648	0.416	0.645	0.414	0.705	0.475	0.699	0.474	0.698	0.47	0.684	0.463	0.668	0.372
313	0.629	0.407	0.627	0.423	0.736	0.437	0.732	0.43	738	0.451	0.73	0.468	0.631	0.384
324	0.631	0.483	0.628	0.493	0.668	0.546	0.674	0.547	0.671	0.546	0.662	0.552	0.649	0.437

* TAG is explained in Table 1.

The air temperature went higher than 29–30 °C during summer, which can potentially cause certain vegetation to become heat stressed [49] and significantly reduce evapotranspiration due to maintaining water balance. Long-term exposure to high heat can cause damage to vegetation, which then reflects further on biological and physical processes [50]. Although vegetation's cooling capacity decreases during winter, it is still recognisably higher than the cooling capacity of urban areas. The cooling ability of agricultural lands (231, 242, 243) depends on the crop's presence and current state. Both metropolitan areas (112) and mixed forests (313) occupy less than 1% of the size of interest (Table 1) but have different cooling capacity values; the cooling capacity of mixed forests is never below the cooling capacity of urban areas.

Table 6. Cooling Capacity Index—average values per surface.

TAG *	14th March	22nd March	18th June	20th July	5th August	6th September	27th December	Avg
112	0.506	0.51	0.576	0.576	0.571	0.577	0.502	0.545
211	0.517	0.521	0.588	0.571	0.557	0.562	0.521	0.548
231	0.525	0.539	0.589	0.591	0.593	0.609	0.537	0.569
242	0.534	0.538	0.609	0.608	0.605	0.613	0.54	0.578
243	0.52	0.523	0.592	0.592	0.588	0.606	0.522	0.563
312	0.533	0.535	0.581	0.578	0.572	0.57	0.512	0.554
313	0.533	0.536	0.61	0.618	0.612	0.618	0.511	0.577
324	0.557	0.559	0.612	0.611	0.609	0.603	0.538	0.584

* TAG is explained in Table 1.

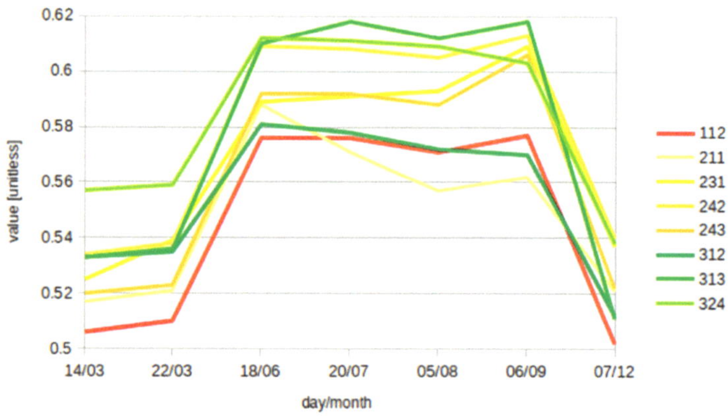

Figure 4. Cooling Capacity values progress through imagining dates.

Non-irrigated arable land (211) is the largest area in the region. Still, it does not have a high cooling potential, even at its maximum during June (Figure 4). It only reaches up to minimal values of the vegetative areas, including those which occupy less than 1% of the region. A sharp decline can be seen (from June to August) during the harvest season due to missing crops in the fields. Areas with higher values of the Cooling Capacity Index are more resilient to higher temperatures connected to Urban Heat Islands [51].

3.2. *Land Surface Temperature*

The measured land surface temperature (*LST*) is processed the same way as the *CCI* (maps, tables, graphs) to understand the extent of landscape cooling better. Values of *LST* range from −2.69 °C to 40.54 °C. Average surface temperature ranges between 13 and 18 °C; lower values tend to be for areas with vegetation. The surface temperature of urban areas (112) is higher than non-urban areas. The largest temperature difference between urban and non-urban areas occurs during August (4.72 °C) and September (5.7 °C) (Table 7).

Overall progress through the year is very similar for every chosen surface in the region. The urban area stays the most heat-accumulating surface type during the year until winter, although the difference between urban and non-urban areas in winter is smaller (Figures 5 and 6). The vegetative areas show very similar temperature flow (Figure 5).

Table 7. Land surface temperature for various surfaces.

* TAG	14th March	22nd March	18th June	20th July	5th August	6th September	27th December	Avg
112	10.23	13.69	27.19	28.42	30.55	14.37	−0.78	17.66
211	9.65	12.77	23.75	26.4	29.16	13.64	−0.92	16.35
231	9.53	12.72	24.12	26.62	27.76	10.85	−0.81	15.83
242	10.09	12.91	25.97	27.77	28.87	8.58	−1.27	16.13
243	9.4	12.73	23.85	26.21	28.04	13.46	−0.74	16.14
312	5.87	9.15	22.95	25.93	26.53	10.16	−1.31	14.18
313	7.54	10.98	22.7	24.97	26.29	10.18	−1.28	14.48
324	5.96	9.06	22.19	25.61	25.83	8.67	−1.25	13.72

* TAG is explained in Table 1.

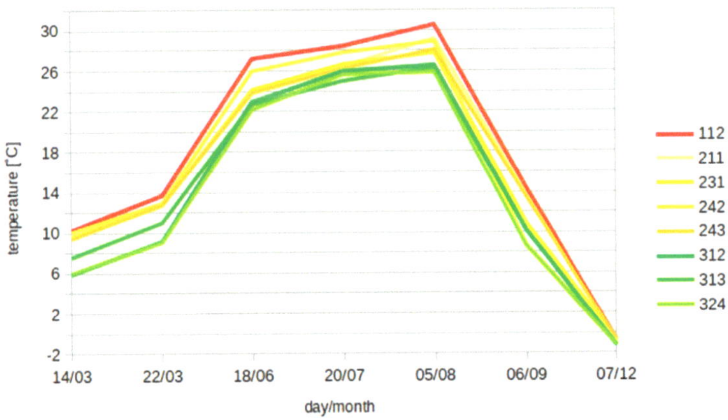

Figure 5. Land surface temperature values progress through imagining dates.

The two types of *LST* and *CCI* maps can then be spatially overlaid to identify locations of greatest vegetation cooling. This can be performed separately for each measured day. By identifying areas with high *CCI* values and low *LST*, we can pinpoint locations where vegetation is most effective in mitigating heat stress. For example, areas with high *CCI* values and low *LST* during summer months are likely to experience lower temperatures and reduced heat-related impacts. Analysing seasonal variations in *LST* and *CCI* can further refine our understanding of vegetation cooling and identify areas that rely heavily on vegetation for cooling during specific periods.

For example, in forested areas, they highlight regions with exceptional cooling potential. Near water bodies, they suggest enhanced cooling benefits.

The proposed approach provides spatiotemporally localised information on the status and evolution of the cooling function of the landscape. The calculation requires only non-contact data (imagery, sensor measurements) that do not damage the landscape (vegetation) in any way (no destructive form of data collection) and allow repeated measurement of the necessary variables at the same location.

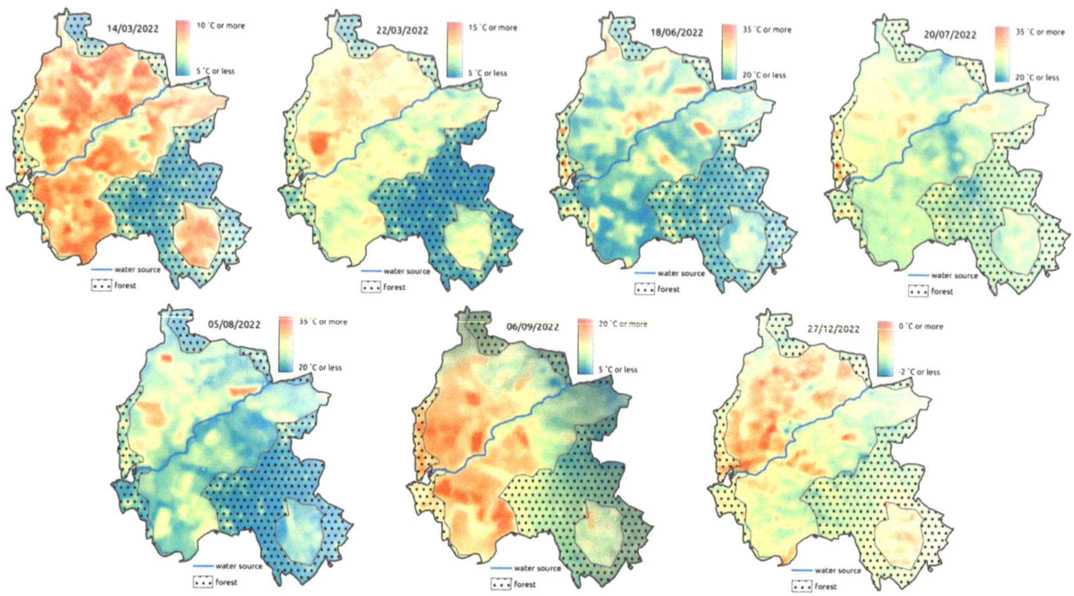

Figure 6. Visualisation development of land surface temperature during the analysed period.

3.3. Validation

Validation of the results was performed using *CCI* calculation through InVEST software—the Urban Cooling model. This model is connected to the Urban Heat Islands effect and is trying to support the assessment of natural green infrastructure in cities and reduce the impacts of UHI [52]. After running the Urban Cooling model, we calculated MAD (Median Absolute Deviation) (Tables 8 and 9) to eliminate the effect of potential data outliers on the sensing day to assess the differences between results obtained by InVEST and our custom calculation. The highest deviation between the two datasets was found in Mixed forest (313) and Transitional woodland/shrubs (324).

Table 8. Biophysical table used for InVEST validation on 20 July 2022.

Green_Area	Albedo	Kc	Shade	LU	TAG
0	0.19	0.560	0.76	discontinuous urban fabric	112
0	0.19	0.477	0.769	non-irrigated arable land	211
0	0.206	0.619	0.765	pastures	231
0	0.199	0.641	0.792	complex cultivations	242
1	0.617	0.617	0.768	mix agriculture and natural vegetation	243
1	0.104	0.561	0.771	coniferous forest	312
1	0.156	0.682	0.797	mixed forest	313
1	0.125	0.719	0.780	transitional woodland shrub	324

* TAG is explained in Table 1.

Table 9. Median absolute deviation of *CCI* on 20 July 2022.

TAG *	20th July
112	0.015
211	0.004
231	0.02
242	0.037
243	0.023
312	0.018
313	0.044
324	0.044

* TAG is explained in Table 1.

InVEST software was filled with the required datasets described in Table 8 below. A biophysical table was created in a comma-separated format. Each surface class was identified by its CLC TAG; other data values (including column names) were averaged for each surface class, and the green area was attributed based on InVEST documentation. The weights for each part of the *CCI* equation were 0.2 (albedo. *ETI*) and 0.6 (shading). The energy valuation and energy consumption options were disabled.

The Green Area Cooling Distance was set to 450 m, as suggested by the literature, with an average air temperature of 23.1 °C, as shown by the metrological station.

The validation results demonstrate the effectiveness of the proposed methodology for assessing vegetation cooling capacity in the study area. The calculated *CCI* values accurately reflect the spatial and temporal variations in cooling potential, aligning with expected patterns based on land use types and surface temperatures. By identifying areas with high *CCI* values and low *LST*, we can pinpoint locations where vegetation is most effective in mitigating heat stress.

4. Discussion

The method is fully transferable even outside the Czech Republic. The primary input is thermal imagery from the Landsat satellite, which covers the whole world, and standardised meteorological variables measured by any weather service or commercially available sensors.

Acquiring a sufficient number of usable satellite images is the most important step in the remote sensing workflow. The success of the calculation depends on having the satellite images with the best spatial and temporal resolution and thermal band. The availability of clear satellite images can be limited due to cloud cover, especially in regions with frequent cloudiness. This can hinder the creation of consistent time series and reduce the accuracy of the *CCI* calculations.

While Landsat 8/9 offers a reasonable spatial and temporal resolution, higher-resolution imagery from satellites such as Sentinel-2 could provide more detailed information on land cover and microclimate variations. However, Sentinel-2 does not include thermal bands for calculating *LST*. *LST* is only available as a pre-calculated band of Sentinel-3 with 1 km/pixel resolution, making it unusable for small-scale studies. For these reasons, Landsat 8/9 was chosen. These two satellites have 16 days of temporal resolution (individually, the combination has eight days). In the presence of clouds, the image is considered unusable. This can be a real problem when acquiring regularly spaced datasets due to the longer temporal resolution. We can partly solve this problem by using an area with fewer cloudy days in a year. Another possibility is to interpolate more images and create a denser time series or to predict the usability of satellite imagery [53]. An alternative solution is to integrate data from the Sentinel-2 and Sentinel-3 satellites. In most cases, neural networks or machine learning techniques are used to merge or interpolate one or more different sensors. The results of both solutions are highly dependent on the skills and abilities of the operator. This greatly reduces the usability of the solution presented [54].

Landsat 8/9 satellites overpass the region of interest between 9:30 AM (GMT) and 10:00 AM (GMT), while the land surface temperature study [37] occurs in the early morning or late afternoon. However, during summer, the surface temperature of urban areas (112) reaches up to reference values (e.g., 9:50 AM (GMT); in August, temperatue reached up to 30.55 °C.

Differences can be found in vegetated areas (312–324) and agricultural areas (211–243) and depend on the season regarding such effects as irrigation or harvesting time. A difference in temperature in vegetated and urban areas is a typical phenomenon connected to Urban Heat Islands. These areas tend to have more built-up surfaces and surfaces accumulating heat (such as asphalt), so they are more absorbing [12]. This difference is especially visible during the night. Cooling Capacity Index or similar indexes (such as the Bowen Ratio) can help locate places needing mitigating heat islands [55]. The relationship between temperature, cooling processes, heat fluxes and other characteristics is complicated, especially with more extreme events during the past years. Research of these relationships as these individual characteristics will be needed in the not-so-distant future as this will help us to understand and mitigate climate change's impacts [56]. The methods shown in this article have proven effective in enabling quick and cost-effective observation and quantification of variables. The advantage of this approach is easy access to most of the data used. The most problematic part regarding data is obtaining Kc (Single Crop Coefficient), as the exact values are derived by laboratory studies performed in the USA, including several crops and plants used in the USA. Kuriata-Potasznik [57] suggests using a constant value of 1.04 and 1.05 according to the crop season or for vegetation not included in the original list of Kc values. InVEST offers a custom-made calculator [19]. However, not all input data were accessible. This problem was aimed to be solved using a substitute vegetation index (*LAI*), as recommended by InVEST. Ultimately, the substitution was performed by an equation proposed by Baastiansen [40] and Brom [41].

By addressing these limitations and exploring future developments, the *CCI* methodology can be further refined and applied to a wider range of case studies, providing valuable insights for urban planning and environmental management.

5. Conclusions

The aim of this study was to test the applicability of the vegetation cooling capacity index and to find a way to determine the index with remotely sensed satellite data, emphasising the maximum usability of satellite imagery. The study area was the town of Černovice in the southwest of the Czech Republic, representing the typical rural landscape in Czechia. Using the contactless, non-destructive index, we calculated and analysed the variability of cooling function regarding specific land use types, which are differentiated by the amount and structure of vegetation and the amount of fulfilling of this function during the growing season.

All data used are non-destructive. We can measure them repeatedly at required locations, and all have spatial characteristics, enabling us to monitor the fulfilment of the cooling function. Most of the data can be obtained using distance methods without field (contact) measurements. One of the advantages of this approach is easy access to most of the data used.

This proved to be a promising approach to assess the cooling function index and, thanks to remotely sensed data, also allows for analysis of spatiotemporal variability.

The results showed vegetation, particularly forests, demonstrated higher cooling capacities throughout the year than urban and non-irrigated agricultural areas.

The Cooling Capacity Index indicated that green spaces, such as parks and forests, can cool the surrounding areas through evapotranspirative processes and shading, even having a significantly smaller size than urban parts of the region of interest.

This method is easily portable and can be used in equal detail at one point, using an area of interest to a large extent. However, this is limited by the spatial and temporal resolution of used satellite imagery. In this way, we can gain a deeper knowledge of the progress of the cooling function of the chosen surface. The process was subsequently successfully applied to several other sites in the Czech Republic, ecologically sensitive areas (e.g., the White Carpathian Protected Landscape Area) or the urban environment of Olomouc.

This method can help predict the impact of heat stress on vegetation and identify potential places needing heat mitigation to establish a more functioning environment. This technique can help us understand heat flows and heat stress over surfaces. These findings highlight the significance of green areas and the need for their preservation, protection and expansion to create resilient and cooler urban environments to ensure living comfort. Overall, urban planning should consider incorporating more green infrastructure to enhance cooling capacities and mitigate the adverse effects of heat stress on vegetation and human populations.

While the method offers several advantages, including non-destructive data acquisition and easy accessibility, it is subject to limitations related to data availability, spatial and temporal resolution, and crop coefficient estimation. To address these challenges, future research should focus on improving data quality, exploring alternative data sources, and refining the *CCI* calculation methodology. By overcoming these limitations, the *CCI* can be further developed and applied to a wider range of case studies, providing valuable insights for urban planning and environmental management.

Author Contributions: Conceptualization, V.P. and T.P.; methodology, V.P. and T.P.; formal analysis, V.P.; investigation, V.P. and T.P.; resources, V.P. and T.P.; data curation, V.P. and T.P.; writing—original draft preparation, T.P.; writing—review and editing, V.P.; visualisation, T.P.; supervision, V.P.; project administration, V.P. All authors have read and agreed to the published version of the manuscript.

Funding: This paper was created within the project "Analysis, modelling, and visualization of spatial phenomena by geoinformation technologies III" (IGA_PrF_2024_018) with the support of the Internal Grant Agency of Palacký University Olomouc).

Data Availability Statement: The satellite data are available through USGS&NASA Earth Explorer (https://earthexplorer.usgs.gov (accessed on 25 August 2024)). Meteorological daily data are published according to Act 123/1998 Coll. as part of the law collection of the Czech Republic (https://www.chmi.cz/historicka-data/pocasi/denni-data/Denni-data-dle-z.-123-1998-Sb (accessed on 25 August 2024).

Conflicts of Interest: The authors declare no conflicts of interest.

Abbreviations

AET	Actual Evapotranspiration
CCI	Cooling Capacity Index
CHMI	Czech Hydrometeorological Institute
CSV	Comma Separated Value
CLC	Corine Land Cover
DN	Digital Number
ET0	Reference Evapotranspiration
ET	Evapotranpiration
ETI	Evapotranspirational Index
GMT	Greenwich Mean Time
inVEST	Integrated Valuation of Ecosystem Services and Tradeoffs
Kc	Crop Coefficient
LAI	Leaf Area Index
LST	Land Surface Temperature
MAD	Median Absolute Deviation
NDVI	Normal Differential Vegetation Index

OLI	Opeational Land Imager
PET	Potential Evapotranspiration
SAVI	Soil Adjusted Vegetation Index
SEBAL	Surface Energy Balance Algorithm for Land
TIRS	Thermal Infrared Sensor
UAV	Unmanned Aerial Vehicle
UHI	Urban Heat Island

References

1. Lim, C.L. Fundamental concepts of human thermoregulation and adaptation to heat: A review in the context of global warming. *Int. J. Environ. Res. Public Health* **2020**, *17*, 7795. [CrossRef]
2. Bolan, S.; Padhye, L.P.; Jasemizad, T.; Govarthanan, M.; Karmegam, N.; Wijesekara, H.; Amarasiri, D.; Hou, D.; Zhou, P.; Biswal, B.K.; et al. Impacts of climate change on the fate of contaminants through extreme weather events. *Sci. Total Environ.* **2024**, *909*, 168388. [CrossRef] [PubMed]
3. Lloret, F.; Escudero, A.; Iriondo, J.M.; Martínez-Vilalta, J.; Valladares, F. Extreme climatic events and vegetation: The role of stabilizing processes. *Glob. Chang. Biol.* **2011**, *18*, 797–805. [CrossRef]
4. Armson, D.; Stringer, P.; Ennos, A.R. The effect of tree shade and grass on surface and globe temperatures in an urban area. *Urban For. Urban Green.* **2012**, *11*, 245–255. [CrossRef]
5. Koc, C.B.; Osmond, P.; Peters, A. Evaluating the cooling effects of green infrastructure: A systematic review of methods, indicators and data sources. *Sol. Energy* **2018**, *166*, 486–508. [CrossRef]
6. Allen, R.G.; Pereira, L.S.; Raes, D.; Smith, M. Crop evapotranspiration—Guidelines for computing crop water requirements. In *FAO Irrigation and Drainage Paper, No. 56*; FAO—Food and Agriculture Organization of the United Nations: Rome, Italy, 1998; Available online: https://www.fao.org/3/x0490e/x0490e00.htm (accessed on 25 August 2024).
7. Lobell, D.B.; Burke, M.B.; Tebaldi, C.; Mastrandrea, M.D.; Falcon, W.P.; Naylor, R.L. Prioritizing Climate Change Adaptation Needs for Food Security in 2030. *Science* **2008**, *319*, 607–610. [CrossRef]
8. Trenberth, K.E. Changes in precipitation with climate change. *Clim. Res.* **2011**, *47*, 123–138. [CrossRef]
9. Silva, C.M.; Gomes, M.G.; Silva, M. Green roofs energy performance in Mediterranean climate. *Energy Build.* **2016**, *116*, 318–325. [CrossRef]
10. Silva, H.R.; Phelan, P.E.; Golden, J.S. Modeling effects of urban heat island mitigation strategies on heat-related morbidity: A case study for Phoenix, Arizona, USA. *Int. J. Biometeorol.* **2009**, *54*, 13–22. [CrossRef]
11. Hayes, A.T.; Jandaghian, Z.; Lacasse, M.A.; Gaur, A.; Lu, H.; Laouadi, A.; Ge, H.; Wang, L. Nature-Based Solutions (NBSs) to Mitigate Urban Heat Island (UHI) Effects in Canadian Cities. *Buildings* **2022**, *12*, 925. [CrossRef]
12. Oke, T.R. The energetic basis of the urban heat island. *Q. J. R. Meteorol. Soc.* **1982**, *108*, 1–24. [CrossRef]
13. Neto, A.D.M.; Fernandes, G.S.T.; Lima, E.D.A.; Lopes, P.M.D.O.; Rodrigues, L.S.; Junior, A.D.S.G.; Lopes, J.R.A.; e Silva, L.L.S.; Da Silva, R.O. Evapotranspiration in the context of land use and land cover changes in MATOPIBA, Brazil. *Rev. Bras. Geogr. Física* **2023**, *16*, 50–62. [CrossRef]
14. Drexler, J.Z.; Snyder, R.L.; Spano, D.; Paw, U. A review of models and micrometeorological methods used to estimate wetland evapotranspiration. *Hydrol. Proc.* **2004**, *18*, 2071–2101. [CrossRef]
15. Rana, G.; Katerji, N. Measurement and estimation of actual evapotranspiration in the field under Mediterranean climate: A review. *Eur. J. Agron.* **2000**, *13*, 125–153. [CrossRef]
16. Carlson, T. An Overview of the "Triangle Method" for Estimating Surface Evapotranspiration and Soil Moisture from Satellite Imagery. *Sensors* **2007**, *7*, 1612–1629. [CrossRef]
17. Euser, T.; Luxemburg, W.M.J.; Everson, C.S.; Mengistu, M.G.; Clulow, A.D.; Bastiaanssen, W.G.M. A new method to measure Bowen ratios using high-resolution vertical dry and wet bulb temperature profiles. *Hydrol. Earth Syst. Sci.* **2014**, *18*, 2021–2032. [CrossRef]
18. Zawadzka, J.; Harris, J.; Corstanje, R. Assessment of heat mitigation capacity of urban greenspaces with the use of InVEST urban cooling model, verified with day-time land surface temperature data. *Landsc. Urban Plan.* **2021**, *214*, 104163. [CrossRef]
19. Miglietta, F.; Gioli, B.; Brunet, Y.; Hutjes, R.W.A.; Matese, A.; Sarrat, C.; Zaldei, A. Sensible and latent heat flux from radiometric surface temperatures at the regional scale: Methodology and evaluation. *Biogeosciences* **2009**, *6*, 1975–1986. [CrossRef]
20. Stisen, S.; Soltani, M.; Mendiguren, G.; Langkilde, H.; Garcia, M.; Koch, J. Spatial Patterns in Actual Evapotranspiration Climatologies for Europe. *Remote Sens.* **2021**, *13*, 2410. [CrossRef]
21. Scanlon, T.M.; Kustas, W.P. Partitioning Evapotranspiration Using an Eddy Covariance-Based Technique: Improved Assessment of Soil Moisture and Land–Atmosphere Exchange Dynamics. *Vadose Zone J.* **2012**, *11*. [CrossRef]
22. Constantinescu, D.; Cheval, S.; Caracaş, G.; Dumitrescu, A. Effective monitoring and warning of Urban Heat Island effect on the indoor thermal risk in Bucharest (Romania). *Energy Build.* **2016**, *127*, 452–468. [CrossRef]
23. Völker, S.; Baumeister, H.; Claßen, T.; Hornberg, C.; Kistemann, T. Evidence for the temperature-mitigating capacity of urban blue space—A health geographic perspective. *Erdkunde* **2013**, *67*, 355–371. [CrossRef]

24. Stanford University, University of Minnesota, Chinese Academy of Sciences, the Nature Conservancy, World Wildlife Fund, and Stockholm Resilience Centre, Natural Capital Project (User's Guide). 2022. Available online: https://storage.googleapis.com/releases.naturalcapitalproject.org/invest-userguide/latest/en/index.html (accessed on 25 August 2024).
25. Quitt, E. *Climatic Regions of Czechoslovakia;* Geografický ústav ČSAV: Brno, Czech Republic, 1971.
26. European Environment Agency (EEA). Corine Land Cover 2018. European Union, Copernicus Land Monitoring Service. 2018. Available online: https://land.copernicus.eu/pan-european/corine-land-cover (accessed on 25 August 2024).
27. United States Geological Survey. Landsat 8 Data Users Handbook, Version 5.0. 2019, pp. 1–114. Available online: https://d9-wret.s3.us-west-2.amazonaws.com/assets/palladium/production/s3fs-public/atoms/files/LSDS-1574_L8_Data_Users_Handbook-v5.0.pdf (accessed on 25 August 2024).
28. U.S. Geological Survey: What Are the Band Designations for the Landsat Satellites? n.d. Available online: https://www.usgs.gov/faqs/what-are-band-designations-landsat-satellites (accessed on 25 August 2024).
29. Czech Hydrometeorological Institute/CHMI. Daily Data according to Act No. 123/1998 Coll., n.d. Available online: https://www.chmi.cz/historicka-data/pocasi/denni-data/Denni-data-dle-z.-123-1998-Sb (accessed on 25 August 2024).
30. Phelan, P.E.; Kaloush, K.; Miner, M.; Golden, J.; Phelan, B.; Silva, H.; Taylor, R.A. Urban Heat Island: Mechanisms, Implications, and Possible Remedies. *Annu. Rev. Environ. Resour.* **2015**, *40*, 285–307. [CrossRef]
31. Zardo, L.; Geneletti, D.; Pérez-Soba, M.; Van Eupen, M. Estimating the cooling capacity of green infrastructures to support urban planning. *Ecosyst. Serv.* **2017**, *26*, 225–235. [CrossRef]
32. Kunapo, J.; Fletcher, T.D.; Ladson, A.R.; Cunningham, L.; Burns, M.J. A spatially explicit framework for climate adaptation. *Urban Water J.* **2018**, *15*, 159–166. [CrossRef]
33. Liang, S. Narrowband to broadband conversions of land surface albedo I. *Remote Sens. Environ.* **2001**, *76*, 213–238. [CrossRef]
34. Sauer, T.J.; Horton, R. Soil Heat Flux. *Micrometeorol. Agric. Syst.* **2005**, *47*, 131–154.
35. Duan, S.-B.; Han, X.-J.; Huang, C.; Li, Z.-L.; Wu, H.; Qian, Y.; Gao, M.; Leng, P. Land Surface Temperature Retrieval from Passive Microwave Satellite Observations: State-of-the-Art and Future Directions. *Remote Sens.* **2020**, *12*, 2573. [CrossRef]
36. Imhoff, M.L.; Zhang, P.; Wolfe, R.E.; Bounoua, L. Remote sensing of the urban heat island effect across biomes in the continental USA. *Remote Sens. Environ.* **2010**, *114*, 504–513. [CrossRef]
37. Hesslerová, P.; Pokorný, J.; Huryna, H.; Seják, J.; Jirka, V. The impacts of greenery on urban climate and the options for use of thermal data in urban areas. *Prog. Plan.* **2021**, *159*, 100545. [CrossRef]
38. Avdan, U.; Jovanovska, G. Algorithm for Automated Mapping of Land Surface Temperature Using LANDSAT 8 Satellite Data. *J. Sens.* **2016**, *2016*, 1480307. [CrossRef]
39. Bréda, N.J.J. Ground-based measurements of leaf area index: A review of methods, instruments and current controversies. *J. Exp. Bot.* **2003**, *54*, 2403–2417. [CrossRef]
40. Bastiaanssen, W. SEBAL-based sensible and latent heat fluxes in the irrigated Gediz Basin, Turkey. *J. Hydrol.* **2000**, *229*, 87–100. [CrossRef]
41. Brom, J. SEBCS for QGIS—Module for Calculation of Energy Balance Features and Vegetation Water Stress Indices. University of South Bohemia in České Budějovice, Faculty of Agriculture and Technology. 2016. Available online: https://github.com/JakubBrom/SEBCS/tree/master (accessed on 25 August 2024).
42. Vani, V.; Mandla, V.R. Comparative Study Of NDVI and *SAVI* Vegetation Indices in Anantapur District Semi-Arid Areas. *Int. J. Civ. Eng. Technol.* **2017**, *8*, 559–566.
43. Cañete-Salinas, P.; Zamudio, F.; Yáñez, M.; Gajardo, J.; Valdés, H.; Espinosa, C.; Venegas, J.; Retamal, L.; Ortega-Farias, S.; Acevedo-Opazo, C. Evaluation of models to determine *LAI* on poplar stands using spectral indices from Sentinel-2 satellite images. *Ecol. Model.* **2020**, *428*, 109058. [CrossRef]
44. Chen, Z.; Jia, K.; Xiao, C.; Wei, D.; Zhao, X.; Lan, J.; Wei, X.; Yao, Y.; Wang, B.; Sun, Y.; et al. Leaf Area Index Estimation Algorithm for GF-5 Hyperspectral Data Based on Different Feature Selection and Machine Learning Methods. *Remote Sens.* **2020**, *12*, 2110. [CrossRef]
45. Yadav, V.P.; Prasad, R.; Bala, R.; Vishwakarma, A.K.; Yadav, S.A.; Singh, S.K. A comparison of machine-learning regression algorithms for the estimation of *LAI* using Landsat 8 satellite data. *Int. Arch. Photogramm. Remote Sens. Spat. Inf. Sci.* **2019**, *XLII-4/W16*, 679–683. [CrossRef]
46. Apolo-Apolo, O.E.; Pérez-Ruiz, M.; Martínez-Guanter, J.; Egea, G. A Mixed Data-Based Deep Neural Network to Estimate Leaf Area Index in Wheat Breeding Trials. *Agronomy* **2020**, *10*, 175. [CrossRef]
47. Ercanlı, i.; Günlü, A.; Şenyurt, M.; Keleş, S. Artificial neural network models predicting the leaf area index: A case study in pure even-aged Crimean pine forests from Turkey. *For. Ecosyst.* **2018**, *5*, 29. [CrossRef]
48. Kristensen, K.J. Actual evapotranspiration in relation to leaf area. *Hydrol. Res.* **1974**, *5*, 173–182. [CrossRef]
49. Wahid, A.; Gelani, S.; Ashraf, M.; Foolad, M.R. Heat tolerance in plants: An overview. *Environ. Exp. Bot.* **2007**, *61*, 199–223. [CrossRef]
50. Gray, S.B.; Brady, S.M. Plant developmental responses to climate change. *Dev. Biol.* **2016**, *419*, 64–77. [CrossRef] [PubMed]
51. Lanxinger, M. Greening Measures to Mitigate Urban Heat Islands during Tropical Nights in Vienna, Austria. Available online: https://studenttheses.uu.nl/bitstream/handle/20.500.12932/42756/Final%20MSc%20Thesis_%20Marina%20Lanxinger.pdf?sequence=1&isAllowed=y (accessed on 25 August 2024).

52. Deilami, K.; Kamruzzaman, M.; Liu, Y. Urban heat island effect: A systematic review of spatio-temporal factors, data, methods, and mitigation measures. *Int. J. Appl. Earth Obs. Geoinf.* **2018**, *67*, 30–42. [CrossRef]
53. Zhu, X.; Helmer, E.H.; Chen, J.; Liu, D. An Automatic System for Reconstructing High-Quality Seasonal Landsat Time-Series. In *Remote Sensing: Time Series Image Processing*; Weng, Q., Ed.; Taylor and Francis Series in Imaging Science; CRC Press: Boca Raton, FL, USA, 2018; 263p.
54. Senty, P.; Guzinski, R.; Grogan, K.; Buitenwerf, R.; Ardö, J.; Eklundh, L.; Koukos, A.; Tagesson, T.; Munk, M. Fast Fusion of Sentinel-2 and Sentinel-3 Time Series over Rangelands. *Remote Sens.* **2024**, *16*, 1833. [CrossRef]
55. Intaraksa, A.; Chunkao, K.; Bualert, S. Bowen Ratio Method for Measuring Heat Transfer on Land Cover Change in Establishing Green Patch in Urban Heat Island of Bangkok. *Mod. Appl. Sci.* **2014**, *8*, 158. [CrossRef]
56. Zhu, J.; Hu, S.; Wang, J.; Zheng, X. Future orientation promotes climate concern and mitigation. *J. Clean. Prod.* **2020**, *262*, 121212. [CrossRef]
57. Kuriata-Potasznik, A.B.; Szymczyk, S. Variability of the water availability in a river lake system—A case study of Lake Symsar. *J. Water Land Dev.* **2016**, *31*, 87–96. [CrossRef]

land

Article

Green Urban Public Spaces Accessibility: A Spatial Analysis for the Urban Area of the 14 Italian Metropolitan Cities Based on SDG Methodology

Angela Cimini [1], Paolo De Fioravante [2], Ines Marinosci [2], Luca Congedo [2,*], Piergiorgio Cipriano [3], Leonardo Dazzi [4], Marco Marchetti [1], Giuseppe Scarascia Mugnozza [5] and Michele Munafò [2]

1 Department of Architecture and Project, University of Rome La Sapienza, Piazza Borghese 9, 00186 Roma, Italy; angela.cimini@uniroma1.it (A.C.); ma.marchetti@uniroma1.it (M.M.)
2 Department of Networks and Environmental Information Systems (SINA), Italian Institute for Environmental Protection and Research (ISPRA), Via Vitaliano Brancati 48, 00144 Rome, Italy; paolo.defioravante@isprambiente.it (P.D.F.); ines.marinosci@isprambiente.it (I.M.); michele.munafo@isprambiente.it (M.M.)
3 Deda Next Srl., legal address Via di Spini 50, 38121 Trento, Italy; piergiorgio.cipriano@dedagroup.it
4 Laboratory GIScience and Drones for Good, Department of Civil, Environmental and Architectural Engineering (ICEA), University of Padua, Via Marzolo 9, 35131 Padua, Italy; leonardo.dazzi@studenti.unipd.it
5 Department for Innovation in Biological, Agro-Food and Forest Systems, University of Tuscia, Via S. Camillo de Lellis, 01100 Viterbo, Italy; gscaras@unitus.it
* Correspondence: luca.congedo@isprambiente.it

Citation: Cimini, A.; De Fioravante, P.; Marinosci, I.; Congedo, L.; Cipriano, P.; Dazzi, L.; Marchetti, M.; Scarascia Mugnozza, G.; Munafò, M. Green Urban Public Spaces Accessibility: A Spatial Analysis for the Urban Area of the 14 Italian Metropolitan Cities Based on SDG Methodology. *Land* **2024**, *13*, 2174. https://doi.org/10.3390/land13122174

Academic Editor: Jianjun Zhang

Received: 28 October 2024
Revised: 2 December 2024
Accepted: 10 December 2024
Published: 13 December 2024

Abstract: Among the most significant impacts related to the spread of settlements and the densification of urban areas, the reduction in the availability of public green spaces plays a central role in the definition of livable cities, in terms of the environment and social cohesion, interaction, and equality. In the framework of target 11.7 of the Sustainable Development Goals (SDG) 11, the United Nations has established the objective of ensuring universal, safe, and inclusive access to public spaces by 2030, for women, children, the elderly, and people with disabilities. This study proposes the evaluation of this objective for the urban area of the 14 Italian metropolitan cities, as defined by EUROSTAT and adopted by the United Nations and the Nature Restoration Law (NRL). A methodology based on open-source data and network analysis tools is tested for the provision of an unprecedented mapping of the availability and accessibility to green urban public spaces, which shows that less than 30% of metropolitan city residents have access to a green space within 300 m on foot, according to OpenStreetMap data (less than one in five for the Urban Atlas data). Furthermore, a critical analysis on the geometric and semantic definition of green urban public spaces adopted by the main European and international tools is carried out, which underlines the strategic role of crowdsourcing but also the need for mapping rules that make the data more consistent with the monitoring objectives set at the institutional level.

Keywords: urban areas; DEGURBA; green urban public spaces; spatial analysis; accessibility; SDG; ecosystem services; 3-30-300 rule; OpenStreetMap; population spatialization

1. Introduction

1.1. State-of-the-Art

The evolution of urban areas in recent decades has been characterized by progressive acceleration and significant evolution, which holds crucial new challenges to ensure sustainability and a good quality of life in cities.

According to estimates by the World Urbanization Prospect, 68% of the global population will live in urban areas by 2050, resulting in a loss of another 1.2 million km^2 of land for the construction of new buildings and infrastructure, often in public places and private green areas [1].

Many studies have shown the beneficial effects of greenery on health, in terms of improved perceived well-being, mental health, reduction in cardiovascular disease, and decreased mortality, but also improvement in air quality, climate mitigation (and reduction in mortality associated with heat waves, which caused between 55,000 and 72,000 deaths in Europe in the years 2003, 2010, and 2022), water infiltration, and landscape and esthetic benefits by improving social cohesion, interaction, and equality [2–7]. The presence of well-managed and fairly large green areas also has an economic benefit, with effects on property values and commercial activities [8]. In spite of that, the European Commission Joint Research Centre estimates that in Europe, only 44% of the urban population has access to a public park within 300 m, with significant variations between Eastern and Mediterranean cities [9]. Several thematic policies and strategies recognize the role of green spaces in urban areas and the need to safeguard them and ensure their accessibility. In Italy, the introduction of urban planning standards (the Ministerial Decree of 2 April 1968 n.1444 establishes that for each new settlement a minimum of 18 m^2 per inhabitant must be foreseen for public spaces, 9 of which are intended for public greenery, excluding the areas pertaining to road infrastructures) has safeguarded urban green spaces from urbanization processes, but this often translates into a merely quantitative application that does not consider the quality of green areas in terms of accessibility and ecosystemic and social functions. This has led to neglect of the positive effects of natural capital in terms of landscape, esthetics, historical–cultural identity, and accessibility, often reducing them to an element of urban furniture in the interstices of the city [10–12]. Tools such as the National Strategy for Urban Green [13] have been developed with the aim of stimulating political and cultural transformation and supporting new forms of planning capable of emphasizing the contribution of green infrastructures and Nature-Based Solutions (NBS) in both urban and rural contexts. The recent Nature Restoration Law (NRL) [14] obliges all European Union member states to eliminate the net loss of urban green areas by 2030 and to ensure their improvement by 2050. Actually, green areas have become crucial indicators to promote the integration of biodiversity values in urban planning.

The growing interest in cities and in settlement and demographic dynamics has highlighted the need for definitions of urban areas, green spaces, and accessibility that are spatially explicit, supported by scientific evidence, and recognized by the main regulatory instruments. The introduction of the United Nations Sustainable Development Goal 11 "Make cities inclusive, safe, resilient and sustainable" has provided standardized definitions and methodologies for mapping urban areas [15] but the definition of green spaces, especially public ones, is not yet univocal. Typically, green public spaces in urban areas are public parks but can also include private gardens, forests, playgrounds, rows of trees along road infrastructures, or blue infrastructures, depending on the objectives of the analysis. At present, there is no universally accepted definition of green urban public spaces (GUPSs), but it varies according to the specific context and objectives [16–21]. At a regulatory level, the Nature Restoration Law [14] introduced a definition of urban greenery to be conserved and implemented immediately. This definition includes all trees, bushes, shrubs, permanent herbaceous plants, lichens and mosses, ponds, and watercourses within cities and small urban agglomerations and excludes all artificial cover and arable land (the measures envisaged for agricultural ecosystems are described in Article 11).

To assess the impacts of green urban areas on human health and well-being, it is important to consider other aspects in addition to the presence of vegetation, such as size, ownership regime, proximity, accessibility, uses, and the functions they host [22]. With reference to size, the 3-30-300 rule [23], adopted by the IUCN, recommends green areas of at least one hectare, as does the study by Matilda Annerstedt Van Den Bosch [24]. The ownership regime allows for evaluation of the function of the green area to promote equality, inclusion, and social interaction, as also supported by SDG 11.7 ("by 2030, provide universal access to safe, inclusive, and accessible green and public spaces, particularly for women and children, older persons, and persons with disabilities"). In the absence of detailed information or mapping of the ownership regime or usability of green areas, the training module of target 11.7.1 suggests the use of the OpenStreetMap (OSM) database [25]. "Accessibility by proximity" [26] considers the theory of "15 min cities", as the possibility for a person to reach relevant daily points of interest (grocery stores, health or recreational areas such as green areas) within 15 min on foot or by bicycle; Deda Next has implemented an algorithm to automatically calculate the proximity index to points of interest [27].

The concept of accessibility has taken on different meanings over time that require combining the normative dimension with the qualitative and quantitative ones [28]. Many studies analyze the accessibility to GUPSs by considering different tools, definitions, study areas, and techniques. There are calculation experiences based on the definition of a buffer around the areas of interest (therefore referring to the distance as the crow flies from the target [29,30]; other studies refer to travel times and others still to the distance referred to the road network [31]). In addition, the assessment is also conditioned by the characteristics of the different elements involved [32]; in fact, to evaluate the accessibility to GUPSs, it is important to appropriately represent the urban areas (the area within which the accessibility to GUPSs is evaluated), the GUPSs (to be accessed) and the relative access points, the distances (to be covered to access the GUPSs), and the population (which can access the GUPSs). The urban area can be identified with respect to different approaches. Poelman and Giuliani consider the footprint of the Urban Atlas [33] Functional Urban Areas (FUA) [34,35] as Della Rosa does too [36], while other experiences refer to the entire municipal territory [27,37]; the UN SDGs introduced the concept of DEGURBA [15] based on the methodology adopted by Eurostat, which represents settlements through the definition of thresholds on the number of inhabitants and population density. This tool is officially adopted by the Nature Restoration Law [14] and the targets related to SDG 11 [25,38]. GUPSs can be identified starting from Urban Atlas classes "1.4—Green urban areas" [36,37] and "3—forest" [34], or by extracting representative categories of green public spaces from OSM by combining land use/cover information with reference to protected areas, agricultural areas, cemeteries, sports areas, green furniture, or even large wooded areas on the edges of the settlement fabric [35,39]. Once GUPSs have been identified, a key aspect for assessing accessibility is the availability of information on their access points, which greatly affects the results of the analysis, especially when using network analysis tools to assess distances, which are more accurate than methodologies based on Euclidean distance [36]. Due to the difficulty in finding the real access points, the centroids of the GUPSs polygons [27,35] or the arrangement of access points at regular intervals along the perimeter of the polygon can be considered [25,34,36,37,40]. Reliable information on the spatial distribution of the population is needed to assess the portion of inhabitants who are near the GUPSs and therefore have access to them. In this sense, it is important to have a spatialized population dataset or to define a criterion for producing it, starting from the most detailed demographic data available. The DEGURBA methodology suggests the use of the Eurostat Population Grid (updated to 2011 for the Italian territory), while in Italy, the ISTAT population data are updated annually with respect to the municipal territory and every 10 years with respect to the census sections (sub-municipal detail, available for 2001, 2011, and 2021) [41].

1.2. Objectives and Structure of the Research

The general objective of this study is the evaluation of accessibility at a national scale using the methodology provided by UN SDG indicator 11.7.1 ("Average share of the built-up area of cities that is open space for public use for all, by sex, age and persons with disabilities"), with reference to green open public spaces, which are one of the three components considered by the indicator metadata, together with gray and blue infrastructures. The focus on green spaces is due to their importance in the current framework of European and international policies—first of all the new Nature Restoration Law, the SDG 11.7.1 itself, or at the national level, the forestation interventions provided by the National Recovery and Resilience Plan (NRRP). In this sense, the accessibility assessment was made in compliance with the definitions and calculation methodologies provided by the regulatory context to obtain spatial data suitable for the decision-making processes, e.g., for the identification of priority areas for urban forestry intervention.

A specific objective concerns the analysis of the technical phases for the calculation of accessibility to GUPSs, with a focus on the input data available for the Italian territory, evaluating their suitability for this purpose in relation to characteristics such as spatial resolution, update frequency, classification system, and spatial coverage. In detail, OSM and UA have been considered, which are among the most detailed data available for mapping urban areas; they are based on different classification systems and mapping and updating logics. UA belongs to the Copernicus Land Monitoring Service (CLMS) and is produced and updated according to a standardized methodology that provides a homogeneous mapping on the main FUAs in terms of minimum mapping unit, classification system, reference year, and accuracy; OSM has higher geometric and temporal heterogeneity, but offers a much wider coverage and is more detailed than UA and more flexible from the thematic point of view, also allows for the introduction of new attributes.

The last specific objective is the assessment of the state of accessibility to GUPSs in the urban areas of the 14 Italian metropolitan cities (delimited through the Eurostat DEGURBA methodology), also with the introduction of two auxiliary indices that describe the presence of GUPSs per capita compared to the total urban green spaces per capita.

This research offers, for the first time, a national-scale assessment of accessibility to GUPSs in compliance with the legal indications of NRL for urban areas monitoring and based on an SDG methodology, allowing us to obtain coherent and comparable estimates between different areas of the Italian territory, all using the most detailed and updated available data (ISTAT population data for census sections and ISPRA LCM) and providing operative ideas that can also be useful for applications in other territorial contexts.

The methodology illustrated in this paper is divided into six phases, whose overview is provided in Section 2.1; in Section 2.2, the choice of the study area is illustrated, i.e., the urban area (according to the DEGURBA methodology) of the 14 Italian metropolitan cities; Section 2.3 describes the allocation of the population and of the starting points in the centers of each cell of a hexagonal grid; Section 2.4 illustrates the classes taken into consideration by the OSM and UA data for the identification of GUPSs and the approach used to identify the access points; Section 2.5 illustrates which elements of the OSM road network that have been considered for the walking routes from the starting points to the access points to the GUPSs, while Section 2.6 describes the calculation of accessibility to GUPSs, considering a maximum walking distance of 300 (threshold indicated by the IUCN rule 3-30-300) and 400 m (indicated by the metadata of SDG 11.7.1).

2. Materials and Methods

2.1. Overview

In this research, indicator 11.7.1 of the United Nations Sustainable Development Goals (UN SDGs) was evaluated for the urban area of the 14 Italian metropolitan cities (MCs), with reference to Green Urban Public Spaces (GUPSs) accessibility.

The urban area of metropolitan cities was delimited using Eurostat's DEGURBA methodology, implemented on Italian territory starting from a 10 m-resolution population grid relating to 2021, produced by the ISPRA on ISTAT data (spatializing the ISTAT population data by census sections on residential buildings of the ISPRA/SNPA National Land Consumption Map).

The accessibility assessment requires the definition of a starting point, an arrival point (to be accessed), and the path between them. Actually, the following operations were carried out:

1. Definition of starting points and spatialization of the resident population of the surrounding area. For the access points, a hexagonal mesh grid was created, considering the centers as the starting points of the routes and on which the spatialized population was allocated;

2. Identification of GUPSs and related access points, with reference to Urban Atlas and OpenStreetMap data. For the OSM access points, those mapped by the "gate" and "entrance" tags were considered; in their absence, the intersection was made between GUPSs and roads or, thirdly, an access was inserted every 100 m along the perimeter of the GUPSs. For UA, it was only possible to consider one access point every 100 m along the perimeter;
3. Definition of the routing network, to be covered on foot, to access the GUPSs from the starting points, defined from the OSM walkable road network;
4. Identification of the shortest route to access from a starting point to the nearest GUPSs. For this purpose, it is considered that the inhabitants located in each hexagon of the grid have access to a green area if the nearest GUPSs can be reached by walking no more than 300 or 400 m. The threshold of 400 m is indicated by the metadata of SDG 11.7.1, while the threshold of 300 m refers to the IUCN 3-30-300 rule, which establishes that every inhabitant must be able to access a park within 300 m, 30 percent of each neighborhood must be wooded, and at least three trees should be visible from every window.

2.2. Study Area

The analysis was carried out on 14 Italian metropolitan cities, considering the portion of their territory classified as an urban area by the DEGURBA methodology (Figure 1), which was adopted by the United Nations Statistical Commission in its 51st session (March 2020) and more recently by the NRL.

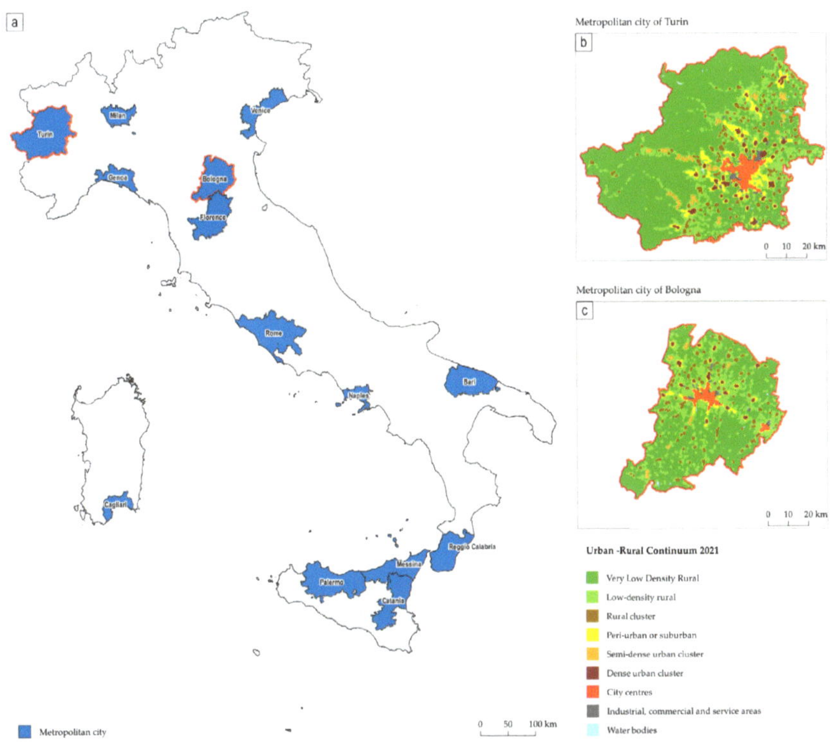

Figure 1. Study area. This research focuses on the 14 Italian metropolitan cities (**a**). On the right, there is an example of the urban–rural continuum for the MCs of Turin (**b**) and Bologna (**c**), used as a reference to delimit the urban area (urban centers and dense urban clusters) on which accessibility was assessed.

The reference to the DEGURBA definition of urban area is required by the metadata of the SDG indicator 11.7.1 (and by other indicators relating to SDG 11, such as 11.3.1 [38]), and is also the official reference for the Nature Restoration Law (for the identification of municipalities subject to legal obligations and to select priority areas of intervention). Furthermore, reference is made to metropolitan cities because they are the subject of significant funding under the NextGenerationEU plan, in particular by mission 2, component 4, investment 3.1 "protection and enhancement of urban and extra-urban greenery" of the NRRP. Such investments are planned to improve the quality of life and well-being of citizens through reforestation interventions that counteract problems related to air pollution, the impact of climate change and the loss of biodiversity. In Italy, the ISPRA represents the urban–rural continuum according to the DEGURBA methodology by spatializing the ISTAT population data with respect to the national land consumption map (LCM) [42] and obtaining a 10 m-resolution product which identifies nine classes based on population density and total population information; in this study, the portion of the metropolitan city that falls into the most populated classes was considered, i.e., the urban centers (population > 50,000 inhabitants and population density $\geq$ 1500 inhabitants per km^2) and dense urban clusters (population $\geq$ 5000 inhabitants and population density $\geq$ 1500 inhabitants per km^2). In the following, "urban area" will mean the surfaces mapped as "urban center" and "dense urban cluster" by the DEGURBA map produced by the ISPRA on 2021 data for the Italian territory. In the first version of the data presented in the previous publication (relating to municipal population data for 2018), the thresholds were adapted to the national context; the version considered in this work (which is based on population data for 2021 by census sections) maintains the population and population density thresholds indicated by DEGURBA. Table 1 summarizes the total areas of MCs and the portion of their territory classified as urban centers and dense urban clusters, taken into consideration for the accessibility; in detail, the urban area occupies 6.3% of the MCs.

Table 1. Total surface area of the 14 Italian metropolitan cities and of the portion of their territory classified as "urban centers" and "dense urban clusters" according to the DEGURBA methodology.

MC	Total Surface	Urban Area	
	ha	ha	%
Tourin	682,973	25,319	3.7
Genoa	183,517	10,855	5.9
Milan	157,674	49,513	31.4
Venice	247,039	8770	3.6
Bologna	370,199	9288	2.5
Florence	351,351	11,937	3.4
Rome	535,581	61,856	11.5
Naples	117,398	50,126	42.7
Bari	382,540	14,670	3.8
Reggio Calabria	318,303	5789	1.8
Palermo	499,302	14,282	2.9
Messina	324,678	6294	1.9
Catania	355,303	15,800	4.4
Cagliari	124,964	6216	5.0
Total	4,650,821	290,716	6.3

The considered MCs represent 15% of the national territory, include more than a third of national "urban centers" and "dense urban clusters", and the main national initiatives for the enhancement of green infrastructures in urban and suburban areas are focused on them [13].

2.3. Identification of Starting Points and Spatialization of the Population

2.3.1. Identification of Starting Points

To define the starting points, a 50 m-apothem hexagonal mesh grid was created. The centroid of each hexagon was taken as a starting point to evaluate accessibility to the nearest GUPSs for the population residing in the hexagon.

The hexagonal grid is widely used for accessibility studies, ensuring a better distribution of sampling points, since it keeps a constant distance between the centers of the meshes compared to the square grid [27,40,43,44].

2.3.2. Population Spatialization

For each hexagon, the resident population was calculated, starting from a population spatialization methodology already described in a previous study by the working group ([42]) and following the steps below (Figure 2):

1. The most detailed population data were identified. In Italy, the ISTAT publishes population data aggregated by census sections (which offer a sub-municipal level of detail) every 10 years, starting from 1991. This study is based on the new ISTAT data updated to 2021.
2. The population was allocated within the census sections by introducing the ISPRA national land consumption map (LCM) and the Copernicus Land Monitoring Service land cover and land use data as ancillary data. The LCM is a 10 m-resolution raster which provides an annually updated consumed land mapping for the Italian national territory. For each census section, the residential population was uniformly distributed across the LCM pixels classified as built-up areas for residential use and which fall under the section itself. In the absence of residential LCM pixels, the population was spread across all the built-up areas of the census section, or, in the absence of built-up pixels, a uniform distribution of the population was maintained across the entire census section. The population density information in the fictitious sections was finally added[1].
3. The previous operation produced a 10 m-resolution raster for the entire study area, which associates a population value to each LCM pixel. It was then crossed with the hexagonal grid to attribute to each hexagon the population of the underlying pixels. To calculate accessibility, the centroid of each hexagon is associated with the population of the entire hexagon.

2.4. Identification of GUPSs and Related Access Points

2.4.1. Identification of GUPSs

GUPSs are the end point against which accessibility is assessed and were identified starting from OSM and UA data.

- From OpenStreetMap (OSM), the green spaces described by the key "Leisure" and the tags "Park" and "Garden" were extracted. Starting from the ITO Map Green space access map (which indicates a range of green spaces, their availability for use, and their access status) and analyzing OSM tags under the "Leisure" key [39], these two tags were identified as the most suitable for the selection of green open public spaces in the Italian context. The "parks" classified as "protected areas" by the tag "boundary" were excluded, since the perimeter of the protected areas is linked to a form of government of the territory and do not necessarily identify an area open to the public (in fact, within them it is also possible to find privately owned agricultural areas).
- From Copernicus Urban Atlas Land Cover/Land Use 2018 (UA), the class "1.4—Artificial non-agricultural vegetated areas" polygons were considered, which includes the subclasses of "1.4.1—Green urban areas" and "1.4.2—Sport and leisure facilities". In detail, only polygons covered by artificial surfaces for less than 20% were selected

(evaluated using the ISPRA LCM).

For both data, only polygons larger than 0.5 ha were considered [23,25].

- The threshold for the percentage of consumed land on UA polygons was defined to exclude areas with a high presence of artificial surfaces, such as sports fields, stadiums, squares, theaters and auditoriums, cemeteries, riding stables, and sports centers.
- On the threshold on the minimum size of polygons, the IUCN rule 3-30-300 suggests considering green areas of at least one hectare; in this research, a threshold of 0.5 ha was defined for the entire study area to preserve many polygons smaller than one hectare (thus enhancing the information content of the input data) while maintaining areas large enough to guarantee the provision of ecosystem services, e.g., in terms of cooling capacity [45].

Among all the polygons with these characteristics, those that fell (entirely or partially) within the urban area of the analyzed MC were considered as green urban public spaces (GUPSs).

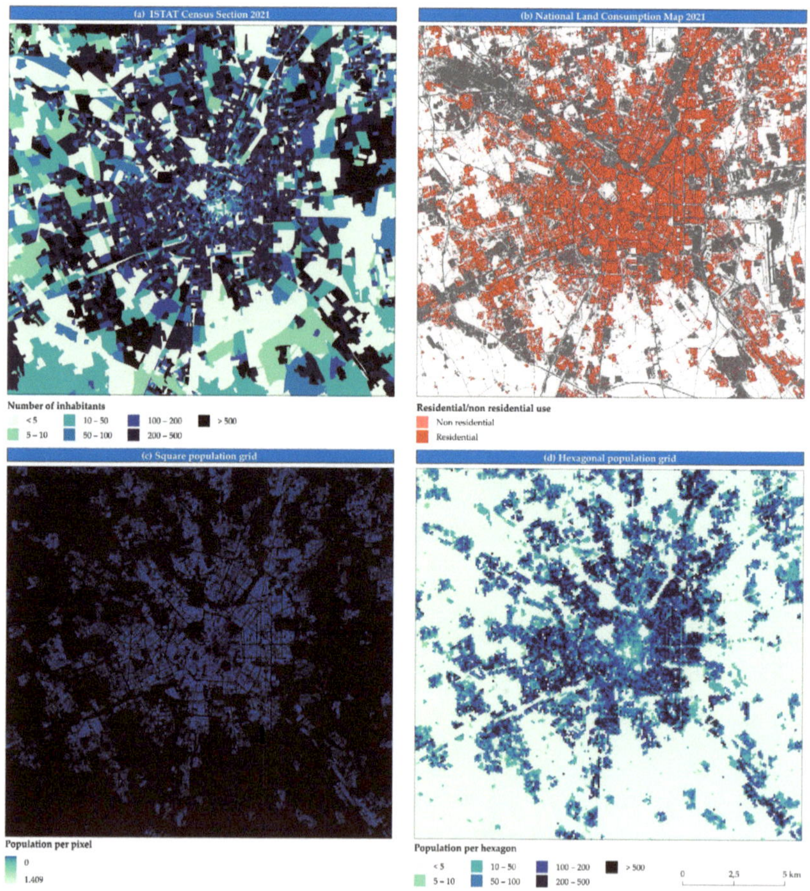

Figure 2. Operational phases for population spatialization: ISTAT census sections for 2021 (**a**), each themed with respect to the resident population. Residential and non-residential built-up areas of the ISPRA LCM (**b**); population spatialized on the built-up area (**c**); spatialized population aggregated with respect to the hexagonal grid, used for the accessibility assessment (**d**).

2.4.2. Identification of GUPS Access Points

The identification of access points to GUPSs requires the availability of detailed information on the edges of the considered green areas, in terms of location of the entrances and their possible limitations.

OSM maps punctual information, called nodes, referring to the accesses to buildings, infrastructures and GUPSs. Actually, OSM GUPS access points were identified by first where the following two conditions are verified:

- OSM nodes classified by the "barrier" key and the "gate[2]" and "entrance[3]" tag, located inside or at the perimeters of the selected green areas can be found;
- Intersection points between OSM streets and OSM GUPS polygons can be derived.

Where neither of the two conditions are met, an access point was defined every 100 m along the perimeter of the GUPSs, according to the indications provided by the SDG indicator 11.7.1 metadata [25].

For the UA GUPSs, only access every 100 m along the perimeter was considered.

2.5. Creation of the Routing Network

To perform a network analysis capable of automatically identifying the best walking route from a starting point to one or more GUPSs in the urban area of interest, the OSM database relating to the road network was used.

Raw OSM data in OSM format were acquired from the Geofabrik download server and transformed into a network dataset using the open-source OSM Editor add-on developed for ESRI ArcMap v.10.4 users, preparing a network configuration xml file. The rules established in the configuration file refer exclusively to the use of walkable road infrastructures, considering the key "highway" and the following tags: "primary", "primary_link", "secondary", "secondary_link", "tertiary", "tertiary_link", "unclassified", "residential", "service", "cycleway", "cyclestreet", "bicycle_road", "footway", "pedestrian", "path", "sidewalk", "steps", "track", "construction", "escape", "bridleway", "living_street". Barriers to movement along the roads were not considered, while, when origins and destinations were not located along the infrastructure network, the distance of each point from the closest network element was considered.

The OD cost Matrix tool implemented in ESRI ArcGis Pro software version 3.2.2 was used to evaluate the routes, considering the distance in meters as a cost attribute and searching for each starting point (the centroids of the hexagons) the closest GUPSs within a maximum distance of 10 km.

The cost matrix provides the "lines" that connect starting points and end points; for each starting point, the shortest path to a GUPS was considered.

2.6. Evaluation of Accessibility to GUPSs

In line with the indications of target 11.7 and with the 3-30-300 rule, this study allowed for evaluation of the total population belonging to the starting point of each hexagon that has access to a GUPS within a certain walking distance. The calculation was carried out both with reference to a distance of 300 m, referring to the IUCN rule 3-30-300, and to a distance of 400 m, defined by the metadata for the calculation of the SDG indicator 11.7.1:

$$GUPS\ Accessibility = \frac{Pop_{wa}}{Pop_{Tua}} * 100$$

where

Pop_{wa} = Total population with access to GUPS (within 300 m or 400 m)
Pop_{Tua} = Total population in urban area

2.7. Evaluation of Supporting Indicators

To evaluate the availability of GUPSs on the total vegetated surface of the urban area, an auxiliary indicator was calculated relating to the urban green spaces (UGSs) surface area per capita (1), which was compared with the GUPSs surface area per capita (2), calculated with respect to both OSM and UA. The UGSs surface area was evaluated starting from the Copernicus CLC+ Backbone 2021 data [46].

$$UGS\ Per\ Capita = \frac{UGS_{Sua}}{Pop_{Tua}} \quad (1)$$

$UGS_{Sua} = Total\ UGS\ Surfaces\ in\ urban\ area\ \left[\mathrm{m}^2\right]\ with\ respect\ to\ UA\ and\ OSM$
$Pop_{Tua} = Total\ population\ in\ urban\ area$

$$GUPS\ Per\ Capita = \frac{GUPS_{Sua}}{Pop_{Tua}} \quad (2)$$

$GUPS_{Sua} = Total\ GUPS\ Surfaces\ in\ urban\ area\ \left[\mathrm{m}^2\right]$
$Pop_{Tua} = Total\ population\ in\ urban\ area$

3. Results

3.1. Identification of Starting Points and Population Spatialization

The spatialization of the population for only the portion of the territory of the 14 MCs classified as urban area by the DEGURBA methodology is shown in Figure 3.

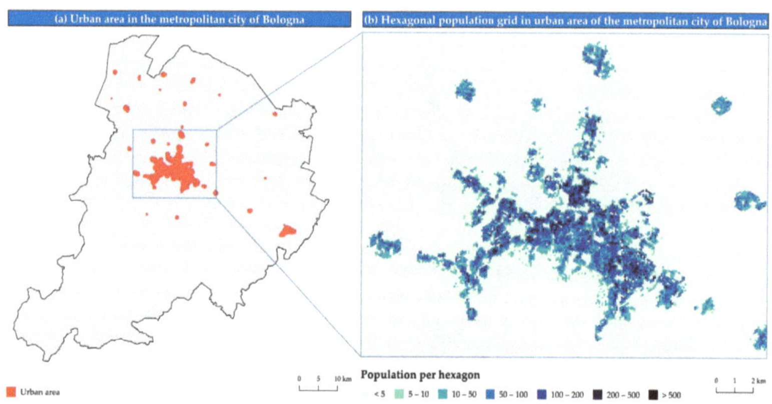

Figure 3. Urban area (**a**) and spatialized population in the urban area (**b**) with reference to the MC of Bologna.

The resident population throughout the metropolitan territory and the urban area of MC are synthesized in Table 2. Over three quarters of the population of the 14 MCs live in urban areas, with minimum values in Venice (44%), Messina (47.1%), and Reggio Calabria (48.3%), while Milan and Naples exceed 92%. The cities that exceed the national average are, in order, Bari (84%), Cagliari (83%), Rome, and Genoa (80%).

Table 2. Total population (number of inhabitants) and population in urban area (number of inhabitants and percentage on the total population) of the 14 Italian MCs.

MC	Total Population	Population in Urban Area	
	Inhabitants	Inhabitants	% of Total Pop.
Tourin	2,208,370	1,556,719	70.5
Genoa	817,402	654,077	80.0
Milan	3,214,630	2,978,997	92.7
Venice	836,916	368,138	44.0
Bologna	1,010,812	562,695	55.7

Table 2. *Cont.*

MC	Total Population	Population in Urban Area	
	Inhabitants	**Inhabitants**	**% of Total Pop.**
Florence	987,260	626,038	63.4
Rome	4,216,874	3,384,763	80.3
Naples	2,931,250	2,710,728	92.5
Bari	1,226,784	1,030,954	84.0
Reggio Calabria	522,127	252,003	48.3
Palermo	1,208,991	868,424	71.8
Messina	603,229	284,205	47.1
Catania	1,077,515	803,756	74.6
Cagliari	421,688	350,559	83.1
Total	21,283,848	16,432,056	77.2

3.2. *Identification of GUPSs and Related Access Points*

3.2.1. Identification of GUPSs

The identification of GUPSs was made with respect to both OSM and UA green spaces (Figure 4).

Figure 4. Example of the municipality of Rome for the procedure of the selection of GUPSs, compared to the OSM data (**a**) and UA (**b**). For OSM, only the polygons classified with the tags "Garden" and "Park" larger than half a hectare were selected. For UA, the polygons classified as "1.4—Artificial non-agricultural vegetated areas" larger than half a hectare and with less than 20% of consumed land were considered.

For OSM, starting from the almost 24,000 polygons classified with the tags "Garden" and "Park", the 5333 larger than half a hectare were selected, for a total surface area of 33,610 hectares.

Regarding UA, from the 13,421 polygons classified as "1.4—Artificial non-agricultural vegetated areas" on the whole territory of the 14 MC, the 4526 larger than half a hectare and with consumed land lower than 20% were selected, for a total of 15,378 hectares. The results for each MC are reported in Appendix A.

The threshold for the maximum percentage of consumed land allowed us to exclude areas not assimilable to GUPSs which belong to the UA class "1.4—Artificial non-agricultural vegetated areas", e.g., cemeteries, sports fields, auditoriums, green areas affected by land take, and other highly artificialized urban public spaces, such as some squares (Figure 5).

Figure 5. Examples of some types of areas mapped by UA as "1.4—Artificial non-agricultural vegetated areas" excluded from the accessibility assessment thanks to the filter on the maximum percentage of consumed land, such as cemeteries (**a**), sports fields (**b**), highly artificialized squares (**c**), and areas affected by land consumption (**d**).

The dimensional threshold of 0.5 ha allowed us to exclude very small areas from both datasets and was particularly effective on OSM, which maps, for example, small polygons of roads pertaining to vegetation, roundabouts, and flower beds because of its variable minimum mappable unit.

The thresholds reduced the differences between the two datasets in terms of number of mapped areas (the remaining differences concern the different polygon shapes), since at the beginning OSM maps many more small-sized areas than UA and UA includes spaces that cannot be assimilated to GUPSs (which can be eliminated from the start by OSM through an appropriate selection of tags).

Considering the green open public spaces larger than half a hectare (for OSM) and with less than 20% CL (for UA) that fall within the urban area, the GUPSs were obtained (Table 3).

Table 3. Portion of green open public spaces larger than half a hectare (for OSM) and with less than 20% of CL (for UA) that fall within the urban area of the 14 MCs in terms of surface and percentage of MC total surface. UA data are not available for Messina.

	GUPSs			
	OSM		UA	
Città Metropolitana	**ha**	**% on Urban Area**	**ha**	**% on Urban Area**
Tourin	1603	6.3	1117	4.4
Genoa	214	2.0	135	1.2
Milan	4208	8.5	3050	6.2
Venice	300	3.4	670	7.6
Bologna	1430	15.4	678	7.3
Florence	776	6.5	510	4.3
Rome	5957	9.6	3222	5.2
Naples	911	1.8	864	1.7
Bari	188	1.3	124	0.8
Reggio Calabria	22	0.4	15	0.3
Palermo	353	2.5	212	1.5
Messina	45	0.7	-	-

Table 3. *Cont.*

	GUPSs			
	OSM		UA	
Città Metropolitana	**ha**	**% on Urban Area**	**ha**	**% on Urban Area**
Catania	88	0.6	64	0.4
Cagliari	264	4.2	153	2.5
Total	16,359	5.6	10,814	3.7

GUPSs larger than half a hectare constitute half of the OSM green open public spaces and approximately two-thirds of those by UA, occupying just under 6% of the urban area for OSM and just under 4% for UA, with a maximum in Bologna in both cases (15.4% for OSM and 7.3% for UA). Accessibility has been assessed in relation to these areas, according to SDG11.7.1 metadata indications.

3.2.2. Identification of GUPS Access Points

The identification of the access points (Figure 6) carried out considering the mapped nodes, the intersection between GUPSs and roads and, in their absence, a point every 100 m, gave the results in Table 4.

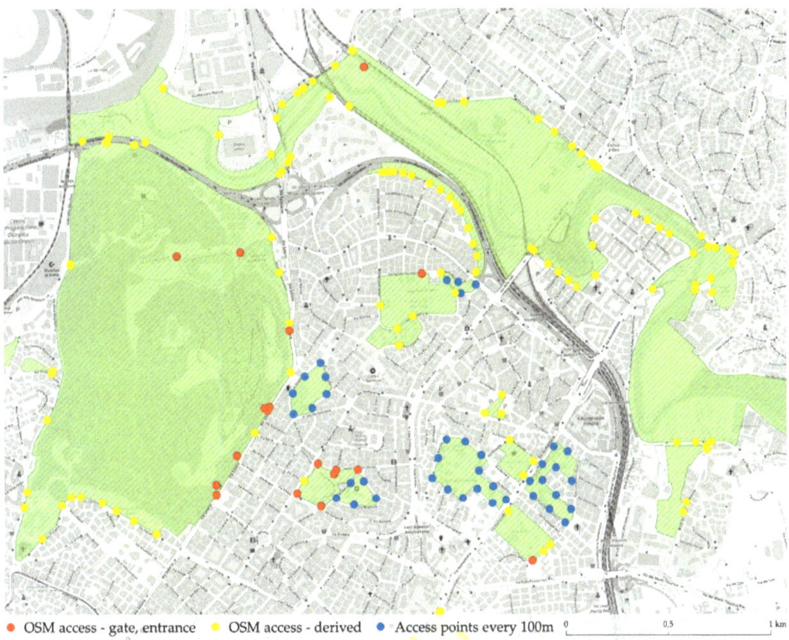

Figure 6. GUPS access points. In red are the access points mapped in OSM, in yellow those obtained by intersection between GUPSs and road network, in blue the points every 100 m along the perimeter. The latter were considered in the absence of the first two for OSM and for all UA GUPSs.

For OSM GUPSs, 35,646 access points were identified, two thirds of which derived from the intersection between GUPS polygons and the road network and less than one in 10 already mapped by OSM as "gate/entrance". For the UA database, only accesses every 100 m along the perimeter were considered.

Table 4. OSM GUPS access points identified considering the mapped points, the intersection between GUPSs and roads, and, in their absence, a point every 100 m.

	Number of Accesses to GUPSs			
MC	**Gate/Entrance**	**Derivati**	**ogni 100 m**	**Total**
Tourin	279	2.718	794	3.791
Genoa	130	569	49	748
Milan	942	8.221	845	10.008
Venice	77	1.263	437	1.777
Bologna	192	4.818	660	5.670
Florence	113	2.147	549	2.809
Rome	729	758	4.429	5.916
Naples	139	1.036	857	2.032
Bari	58	431	258	747
Reggio Calabria	9	129	51	189
Palermo	36	233	139	408
Messina	20	68	81	169
Catania	61	200	201	462
Cagliari	66	674	180	920
Total	2.851	23.265	9.530	35.646

3.3. Evaluation of Accessibility to GUPSs

The evaluation of accessibility to GUPSs starting from the centers of the hexagons in the urban areas produced the results in Figure 7 and Appendix B.

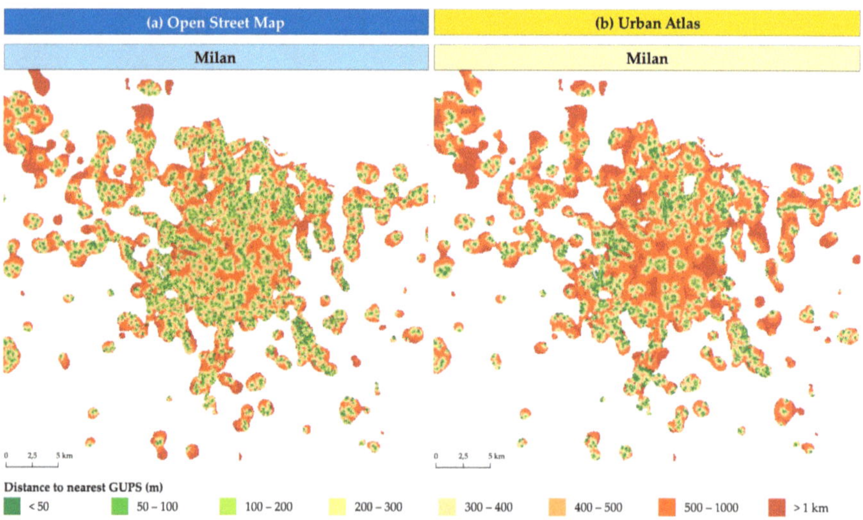

Figure 7. Example of the result of the calculation of accessibility to GUPSs compared to OSM (**a**) and UA (**b**) on the city of Milan. The accessibility mapping for all 14 MCs is reported in Appendix B.

By evaluating the resident population of the hexagons that have access to a GUPS within 300 m (considering the IUCN threshold) and within 400 m (considering the SDG 11.7.1 indicator metadata threshold), the results of Table 5 were obtained.

Considering OSM, approximately 28% of the population living in urban areas has access to a GUPS within 300 m, with values below the national average in Genoa and in the MCs of the south and the islands; the value rises to 40% considering a radius of 400 m. Milan and Bologna guarantee access to an OSM GUPS within 300 m on foot to more than half of the population; to these Turin and Florence are added when considering a maximum distance of 400 m.

Table 5. Number of inhabitants with access to a GUPS within 300 or 400 m from the starting point, with reference to the OSM and UA databases.

	Population That Has Access to a GUPS							
	OSM				UA			
	Within 300 m		Within 400 m		Within 300 m		Within 400 m	
MC	Number of Inhab.	% *	Number of Inhab.	% *	Number of Inhab.	% *	Number of Inhab.	% *
Tourin	590,640	37.9	839,799	53.9	279,268	17.9	420,090	27
Genoa	124,725	19.1	192,809	29.5	103,234	15.8	159,562	24.4
Milan	1,484,951	49.8	1,978,162	66.4	891,384	29.9	1,283,206	43.1
Venice	104,030	28.3	155,779	42.3	75,664	20.6	114,489	31.1
Bologna	318,255	56.6	411,987	73.2	217,860	38.7	289,490	51.4
Florence	247,748	39.6	347,600	55.5	163,729	26.2	239,398	38.2
Rome	994,934	29.4	1,427,518	42.2	738,062	21.8	1,031,376	30.5
Naples	333,286	12.3	514,773	19	469,901	17.3	677,108	25
Bari	193,773	18.8	308,033	29.9	56,421	5.5	86,897	8.4
Reggio Calabria	15,958	6.3	25,114	10	6830	2.7	10,101	4
Palermo	85,910	9.9	133,860	15.4	108,735	12.5	166,965	19.2
Messina	19,836	7	32,444	11.4	-	-	-	-
Catania	58,901	7.3	99,027	12.3	33,929	4.2	54,310	6.8
Cagliari	67,043	19.1	107,180	30.6	77,038	22	117,162	33.4
Total	4,639,989	28.2	6,574,084	40	3,222,055	19.6	4,650,153	28.3

(*) Percentage of population that has access, compared to the total population of the urban area.

The values obtained by considering the UA GUPSs are about 10 percentage points lower than OSM ones, for both 300 m and 400 m radius. Less than a fifth of the population has access to a UA GUPS within 300 m, and even in this case Genoa and the metropolitan cities of the south and the islands (except for Cagliari) show values below the national average. The value reaches just under a third considering a maximum distance of 400 m and only the metropolitan city of Bologna ensures access to a GUPS for more than half of the population.

3.4. Evaluation of GUPSs per Capita

The CLC Plus Backbone total green urban area per capita ranges from 53.2 m^2/inhabitant (Turin) to 91.9 m^2/inhabitant (Rome), with a national average value of just under 75 m^2/inhabitant (Table 6).

Table 6. Total green areas per capita (from CLC Plus Backbone) and public green areas per capita (from OSM and UA) in urban areas.

MC	Urban Green Spaces	UGS per Capita	OSM GUPSs per Capita	UA GUPSs per Capita
	[ha]	[m^2/ab]	[m^2/ab]	[m^2/ab]
Tourin	8288	53.2	10.3	7.2
Genoa	5185	79.3	3.3	2.1
Milan	18,534	62.2	14.1	10.2
Venice	3035	82.4	8.1	18.2
Bologna	3862	68.6	25.4	12.0
Florence	5108	81.6	12.4	8.1
Rome	31,110	91.9	17.6	9.5
Naples	21,808	80.4	3.4	3.2

Table 6. *Cont.*

MC	Urban Green Spaces	UGS per Capita	OSM GUPSs per Capita	UA GUPSs per Capita
	[ha]	[m²/ab]	[m²/ab]	[m²/ab]
Bari	5117	49.6	1.8	1.2
Reggio di Calabria	2091	83.0	0.9	0.6
Palermo	5882	67.7	4.1	2.4
Messina	2709	95.3	1.6	-
Catania	6916	86.0	1.1	0.8
Cagliari	2441	69.6	7.5	4.4
Total	122,087	74.3	10.0	6.6

Regarding the availability of green urban spaces for public use, OSM shows values 7 times lower (10.0 m^2/inhabitant), with UA over 10 times lower (6.6 m^2/inhabitant) compared to the total urban greenery.

Bologna has the highest value according to OSM data (25.4 m^2/inhabitant) and it is also the MC with the second highest value for UA data (12.0 m^2/inhabitant), while for UA, Venice is the city with the highest value (18.2 m^2/inhabitant). Only a third of the MCs offer a GUPS per capita value in line with the limit of 9 m^2/inhabitant set as the national urban planning standard, with a minimum in Reggio Calabria and Catania for both OSM and UA.

4. Discussion

This study presents for the first time a large-scale assessment of GUPS accessibility in Italy based on the official UN methodology for SDG indicator 11.7.1. The evaluation is strictly dependent on the availability of adequate input data on starting points, road networks (accessible on foot), green urban areas (to be accessed), and the related access points.

With reference to the starting points, the regular hexagonal mesh grid allows for good coverage of the study area and is easily updatable, replicable, and interpretable. It also offers a more homogeneous representation of the starting points and a more realistic distribution of the population in the urban area, compared to other studies that consider, for example, the centroids of the census section polygons [36]. A more detailed representation would require information on the location of individual residential buildings and their access points, which are however often not easily available or regularly updated. In Italy, data about residential buildings are collected in cadastral databases or in the "national synthesis database" (DBSN), both of which are not always updated and with differences from region to region. As an alternative, the OSM database can be used, but by considering the inhomogeneity of the update of this dataset compared with other official data such as the LCM.

The OSM road network on the 14 MCs is satisfactory except for a few inhomogeneities and omissions that affect the calculation of accessibility in the corresponding hexagons. Actually, the participatory approach underlying the mapping activity in OSM allows for correcting errors in the dataset, which are then validated by the OSM volunteer community [47].

The identification of GUPSs and the related access points are the most critical aspects, both from a geometric and semantic point of view, both for UA and OSM.

From a semantic point of view, the identification of GUPSs requires a good knowledge of the study area in terms of LC and LU. LC information allows for the identification of the vegetated areas and is easily obtainable through remote sensing or existing LC data, while the LU information is necessary to discriminate public and private uses, e.g., internal courtyards of buildings, private gardens, and the green components of street furniture.

The two input datasets show different results in terms of the number and extension of mapped GUPSs; the UA database has temporal, spatial, and semantic homogeneity, ensuring a uniform and standard mapping of green urban areas on the main FUAs, but also includes areas with nothing in common with GUPSs (cemeteries, auditoriums, highly artificial squares, theaters, sports fields, villas, etc.) or previously natural areas where land take has occurred since 2018 (the reference year of the most recent version of the data) and that need to be filtered. For this purpose, the threshold for the percentage of consumed land was introduced. The threshold, defined in this phase for the entire study area, can lead in some cases to an underestimation of accessibility. However, this result, although conservative, can be improved by considering different thresholds in relation to the specificities of the territorial contexts or by introducing ancillary data, such as the census provided by municipalities (which, however, are often not available) according to law 10/2013 "Regulations for the development of urban green spaces" (which was the starting point for valorizing the role of green spaces in cities not only from an environmental point of view but also from a social and cultural one). This activity will be one of the main future developments of this research. OSM is a very powerful tool to support studies that require LU information, especially at a detailed scale and for large territories, thanks to its great versatility, the valorization of information from those who know the territory, and the possibility of implementing even further new information. It also offers higher margins of improvement and refinement than UA in the definition of specific semantics for the assessment of accessibility, thanks to the open approach.

The OSM structure offers many possibilities for representing objects through keys and tags, but currently the description of GUPSs is often too incomplete to be fully used for accessibility assessment. Additional information on the usability of green urban areas should be implemented, e.g., to distinguish open areas with free access from private areas or with restricted, occasional, or paid access. In this sense, the compilation of the "access" and "fee" tags, as already foreseen in the OSM mapping system [39], should be encouraged. In addition, OSM can support the evaluation of cultural ecosystem services provided by GUPSs, since the mappers' in-depth knowledge of the territory provides information on the usability and potential of an area such as esthetic, spiritual, educational, and recreational values. These aspects are linked to the experiential sphere and cannot be deduced through remote sensing or other similar monitoring tools. An example is the activity carried out in the municipality of Rome to evaluate the appreciation of public green spaces by the population through information extracted from social networks [37]. Actually, these considerations would also be useful for the description of the road network, to integrate the mapping criteria by introducing attributes related to the usability of the route, e.g., in terms of perceived sense of safety, presence of obstacles or architectural barriers or slope, an example is the Bike Improver Day initiative in Trento, to collect information on the cycling network in terms of user satisfaction and perception of risk and/or degradation of the infrastructure.

However, the crowdsourcing population of the OSM database introduces critical issues related to the homogeneity of the mapping, from a thematic and geometric point of view, to be considered when reading the data. The variable minimum mapping unit and the GUPS definitions affect the shape of the polygons and the attribution of tags, which depend greatly on the sensitivity of the operator, leading to results that are often inconsistent or misleading.

Figure 8 shows some of the main critical issues encountered in the analysis of the OSM "park" and "garden" tags, which affect the accessibility assessment. In some cases, roundabouts or avenues are mapped as "parks" (Figure 8a,b) that have trees but do not have characteristics that would make them GUPSs. In other cases, both areas open to the public and private gardens or spaces with public access restrictions are mapped with the same "tag" (Figure 8c,d). A further critical issue is the lack of homogeneity in the mapping of some green areas, which are partially or incompletely delimited.

The choice of selecting polygons with a minimum size of 0.5 hectares allows for the exclusion of areas from the calculation that generally are not GUPSs (especially roundabouts, green furnishing, or green courts); this threshold had a greater effect on OSM, while it had less impact on UA, which has a MMU similar to the threshold.

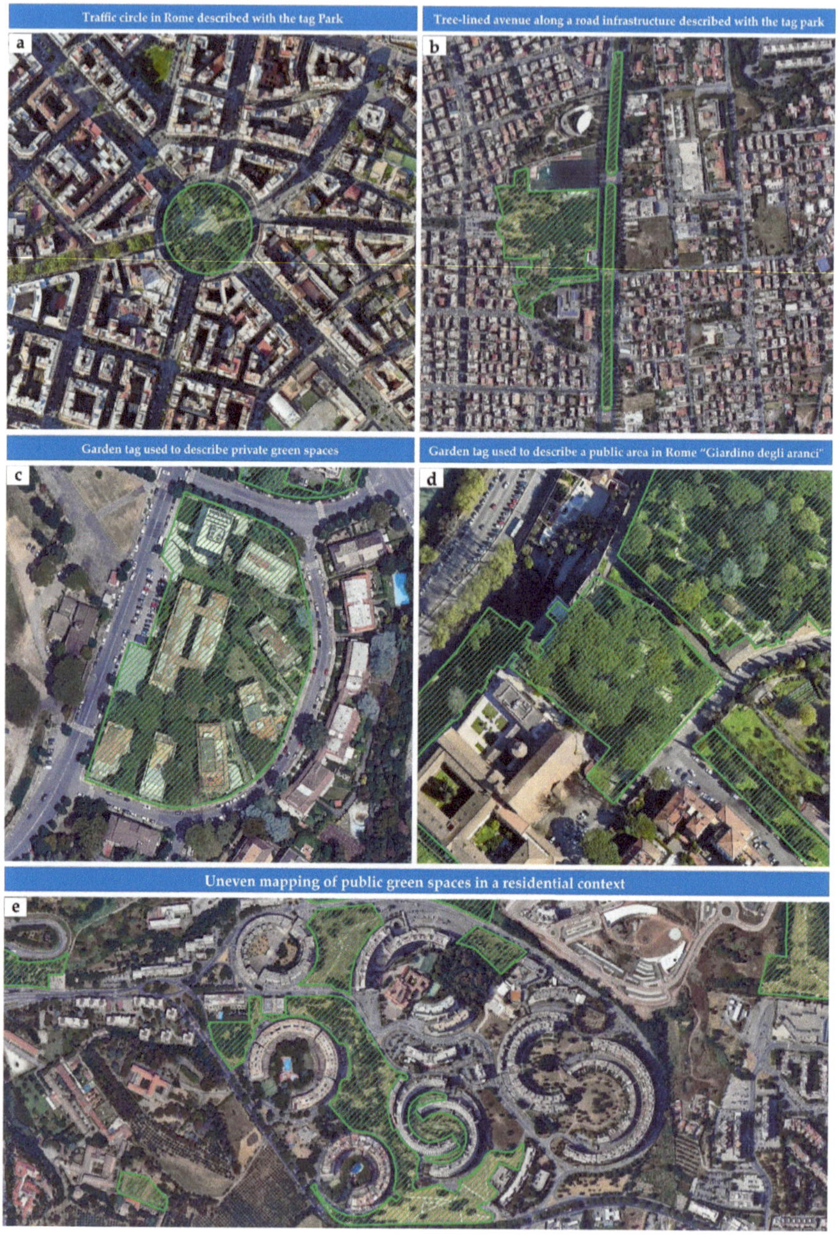

Figure 8. Examples of mapping in the OSM data (identified in the figures with green polygons) that influence the accessibility assessment.

Overall, both datasets show limitations and potential for GUPS identification, although OSM appears more promising, especially in identifying access points and in evaluating the actual possibility of access. The homogeneity and standardization of UA make it appear as a perfect dataset for the assessment of accessibility; however, the rigid definition of green urban areas provided by the class "1.4—Artificial non-agricultural vegetated areas" (which forces the identification of a strategy and/or ancillary data to filter non-GUPS areas, such as cemeteries and sports fields), the reduced update frequency (which causes losing track of areas that have changed in the meantime, for example, due to land take), the limited spatial coverage (available only for European FUAs with at least 50,000 inhabitants), and the absence of information on accesses (which forces the definition of fictitious accesses) strongly condition the identification of GUPSs. OSM, despite having greater heterogeneity in the quality of the mapping, offers a much wider spatial coverage and greater possibilities of interaction with the data thanks to the open approach, with the opportunity to introduce tailor-made mapping criteria for the needs of accessibility assessment.

The information on access points is partial on OSM and not provided by UA. The introduction of derived points has allowed us to reduce the number of polygons without accesses; however, the mapping of these entities should be encouraged, and are essential to evaluate the accessibility of a GUPS.

In the case of UA, the introduction of fictitious accesses every 100 m may have led to an overestimation of accessibility, especially in green areas with a limited number of accesses (for example where the perimeters of green areas have fences or surrounding walls). On OSM, the overestimation of accessibility is more limited and may only affect polygons for which access information is not natively provided or where such information could not be derived by cross-referencing the GUPSs with the road network.

The availability of adequately spatialized and updated demographic data is a further element that greatly affects the evaluation of accessibility, since it determines the distribution of the population around the GUPSs. In Italy, the ISTAT census sections data are a valid support for the spatialization of the population, but the update frequency (10 years) conditions the monitoring. The municipal population data are updated annually and are spatializable (on the residential buildings of the LCM), even if they do not allow for consideration the uneven distribution of the population within the municipality. To improve the distinction of residential buildings from the rest of the construction, ISPRA has started a classification activity of the land use of consumed land [48,49], allowing for the methodology to free itself from input data coming from heterogeneous sources and valorizing the knowledge of the territory of the regional agencies (ARPA).

The tools analyzed in this research are an important starting point for conducting assessments related to the accessibility of open public spaces in urban areas, which is essential, especially considering the obtained results. In this sense, developing monitoring tools to observe territorial dynamics in detail is an essential prerequisite for defining effective actions and initiatives. Furthermore, tools of this type are essential to respond to institutional requests in a technically adequate way, first and foremost the monitoring of accessibility to public urban greenery envisaged by the UN and the 2030 Agenda for Sustainability.

Analyzing the results, in the urban areas of the 14 MCs, less than one person in three has access to a GUPS within a 300 m walking distance (less than one in five for the UA data), and in 9 of the 14 metropolitan cities, the availability of GUPSs per capita is less than the 9 m^2/inhabitant required by the Urban Standards Law [50]; in fact, even if the total greenery in the urban area is an order of magnitude higher than the GUPSs, this also includes non-usable areas, such as roundabouts, green areas pertaining to infrastructures or street furniture, private agricultural areas, and green areas of private gardens. Moreover, significant differences are noted between cities in the north and center compared to those in the south and on the islands, which show the lowest values of availability and accessibility to GUPSs regardless of the input data (OSM or UA).

The results obtained on the availability of GUPSs are consistent with what emerges from the European plan NextGenerationEU, which allocates more than half of the funds foreseen for Italy for new forestation interventions to the MC of the south and the islands, of which implementation in urban areas can increase the presence of configured public areas (similar to GUPSs) and which can contribute to improving accessibility and the provision of multiple other recreational and regulatory ecosystem services in urban context.

In this sense, this research aims to have practical value, and to offer a series of useful tools to support policy makers in achieving the legal objectives imposed by the UN SDGs or in managing the financing of forestry projects envisaged by the PNRR, e.g., in identifying priority areas for new interventions, but also to support the development of restoration plans envisaged by NRL. For this purpose, methodologies officially adopted by the main regulatory instruments that deal with urban areas and urban greenery (NRL, the SDGs, the PNRR) have been considered, operating on two fronts: firstly, by providing an accessibility mapping in line with the legislation, directly usable for the identification of critical areas or priority areas of intervention; then, analyzing and evaluating the limits of the data currently available to achieve a representation closer to reality.

These limitations have mainly affected the mapping of GUPSs and their access points while the mapping of the road network is satisfactory, and the new population data for census sections offer great added value to producing detailed spatialized demographic data for Italy, useful for the representation of the urban area in line with the legal indications.

The reflections reported in this work are, however, perfectly valid in all areas covered by OSM and UA. This research concerned the urban area of the 14 Italian MCs, but the evaluation can be extended to other areas of the Italian territory or to the entire national territory, thanks to the coverage of the population grid (spatialized by the ISPRA starting from ISTAT 2021 data) and the OSM data (relating to the representation of the road network and green areas). The global coverage of OSM data makes it possible to extend the considerations presented in this work relating to the road network, green areas, and access points to even outside of the European territory, while the spatialization of the population (and, consequently, the representation of the urban area according to DEGURBA) require the availability of reliable and updated demographic and residential building data. In this sense, the CLMS high-resolution layer data allow for the exact replication of the spatialization of the population on any area of the European territory for which population data are available. Alternatively, for non-European areas, it is possible to use the world population grid or the Eurostat DEGURBA layer, which offer global coverage, albeit with a spatial resolution of 1 km.

5. Conclusions

The availability of timely and detailed information on the presence, characteristics, and accessibility of GUPSs is a crucial element in defining policies and initiatives for the management of urban areas for the near future [51]; this is with a view to increasing the resilience of settlements to extreme climate events, improving the quality of life of citizens, and maximizing the provision of ecosystem services provided by GUPSs.

The crucial importance of the correct management of urban areas is also evident considering the recent approval of the NRL, which establishes a halt to land consumption in urban areas and requires EU member states to develop a monitoring plan for urban green areas based on Copernicus or national data. Urban greenery is also at the center of national initiatives, such as the VeBS project (the good use of green and blue spaces for the promotion of health and well-being) [52], supported by the Ministry of Health and coordinated by the Istituto Superiore di Sanità with the participation of the ISPRA, which promotes the use and benefits of green and blue infrastructures in urban areas and protected areas.

This research is a part of the ISPRA activities on land consumption monitoring and urbanization dynamics in Italy [42]. It explores, for the first time, the feasibility of a large-scale evaluation of accessibility to GUPSs, based on freely accessible data at European (UA) and global (OSM) scales and following the UN methodology for the calculation of SDG indicators. Actually, the accessibility to GUPSs shown in this work is one of the three components of the SDG indicator 11.7.1, together with access to gray and blue infrastructures. The complete calculation of the indicator is one of the main and upcoming developments of this research activity, which is, however, an important step in monitoring the sustainable development goals required by the 2030 agenda.

This study proposes a technical analysis on spatial data available for the entire European territory (UA) and at global level (OSM), considering their aptitude to identify GUPSs, their access points, and road infrastructure. It therefore constitutes an important starting point to support further studies of this nature on different territorial realities and to support the improvement of currently available crowdsourcing products, such as OSM. The workflow is scalable and replicable on all the main national and international urban contexts, and the intermediate products, such as the population grid or the perimeter of urban areas [15,42], are freely accessible and usable for conducting further studies.

For example, the population grid finds application in the improvement of the methodology developed by Deda Next within the Horizon Europe project "USAGE", which works on datasets at different scales, from the local (city) to the national and European level. In addition to the introduction of orography in the calculations of walkability through digital terrain models (DTMs) from different sources and with different accuracies, the national population dataset is halfway between local-level applications (successfully carried out on the Municipality of Florence), and implementations on global-level data (on a GHSL basis for the population and OSM for addresses and buildings).

The tools and observations developed in this research can also support other types of analyses, such as the link between access to greenery, climate comfort, and risks associated with the UHI phenomenon or territorial inequalities between urban areas with different income capacities. The analysis of the representation and classification of GUPSs also provides interesting insights to support the monitoring of these areas even beyond the assessment of accessibility, for example, as an ancillary indicator for compliance with the regulatory obligations of the Nature Restoration Law on the protection of green areas in urban contexts.

Author Contributions: Conceptualization, A.C., P.D.F. and M.M. (Michele Munafò); methodology, A.C., P.D.F. and M.M. (Michele Munafò); software, A.C.; validation, A.C., P.C. and L.D.; formal analysis, A.C., P.D.F. and L.C.; investigation, A.C., P.C. and I.M.; resources, M.M. (Michele Munafò), M.M. (Marco Marchetti) and G.S.M.; data curation, A.C. and P.D.F.; writing—original draft preparation, A.C. and P.D.F.; writing—review and editing, A.C., P.D.F. and P.C.; visualization, A.C. and P.D.F.; supervision, M.M. (Michele Munafò), M.M. (Marco Marchetti) and G.S.M.; project administration, L.C., M.M. (Michele Munafò), M.M. (Marco Marchetti) and G.S.M.; funding acquisition, L.C. and M.M. (Michele Munafò). All authors have read and agreed to the published version of the manuscript.

Funding: This research was funded by the Italian Institute for Environmental Protection and Research (ISPRA) structural funds.

Data Availability Statement: The datasets presented in this article are not readily available because the data are part of an ongoing study. Requests to access the datasets should be directed to the corresponding author.

Conflicts of Interest: Author Piergiorgio Cipriano was employed by the company Deda Next Srl. The remaining authors declare that the research was conducted in the absence of any commercial or financial relationships that could be construed as a potential conflict of interest.

Appendix A Selection of Suitable Green Spaces

Table A1. For the territory of each of the 14 MCs, the following are shown: the areas mapped with the "Garden" and "Park" tags by OSM (in terms of total number of patches, number of patches with an area greater than half a hectare, and their total surface area) and areas classified as "1.4—Artificial non-agricultural vegetated areas" by UA (in terms of total number of patches, number of patches with consumed land <20%, and their total area). OSM = OpenStreetMap, UA = Urban Atlas, CL = consumed land.

MC	OSM Patches			UA Patches		
	Count (Total)	Count (>0.5 ha)	Surface (ha) (>0.5 ha)	Count (Total)	Count (CL < 20%, >0.5 ha)	Surface (ha) (CL < 20%, >0.5 ha)
Tourin	2990	624	2089	1808	594	1865
Genoa	636	108	235	442	129	248
Milan	4449	1380	4721	2217	877	3813
Venice	1238	336	854	981	256	670
Bologna	1913	1002	3623	787	411	1151
Florence	926	273	776	801	305	763
Rome	2924	765	5957	2847	1000	4892
Naples	3620	340	6694	1455	468	985
Bari	1114	151	281	624	104	200
Reggio Calabria	397	41	55	178	21	64
Palermo	436	74	425	551	137	302
Messina	497	41	96	-	-	-
Catania	2306	85	288	365	78	111
Cagliari	513	113	7516	365	146	314
Total	23,959	5333	33,610	13,421	4526	15,378

Appendix B Accessibility Maps to GUPSs in MC According to OSM and UA

From Figures A1–A4. The results of the accessibility evaluation with respect to the GUPSs mapped by OSM and UA are reported for the remaining 10 MCs not shown in Figure 7.

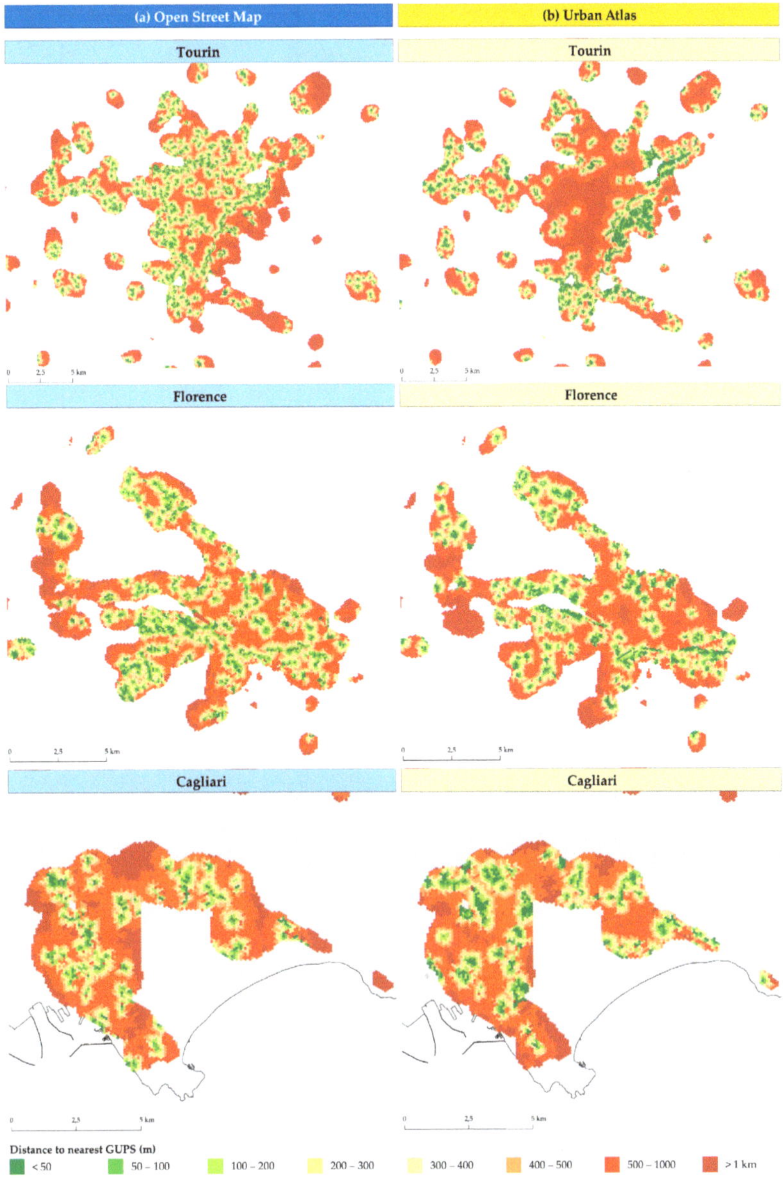

Figure A1. Accessibility to GUPSs compared to OSM (**a**) and UA (**b**) in the cities of Tourin, Florence, and Cagliari.

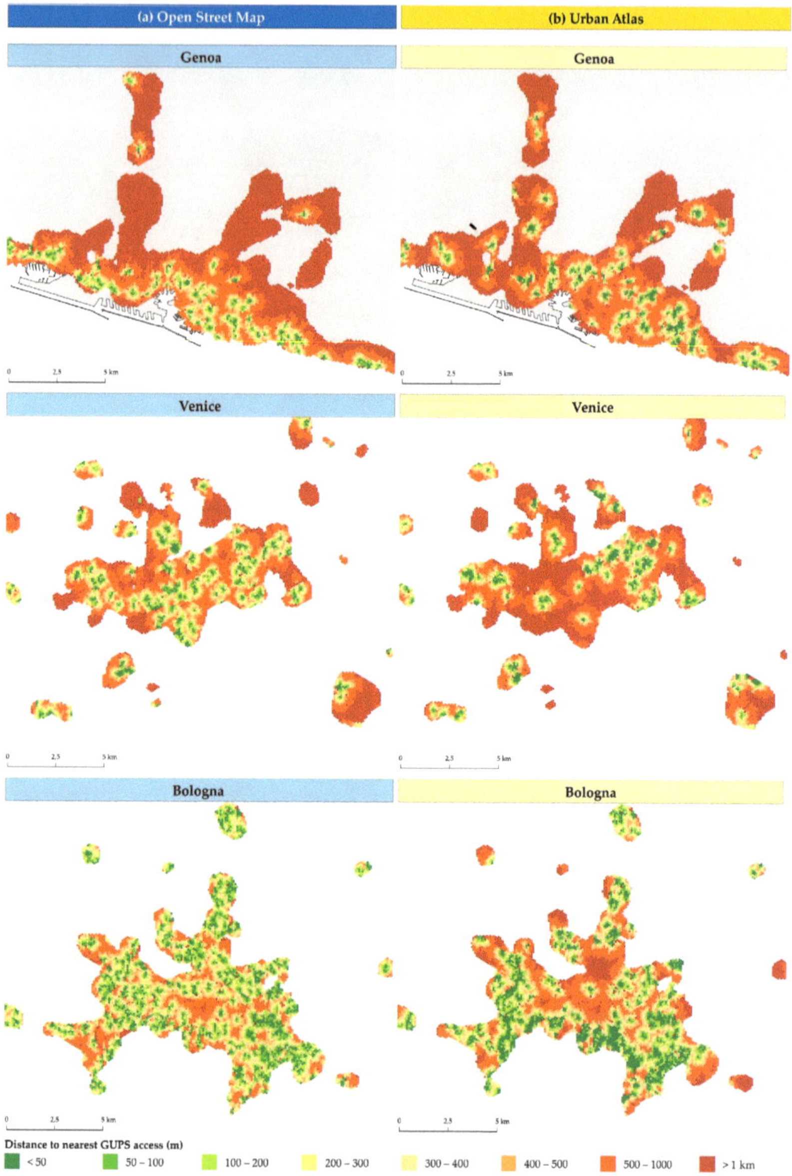

Figure A2. Accessibility to GUPSs compared to OSM (**a**) and UA (**b**) in the cities of Genoa, Venice, and Bologna.

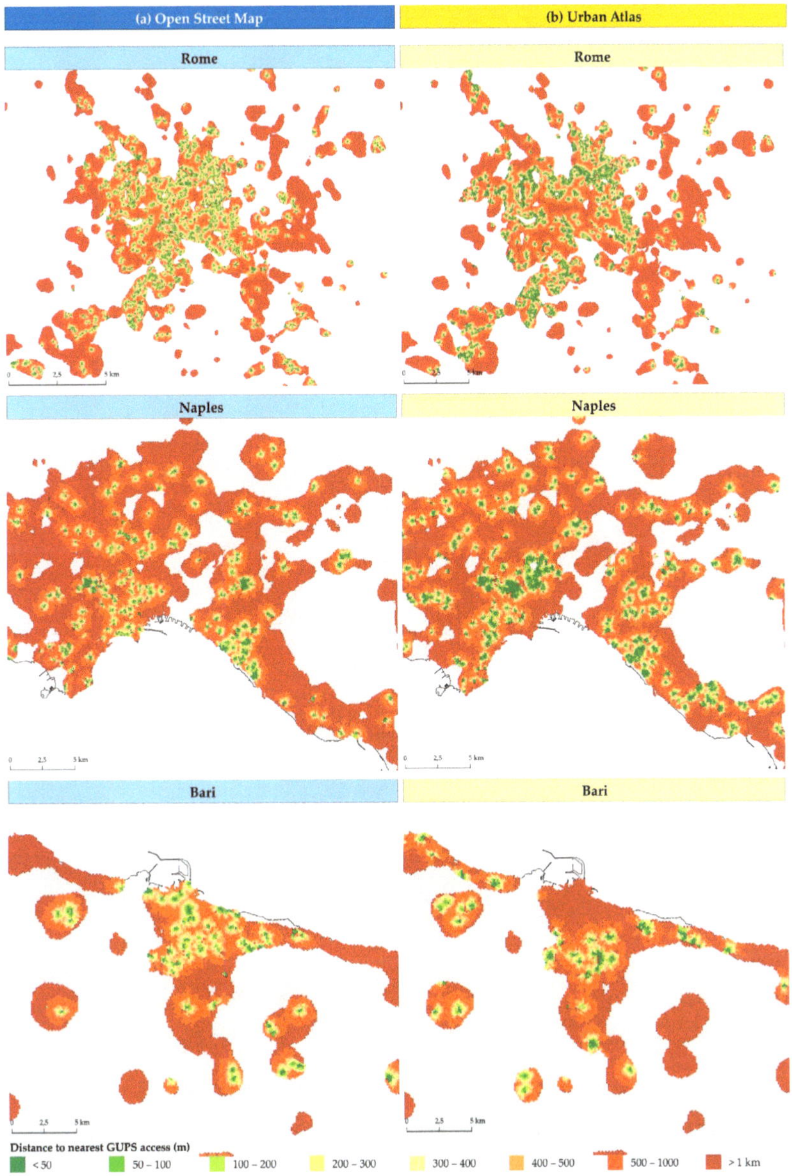

Figure A3. Accessibility to GUPSs compared to OSM (**a**) and UA (**b**) in the cities of Rome, Naples, and Bari.

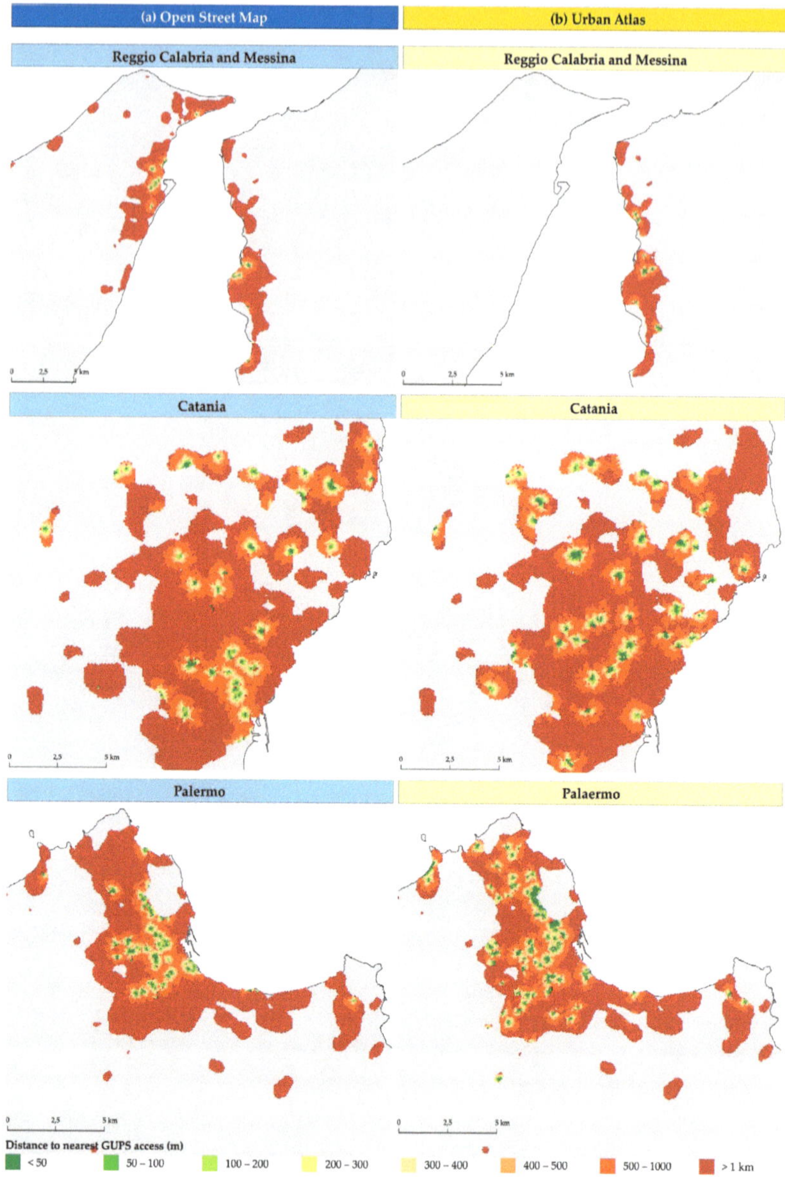

Figure A4. Accessibility to GUPSs compared to OSM (**a**) and UA (**b**) in the cities of Reggio Calabria, Messina, Catania, and Palermo.

Notes

1 The fictitious sections are introduced to allocate people without an official address, such as homeless people registered in the registry and allocated to a conventional address established by the municipality, individuals registered in the registry at associations or reception facilities, or residents in municipalities affected by seismic events.

2 The "Gate" tag indicates the presence of physical barriers such as doors or gates.

3 The "Entrance" tag is used to describe the point at which it is possible to enter a building or an enclosed area, in the absence of a physical barrier.

References

1. Department of Economic and Social Affairs, United Nations. *Population Division World Urbanization Prospects: The 2018 Revision*; ST/ESA/SER.A/420; United Nations: New York, NY, USA, 2018.
2. Yeager, R.A.; Smith, T.R.; Bhatnagar, A. Green Environments and Cardiovascular Health. *Trends Cardiovasc. Med.* **2020**, *30*, 241–246. [CrossRef]
3. Wong, N.H.; Tan, C.L.; Kolokotsa, D.D.; Takebayashi, H. Greenery as a Mitigation and Adaptation Strategy to Urban Heat. *Nat. Rev. Earth Environ* **2021**, *2*, 166–181. [CrossRef]
4. Alkama, R.; Forzieri, G.; Duveiller, G.; Grassi, G.; Liang, S.; Cescatti, A. Vegetation-Based Climate Mitigation in a Warmer and Greener World. *Nat. Commun.* **2022**, *13*, 606. [CrossRef]
5. Cohen-Cline, H.; Turkheimer, E.; Duncan, G.E. Access to Green Space, Physical Activity and Mental Health: A Twin Study. *J. Epidemiol. Community Health (1978)* **2015**, *69*, 523–529. [CrossRef]
6. World Meteorological Organization (Ed.) *State of the Global Climate 2023*, 2024th ed.; Chair, Publications Board: Geneva, Switzerland, 2024; Volume 1347, ISBN 978-92-63-11347-4.
7. Bočkarjova, M.; Kačalová, A. Greening of European Cities: Social Benefits of Urban Nature for Urban Air Quality. *Eur. Stud. Rev. Eur. Law Econ. Politics* **2021**, *8*, 177–204. [CrossRef]
8. Piaggio, M. The Value of Public Urban Green Spaces: Measuring the Effects of Proximity to and Size of Urban Green Spaces on Housing Market Values in San José, Costa Rica. *Land Use Policy* **2021**, *109*, 105656. [CrossRef]
9. Vandecasteele, I.; Baranzelli, C.; Siragusa, A.; Aurambout, J.P.; Alberti, V.; Alonso Raposo, M.; Attardo, C.; Auteri, D.; Barranco, R.; Batista e Silva, F.; et al. *The Future of Cities—Opportunities, Challanges and the Way Forward*; EUR (Luxembourg. Online); Publications Office: Luxembourg, 2019.
10. Lanzani, A. L'Italia Degli Standard Urbanistici. Che Fare, Oggi? *TERRITORIO* **2020**, 77–83. [CrossRef]
11. Baioni, M.; Basso, S.; Caudo, G.; Franzese, A.; Marchigiani, E.; Munarin, S.; Renzoni, C.; Savoldi, P.; Tosi, M.C.; Vazzoler, N. *Diritti in Città: Gli Standard Urbanistici in Italia Dal 1968 a Oggi*; Donzelli: Roma, Italy, 2021; Volume 1, ISBN 8855220470.
12. Bottalico, G.; Brini, S.; Vitali, W. *Infrastrutture Verdi Urbane e Periurbane*; ASviS, Ed.; ASviS: Roma, Italy, 2022.
13. Atelli, M.; Blasi, C.; Boldini, G.; Cignini, B.; Cosenza, G.; Emiliani, V.; Marchetti, M.; Maria Maggiore, A.; Pericoli, T.; Ricciardi, A.; et al. Strategia Nazionale Del Verde Urbano. Foreste Resilienti Ed Eterogenee per La Salute e Il Benessere Dei Cittadini. 2018. Available online: https://www.mase.gov.it/sites/default/files/archivio/allegati/comitato%20verde%20pubblico/strategia_verde_urbano.pdf (accessed on 5 June 2024).
14. Council of the European Union. *On Nature Restoration and Amending Regulation*; Office Journal of the European Union: Luxembourg, 2024; pp. 1–93. Available online: https://eur-lex.europa.eu/legal-content/EN/TXT/PDF/?uri=OJ:L_202401991 (accessed on 18 July 2024).
15. Eurostat. *Applying the Degree of Urbanisation: A Methodological Manual to Define Cities, Towns and Rural Areas for International Comparisons*, 2021st ed.; Publications Office of the European Union: Luxembourg, 2021; ISBN 9789276203063.
16. Taylor, L.; Hochuli, D.F. Defining Greenspace: Multiple Uses across Multiple Disciplines. *Landsc Urban Plan* **2017**, *158*, 25–38. [CrossRef]
17. World Health Organization. *Urban Green Spaces: A Brief for Action*; WHO: Copenhagen, Denmark, 2017.
18. Sangwan, A.; Saraswat, A.; Kumar, N.; Pipralia, S.; Kumar, A. *Urban Green Spaces Prospects and Retrospect's*; Castanho, R.A., Fernández, J.C., Eds.; Urban Green Spaces; IntechOpen: London, UK, 2022.
19. Huang, B.X.; Li, W.Y.; Ma, W.J.; Xiao, H. Space Accessibility and Equity of Urban Green Space. *Land* **2023**, *12*, 766. [CrossRef]
20. Przewoźna, P.; Inglot, A.; Mielewczyk, M.; Maczka, K.; Matczak, P. Accessibility to Urban Green Spaces: A Critical Review of WHO Recommendations in the Light of Tree-Covered Areas Assessment. *Ecol. Indic.* **2024**, *166*, 112548. [CrossRef]
21. Marinosci, I. *La Classificazione Del Verde Urbano: Una Proposta Metodologica*; Institute for Environmental Protection and Research (ISPRA): Roma, Italy, 2009; Volume 96, ISBN 9788844803032.
22. Cardinali, M.; Beenackers, M.A.; van Timmeren, A.; Pottgiesser, U. The Relation between Proximity to and Characteristics of Green Spaces to Physical Activity and Health: A Multi-Dimensional Sensitivity Analysis in Four European Cities. *Environ. Res.* **2024**, *241*, 117605. [CrossRef]
23. Konijnendijk, C.C. Evidence-Based Guidelines for Greener, Healthier, More Resilient Neighbourhoods: Introducing the 3–30–300 Rule. *J. For. Res.* **2023**, *34*, 821–830. [CrossRef]
24. Annerstedt Van Den Bosch, M.; Egorov, A.I.; Mudu, P.; Uscila, V.; Barrdahl, M.; Kruize, H.; Kulinkina, A.; Staatsen, B.; Swart, W.; Zurlyte, I. Development of an Urban Green Space Indicator and the Public Health Rationale. *Scand. J. Public Health* **2016**, *44*, 159–167. [CrossRef]
25. UN Habitat Metadata on SDGs Indicator 11.7.1 Indicator Category: Tier II. Available online: https://unhabitat.org/sites/default/files/2020/07/metadata_on_sdg_indicator_11.7.1.pdf (accessed on 10 January 2024).
26. Silva, C.; Larsson, A. *Challenges for Accessibility Planning and Research in the Context of Sustainable Mobility*; OECD: Paris, France, 2018.
27. Olivari, B.; Cipriano, P.; Napolitano, M.; Giovannini, L. Are Italian Cities Already 15-Minute? Presenting the Next Proximity Index: A Novel and Scalable Way to Measure It, Based on Open Data. *J. Urban Mobil.* **2023**, *4*, 100057. [CrossRef]
28. Levine, J. A Century of Evolution of the Accessibility Concept. *Transp. Res. D Transp. Environ.* **2020**, *83*, 102309. [CrossRef]

29. Kabisch, N.; Strohbach, M.; Haase, D.; Kronenberg, J. Urban Green Space Availability in European Cities. *Ecol. Indic.* **2016**, *70*, 586–596. [CrossRef]
30. De La Barrera, F.; Reyes-Paecke, S.; Banzhaf, E. Indicators for Green Spaces in Contrasting Urban Settings. *Ecol. Indic.* **2016**, *62*, 212–219. [CrossRef]
31. Fang, D.; Liu, D.; Kwan, M.-P. Evaluating Spatial Variation of Accessibility to Urban Green Spaces and Its Inequity in Chicago: Perspectives from Multi-Types of Travel Modes and Travel Time. *Urban For. Urban Green* **2025**, *104*, 128593. [CrossRef]
32. Ekkel, E.D.; de Vries, S. Nearby Green Space and Human Health: Evaluating Accessibility Metrics. *Landsc. Urban Plan* **2017**, *157*, 214–220. [CrossRef]
33. EC. *Mapping Guide v6.2 for a European Urban Atlas Regional Policy*; EC: Brussels, Belgium, 2020.
34. Poelman, H. *A Walk to the Park? Assessing Access to Green Areas in Europe's Cities*; Regional and Urban Policy: Brussels, Belgium, 2018.
35. Giuliani, G.; Petri, E.; Interwies, E.; Vysna, V.; Guigoz, Y.; Ray, N.; Dickie, I. Modelling Accessibility to Urban Green Areas Using Open Earth Observations Data: A Novel Approach to Support the Urban SDG in Four European Cities. *Remote Sens.* **2021**, *13*, 422. [CrossRef]
36. La Rosa, D. Accessibility to Greenspaces: GIS Based Indicators for Sustainable Planning in a Dense Urban Context. *Ecol. Indic.* **2014**, *42*, 122–134. [CrossRef]
37. Benati, G.; Calcagni, F.; Matellozzo, F.; Ghermandi, A.; Langemeyer, J. Unequal Access to Cultural Ecosystem Services of Green Spaces within the City of Rome—A Spatial Social Media-Based Analysis. *Ecosyst. Serv.* **2024**, *66*, 101594. [CrossRef]
38. UN Habitat. Metadata on SDGs Indicator 11.3.1. Indicator Category: Tier II. Available online: https://unhabitat.org/sites/default/files/2020/07/metadata_on_sdg_indicator_11.3.1.pdf (accessed on 28 September 2023).
39. Ito World Green Space Access ITO Map. Available online: https://wiki.openstreetmap.org/wiki/Ito_Map (accessed on 22 October 2024).
40. Burdziej, J. Using Hexagonal Grids and Network Analysis for Spatial Accessibility Assessment in Urban Environments—A Case Study of Public Amenities in Toruń. *Misc. Geogr.* **2019**, *23*, 99–110. [CrossRef]
41. Istituto Nazionale di Statistica (ISTAT). *Basi Territoriali Anni 1991, 2001, 2011 e 2021*; ISTAT: Roma, Italy, 2024.
42. Cimini, A.; De Fioravante, P.; Riitano, N.; Dichicco, P.; Calò, A.; Scarascia Mugnozza, G.; Marchetti, M.; Munafò, M. Land Consumption Dynamics and Urban–Rural Continuum Mapping in Italy for SDG 11.3.1 Indicator Assessment. *Land* **2023**, *12*, 155. [CrossRef]
43. Wang, W.; Zhou, L.; Zhu, A.-X.; Lv, G. Isoline Extraction Based on a Global Hexagonal Grid. *Cartogr. Geogr. Inf. Sci.* **2024**, 1–15. [CrossRef]
44. Burdziej, J. A Web-Based Spatial Decision Support System for Accessibility Analysis-Concepts and Methods. *Appl. Geomat.* **2012**, *4*, 75–84. [CrossRef]
45. Yu, Z.; Yang, G.; Zuo, S.; Jørgensen, G.; Koga, M.; Vejre, H. Critical Review on the Cooling Effect of Urban Blue-Green Space: A Threshold-Size Perspective. *Urban Urban Green* **2020**, *49*, 126630. [CrossRef]
46. European Environment Agency CLC+ Backbone Product Specification and User Manual for CLC+ Backbone Raster Product 2021. Available online: https://land.copernicus.eu/en/products/clc-backbone/clc-backbone-2021 (accessed on 28 July 2023).
47. OpenStreetMap Community. Humanitarian OpenStreetMap Team LearnOSM. Available online: https://learnosm.org/en/contribute/ (accessed on 22 October 2023).
48. Munafò, M. Consumo Di Suolo, Dinamiche Territoriali e Servizi Ecosistemici. *Edizione* **2023**, *37*, 2023.
49. Albanese, A.; Cecili, G.; Cimini, A.; Congedo, L.; D'Agata, A.; De Fioravante, P.; Dichicco, P.; Falanga, V.; Marinosci, I.; Mariani, L.; et al. Linee Guida per Il Monitoraggio Del Consumo Di Suolo Nell'ambito Delle Attività Del SNPA 50. 2024. Available online: https://www.snpambiente.it/wp-content/uploads/2024/05/Linee-Guida-SNPA-50_24.pdf (accessed on 20 September 2024).
50. Decreto Ministeriale 2 Aprile 1968. Limiti Inderogabili Di Densità Edilizia, Di Altezza, Di Distanza Tra i Fabbricati e Rapporti Massimi Tra Spazi Destinati Agli Insediamenti Residenziali e Produttivi e Spazi Pubblici o Riservati Alle Attività Collettive Ai Fini Della Formazione Dei Nuovi Strumenti Urbanistici o Della Revisione Di Quelli Esistenti, Ai Sensi Dell'art.17 Della Legge 6 Agosto 1967, n.765; pp. 2340–2342. Available online: https://www.gazzettaufficiale.it/eli/id/1968/04/16/1288Q004/sg (accessed on 2 April 2024).
51. van Dillen, S.M.E.; de Vries, S.; Groenewegen, P.P.; Spreeuwenberg, P. Greenspace in Urban Neighbourhoods and Residents' Health: Adding Quality to Quantity. *J. Epidemiol. Community Health (1978)* **2012**, *66*, e8. [CrossRef]
52. ARTA Abruzzo VeBS Il Buon Uso Degli Spazi Verdi e Blu per La Promozione Del Benessere e Della Salute. Available online: https://vebs.it/ (accessed on 22 May 2024).

land

MDPI

Article

Urbanization and Carbon Storage Dynamics: Spatiotemporal Patterns and Socioeconomic Drivers in Shanghai

Hao Wu [1], Caihua Yang [1], Anze Liang [2], Yifeng Qin [1], Dobri Dunchev [3], Boryana Ivanova [3] and Shengquan Che [1,*]

[1] School of Design, Shanghai Jiao Tong University, Shanghai 200240, China; wuhao1101@sjtu.edu.cn (H.W.); yangcaihua.bang.sjtu@sjtu.edu.cn (C.Y.); qyf0162@sjtu.edu.cn (Y.Q.)

[2] School of Agriculture and Biology, Shanghai Jiao Tong University, Shanghai 200240, China; 018150210001@sjtu.edu.cn

[3] Department of Economics, Faculty of Economics, Agricultural University Plovdiv, 12 Mendeleev Blvd., 4000 Plovdiv, Bulgaria; d_dunchev@au-plovdiv.bg (D.D.); bivanova@au-plovdiv.bg (B.I.)

* Correspondence: chsq@sjtu.edu.cn

Abstract: Combating climate change by increasing urban carbon storage is one of the critical issues which urban policymakers must address. Understanding the characteristics and driving factors of carbon storage changes during urbanization can assist urban managers in formulating responsive land use policies. This study employs the INVEST model to evaluate carbon storage in Shanghai from 2000 to 2020, analyzing land use changes and their carbon impacts. It analyzes the transformation of land use in Shanghai during the same period and its impact on carbon storage. Using a 1 km grid for sampling, this study examines the spatiotemporal distribution patterns of carbon storage in Shanghai. Furthermore, it employs linear regression to discuss the social and economic drivers influencing carbon storage in the city. Carbon storage in Shanghai, predominantly from cultivated land and artificial surfaces, increased from 16.78 Mt in 2000 to 18.40 Mt in 2020, with an annual rise of 0.81 Mt. The spatial distribution of carbon storage exhibited a stable southeast-northwest pattern, with variations in dispersion between the north-south and east-west directions. The distribution of carbon storage shifted from a bimodal to a unimodal pattern, indicating an overall increase. There was a significant positive correlation between carbon storage and both the per capita green space area and the industrial output value, which can be attributed to Shanghai's policies on green industrial development. This research aids in formulating land use policies to enhance urban carbon storage.

Keywords: carbon storage; urbanization; INVEST; socioeconomic drivers

Citation: Wu, H.; Yang, C.; Liang, A.; Qin, Y.; Dunchev, D.; Ivanova, B.; Che, S. Urbanization and Carbon Storage Dynamics: Spatiotemporal Patterns and Socioeconomic Drivers in Shanghai. *Land* **2024**, *13*, 2098. https://doi.org/10.3390/land13122098

Academic Editors: Michele Munafò, Luca Congedo and Francesca Assennato

Received: 6 November 2024
Revised: 29 November 2024
Accepted: 3 December 2024
Published: 5 December 2024

1. Introduction

Since the Industrial Revolution, anthropogenic activities have led to the excessive emission of carbon dioxide, which is the primary cause of the greenhouse effect and global warming [1]. Global warming has been proven to alter global climate patterns over the past few decades, intensifying the duration and frequency of extreme weather events, and it continues to pose a severe threat to human survival and well-being [2–4]. To mitigate climate change and achieve sustainable development in human society, the global average temperature rise must not exceed 2 °C above pre-industrial levels [5]. How to absorb the carbon dioxide emitted by human activities and achieve the goal of carbon neutrality has become a target for many governments and institutions. The Paris Agreement advocates for capping the rise in the global mean temperature at a level significantly below 2 °C above pre-industrial benchmarks, while also striving to confine the temperature increase to 1.5 °C above pre-industrial benchmarks. The Chinese government proposed striving for carbon neutrality by 2060, which is a formidable task, as urbanization has been rapidly progressing globally over the past two decades, which has led to significant infrastructure construction and land use conversion, resulting in substantial carbon dioxide emissions in the atmosphere. According to the United Nations, by 2050, global urbanization will further develop, and how to reduce carbon dioxide emissions during the urbanization process has become a hot topic of concern for scholars and policymakers. Studies have shown that terrestrial ecosystems play a crucial role in the global carbon cycle, absorbing 29% of carbon dioxide emissions through various ecological processes over the past decade [6]. Carbon storage has become an important indicator of sustainable development and climate change mitigation in terrestrial ecosystems [7,8]. Areas undergoing urbanization, influenced by human activities, often experience drastic changes in natural conditions, such as the conversion of forests, farmlands, wetlands, and rivers into impervious surfaces. This process affects the original carbon cycle, alters the carbon stock in the soil, and has a significant impact on carbon storage [9,10]. Over the past century, land use changes due to urban construction have accounted for 35% of anthropogenic carbon dioxide emissions [11]. Therefore, studying the impact of land use changes caused by urbanization on carbon storage is of great significance for both carbon emissions and carbon storage, providing important theoretical support for urban policymakers.

Carbon storage in terrestrial ecosystems is one of the most critical ecosystem services [12]. Many studies have investigated the impact of land use changes on terrestrial carbon storage. Urbanization in China from 1980 to 1990 resulted in an average reduction in carbon storage of 0.72 TgC/y, and from 2000 to 2010, this figure rose to 8.72 TgC/y. Based on this trend, it is projected that the rate of carbon storage reduction will range from 9.31 TgC/y to 12.94 TgC/y from 2010 to 2050 [13]. In contrast, afforestation in China from 1990 to 2020 increased carbon storage by an average of approximately 40 TgC/y [14]. This suggests that urbanization could offset about 25% of the increase in carbon storage attributable to afforestation. Studying the impact of urban land use changes on carbon storage is of great significance for the achievement of China's carbon neutrality goals. Compared with the urbanization process of the previous decade, China has established more ambitious goals and requirements for urbanization in the coming decade, implying that mitigating carbon storage loss during urbanization—or even enhancing carbon storage in urban areas—is of great importance. Research indicates a strong correlation between urban land use and carbon storage, and thus the focus of research has shifted from how to increase the carbon density of land use to the impact of land use transitions on the trends in carbon storage changes [15,16]. Carbon storage assessment models based on land use have been widely applied in studies across various spatial and temporal scales [9,17]. These models, when integrated with land use prediction models, are used to forecast changes in carbon storage under various development scenarios [18–20]. The increase in impervious surface area during urbanization reduces vegetation cover, which significantly decreases plant carbon storage. Studies have also shown that impervious surfaces isolate the soil from the atmosphere, thereby partially increasing the stability of soil organic carbon [21]. However, over time, the continuous lack of organic carbon input may lead to a decrease in soil carbon storage [22]. In addition to the impact of impervious surface area expansion, urban carbon storage is also influenced by climatic factors, particularly temperature and precipitation [23,24]. Temperature and precipitation can affect the growth of urban vegetation and influence the input and decomposition of organic carbon in the soil [25,26]. Furthermore, urban carbon storage is affected by socioeconomic factors. For instance, high-quality economic development is believed to lead to higher-quality green space construction and management, which can increase urban carbon storage, and a lower population density can allow for more green spaces in cities, thereby increasing carbon storage [27,28]. The spatial heterogeneity and high human activity in cities introduce significant uncertainty into the relationship between urban carbon storage and urbanization [28], and this relationship is also challenging to apply across different spatial and temporal scales. Current research on urban carbon storage primarily focuses on the national scale [19,29] and the city scale [28,30]. However, at smaller scales, such as the county and community levels, the relationship between urban carbon storage and urbanization has not been adequately studied [31]. Practical management is directly applied at the county level and smaller scales, and the mismatch between urban carbon storage and urbanization could potentially have a detrimental impact on the increase in urban carbon storage.

Shanghai is one of the most urbanized cities in China, with an urbanization rate exceeding 89%, which is significantly higher than the national average of 66%. The rapid urbanization over the past two decades has resulted in a dramatic expansion of impervious surface areas in Shanghai [32]. Due to varying economic levels and development patterns across different districts in Shanghai, there are noticeable differences in carbon storage between the central urban areas and the surrounding regions [33]. In 2022, the Shanghai Municipal People's Government issued the "Shanghai Carbon Peak Implementation Plan", which set achieving carbon neutrality as an important goal in the urban development of Shanghai. This study selected Shanghai as the research subject for several reasons. (1) Shanghai has a high level and long history of urban development, and analyzing the characteristics of carbon storage in Shanghai during different development periods can provide a reference for other cities. (2) Also, being located in the Yangtze River Delta, Shanghai has flat terrain and a warm and humid climate, making the impact of natural conditions on carbon storage relatively balanced and allowing this study to focus more on the effects of land use changes caused by human activities on carbon storage. The main objectives of this study are (1) to analyze the spatiotemporal characteristics of carbon storage in Shanghai from 2000 to 2020; (2) to examine the changes in land use and carbon storage in Shanghai from 2000 to 2020; (3) to evaluate the distribution patterns of carbon storage across different districts in Shanghai from 2000 to 2020; and (4) to provide effective recommendations and strategies for each district in Shanghai.

2. Materials and Methods

2.1. Research Area and Data Resource

Shanghai is characterized by a subtropical monsoon climate, with distinct seasons, ample sunshine, and abundant rainfall. The spring and autumn seasons are relatively short, whereas the winter and summer seasons are longer. The terrain of Shanghai is flat, predominantly consisting of the alluvial plain of the Yangtze River Delta, with an average elevation of about 4 m. The land slopes slightly from east to west. Apart from a few hills and mountains in the southwest, the region is generally a vast, low-lying plain. The highest point within Shanghai's territory is Dajinshan, with an elevation of 103.4 m. Shanghai governs three islands—Chongming, Changxing, and Hengsha—with Chongming Island being the third-largest island in China. As of 2023, Shanghai's gross domestic product (GDP) reached CNY 4721.866 billion, a 5.0% increase from the previous year. The value added by the primary industries was CNY 9.609 billion; the secondary industries contributed CNY 1161.297 billion; and the tertiary industries accounted for CNY 3550.960 billion. The permanent population of the city is approximately 24.8745 million, making it one of the most urbanized cities in China.

The research area is shown in Figure 1, and the data utilized in this study are listed in Table 1. The carbon density data for different land uses in this paper were partly derived from field surveys, with missing data extracted from existing databases [34]. Land use data in this study were sourced from GlobeLand30, and land use changes were derived from spatial calculations conducted using the ARCGIS 10.4 platform based on the land use data.

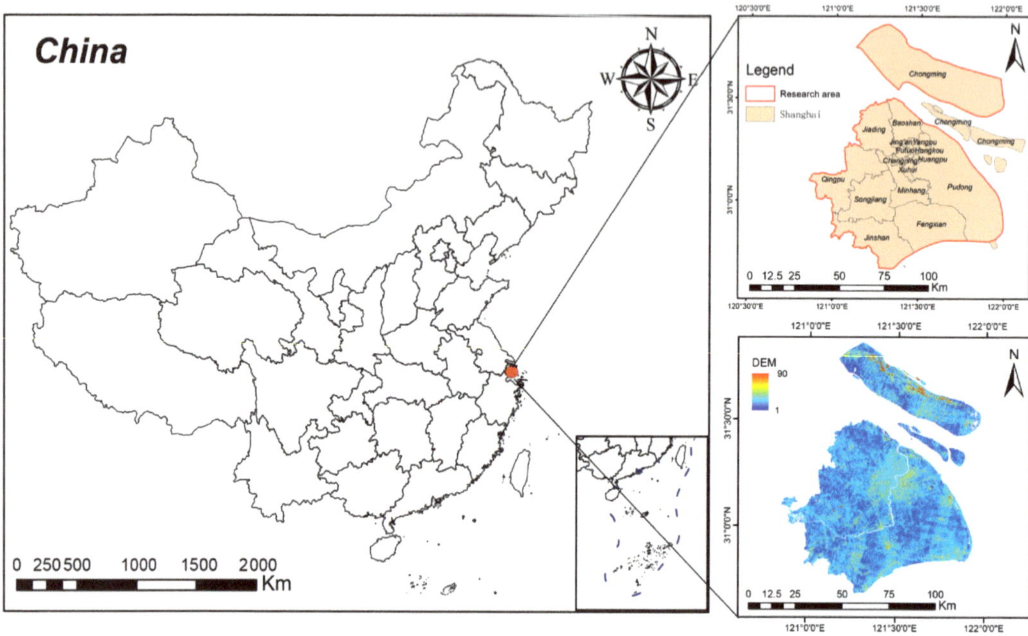

Figure 1. Research area.

Figure 2 shows the methodical framework for this research.

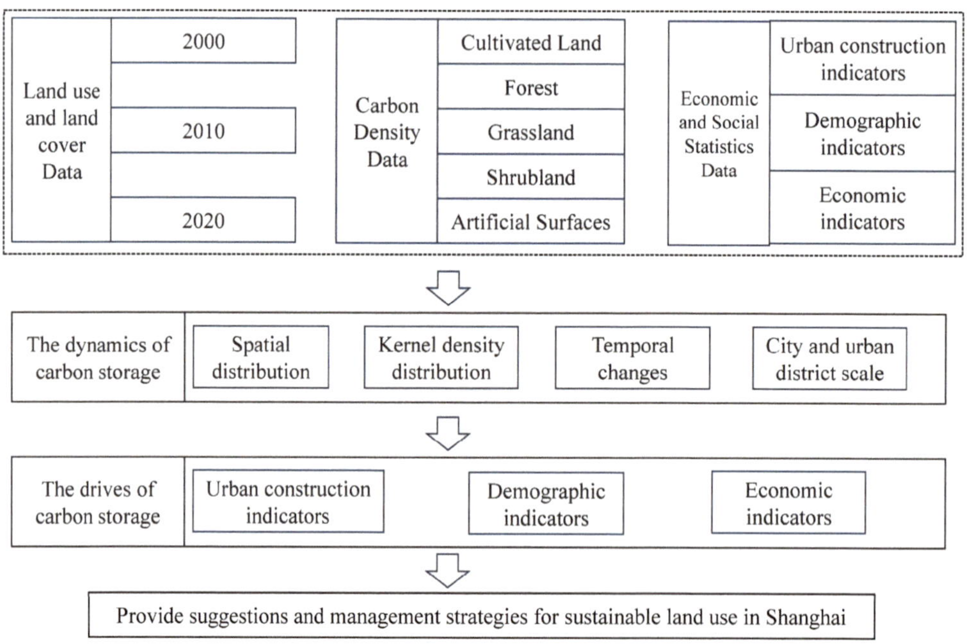

Figure 2. Research framework.

Table 1. Data used in this study.

Data	Resolution	Time	Data Sources
Land Use and Land Cover	30 m	2000, 2010, 2020	GlobeLand30 (https://www.webmap.cn/commres.do?method=dataDownload, accessed on 6 August 2024)
DEM	12.5 m	\	EARTHDATA (https://search.asf.alaska.edu/, accessed on 6 August 2024)
Carbon Density	\	2010s	Field Investigation and Dataset [34]
Economic and Social Statistics Data	\	2010, 2020	ShangHai Statistical Yearbook 2020 (https://tjj.sh.gov.cn/tjnj/20210303/2abf188275224739bd5bce9bf128aca8.html, accessed on 6 August 2024), ShangHai Statistical Yearbook 2010 (https://tjj.sh.gov.cn/tjnj/20210303/2abf188275224739bd5bce9bf128aca8.html, accessed on 15 August 2024)

2.2. Carbon Storage

This study calculated the carbon storage from aboveground biomass, belowground biomass, soil, and dead matter:

$$C_{total} = C_{above} + C_{below} + C_{soil} + C_{dead} \quad (1)$$

where C_{total} represents the total carbon storage, C_{above} denotes the carbon from aboveground biomass, C_{below} signifies the carbon from belowground biomass, C_{soil} refers to the carbon from soil, and C_{dead} refers to the carbon from dead matter.

The carbon storage calculation was performed using the INVEST model. INVEST is an open-source modeling tool designed for evaluating the efficacy of ecosystem services, including carbon storage. INVEST aids in the management of ecosystems and informs decision-making processes. The carbon densities of different land use types for this study are displayed in Table 2. The data presented in Table 2 were derived from field surveys and existing databases [34], which provide the carbon storage per unit area (1 ha) for different land use types in Shanghai. These data were utilized as inputs for the INVEST model to calculate the carbon storage values across different regions of Shanghai. In this study, the scope of the carbon storage calculation was defined to include surface and subsurface depths not exceeding 100 cm. For water bodies exceeding 100 cm in depth, carbon storage was not accounted for in the model. For wetlands, the aboveground biomass and soil carbon were calculated, while the calculation parameters for waterbodies and the ocean were set to zero. The carbon storage of artificial surfaces represents the carbon storage in urban built-up areas. These areas primarily consist of buildings, roads, squares, and other artificial facilities, with vegetation mainly being artificial greenery. Therefore, in the calculation of carbon storage for artificial surfaces, the carbon in this portion of biomass was not accounted for, with c_above, c_below, and c_dead all being set to 0. However, the carbon in the soil could not be overlooked. Based on field surveys and data from the database [34], we calculated the carbon storage for artificial surfaces.

Table 2. Carbon density values of different land use types for INVEST model (t/ha).

Land Use Type	c_above	c_below	c_soil	c_dead
Cultivated Land	5	1	25.6	0
Forest	47.8	9.94	120.8	40
Grassland	0.25	1.11	18.2	5.2
Shrubland	9.303	2	25.6	3
Wetland	1	0	33	0
Water Bodies	0	0	0	0
Artificial Surfaces	0	0	25.3	0
Ocean	0	0	0	0

2.3. Standard Deviation Ellipse Analysis

Standard deviation ellipse analysis is one of the most commonly used tools in spatial data analysis for examining the orientation of data distribution. This study employed standard deviation ellipse analysis to examine the spatial distribution of urban carbon storage in Shanghai for 2000, 2010, and 2020. Standard deviation ellipse analysis requires determining the center of the ellipse, the rotation angle, and the lengths of the X and Y axes. The formula for determining the center is as follows:

$$SDE_x = \sqrt{\frac{\sum_{i=1}^{n}(x_i - \overline{X})}{n}} \tag{2}$$

$$SDE_y = \sqrt{\frac{\sum_{i=1}^{n}(y_i - \overline{Y})}{n}} \tag{3}$$

where x_i and y_i represent the spatial coordinates of each feature, $\overline{X}$ and $\overline{Y}$ are the arithmetic mean centers, and SDE_x and SDE_y are the coordinates of the ellipse's center. We also have

$$\tan\theta = \frac{A+B}{C} \tag{4}$$

$$A = \left(\sum_{1}^{n}\tilde{x}_i^{2} - \sum_{1}^{n}\tilde{y}_i^{2}\right) \tag{5}$$

$$B = \sqrt{\left(\sum_{1}^{n}\tilde{x}_i^{2} - \sum_{1}^{n}\tilde{y}_i^{2}\right)^2 + 4\left(\sum_{1}^{n}\tilde{x}_i\tilde{y}_i\right)^2} \tag{6}$$

$$C = 2\sum_{1}^{n}\tilde{x}_i\tilde{y}_i \tag{7}$$

where $\tilde{x}_i$ and $\tilde{y}_i$ are the differences between the arithmetic mean center and the x and y coordinates, respectively, while

$$\sigma_x = \sqrt{2}\sqrt{\frac{\sum_{i=1}^{n}\left(\tilde{x}_i\cos\theta - \tilde{y}_i\sin\theta\right)^2}{n}} \tag{8}$$

$$\sigma_y = \sqrt{2}\sqrt{\frac{\sum_{i=1}^{n}\left(\tilde{x}_i\cos\theta + \tilde{y}_i\sin\theta\right)^2}{n}} \tag{9}$$

where σ_x and σ_y represent the lengths of the ellipse's X axis and Y axis, respectively.

2.4. Kernel Density Estimation

The kernel density curves of carbon storage for each district in Shanghai were calculated to analyze the distribution characteristics of carbon storage across different districts, identify local distribution patterns of carbon storage within each district, and discern the temporal trends in the distribution patterns of carbon storage across different districts.

The calculation formula for the kernel density curves is as follows:

$$\hat{f}_h(x) = \frac{1}{n}\sum_{i=1}^{n} K_h(x - x_i) = \frac{1}{nh}\sum_{i=1}^{n} K\left(\frac{x - x_i}{h}\right) \tag{10}$$

where $x_1, x_1, \cdots, x_i$ are n independent and identically distributed samples, h is the bandwidth, and K is the density function of any arbitrary distribution.

2.5. Linear Regression

This study employed linear fitting to investigate the relationship between carbon storage and various influencing factors across different districts of Shanghai:

$$y = \beta_0 + \beta_1 x \tag{11}$$

where β_0 represents the intercept and β_1 denotes the slope.

The explanatory variables were obtained through interviews with city managers and experts, selecting economic indicators, urban construction indicators, and demographic indicators. For the economic indicators, the per capita GDP and per capita industrial output value were chosen. For the urban construction indicators, the per capita public expenditure, per capita green space area, and per capita housing completion area were selected. For the demographic indicators, population density was chosen.

This study employed the ARCGIS 10.7 platform for spatial analysis and cartography, while data analysis and plotting were conducted using Python 3.0. All data were sampled using a 1 km grid for subsequent analysis.

3. Results

3.1. Carbon Storage of Shanghai

Table 3 presents the total carbon storage for different land use types in Shanghai for the years 2000, 2010, and 2020. From 2000 to 2010, the carbon storage in Shanghai increased from 16.78 Mt to 17.07 Mt and further rose to 18.40 Mt in 2020, with an increment of 0.81 Mt/year. The carbon storage in Shanghai was primarily provided by cultivated land and artificial surfaces, constituting 97.18%, 94.85%, and 84.48% of the total carbon storage in the years 2000, 2010, and 2020, respectively. Except for cultivated land and water bodies, the carbon storage of other land use types increased, with the most significant increase observed in forests, which grew from 0.04 Mt to 1.98 Mt. Overall, the carbon storage in Shanghai exhibited an upward trend, with the carbon storage of ecological land uses such as forests, wetlands, and grasslands increasing and their proportion in the total carbon storage also gradually increasing (Figure 3).

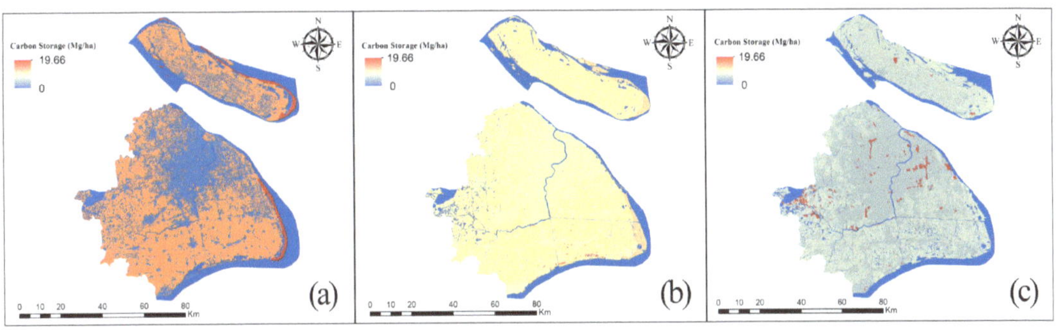

Figure 3. Carbon storage of Shanghai from 2000 to 2020: (**a**) 2000, (**b**) 2010, and (**c**) 2020.

Table 3. Carbon storage of different land use types in Shanghai.

Land Use	2000 (Mt)	2010 (Mt)	2020 (Mt)
Cultivated Land	12.23	10.05	8.52
Artificial Surfaces	4.08	6.14	7.03
Water Bodies	0.00	0.00	0.00
Ocean	0.00	0.00	0.00

Table 3. *Cont.*

Land Use	2000 (Mt)	2010 (Mt)	2020 (Mt)
Wetland	0.39	0.63	0.63
Grassland	0.04	0.05	0.25
Forest	0.04	0.21	1.98
Shrubland	\	0.00	0.00
Total	16.78	17.07	18.40

3.2. Standard Deviation Ellipse Analysis Results

Due to Chongming District in Shanghai being an island and not directly adjacent to other areas, standard deviation ellipse analysis was conducted while including only the mainland urban areas of Shanghai. The analysis results indicate that from 2000 to 2020, the spatial distribution of carbon storage in the mainland urban areas of Shanghai exhibited a southeast-northwest orientation (Figure 4). The direction of carbon storage in the mainland urban areas of Shanghai changed little from 2000 to 2010, and then it rotated to a north-south direction by 97.00° from 2010 to 2020 (Figure 5). The total carbon storage within the ellipse accounted for 68% of the total carbon sink in the mainland urban areas of Shanghai. In 2000, the ellipse area accounted for approximately 48.04%; in 2010, it accounted for 49.91%; and in 2020, it accounted for about 49.71% (Table 4). From 2000 to 2020, the concentration of carbon storage in the mainland urban areas of Shanghai first increased and then stabilized. Overall, the carbon storage in the mainland urban areas of Shanghai is primarily concentrated in the central western part of the mainland, including the central urban district, southern Baoshan, southeastern Jiading, northeastern Qingpu, central northern Songjiang, northeastern Jinshan, northwestern Fengxian, western Pudong, and Minhang District. The overall pattern of carbon storage from 2000 to 2020 has not changed significantly.

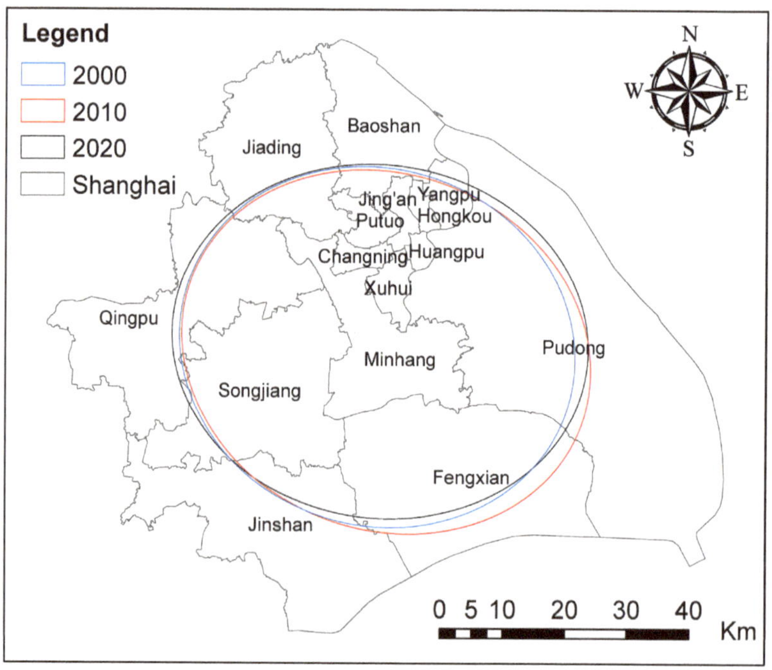

Figure 4. Standard deviation ellipse analysis diagram.

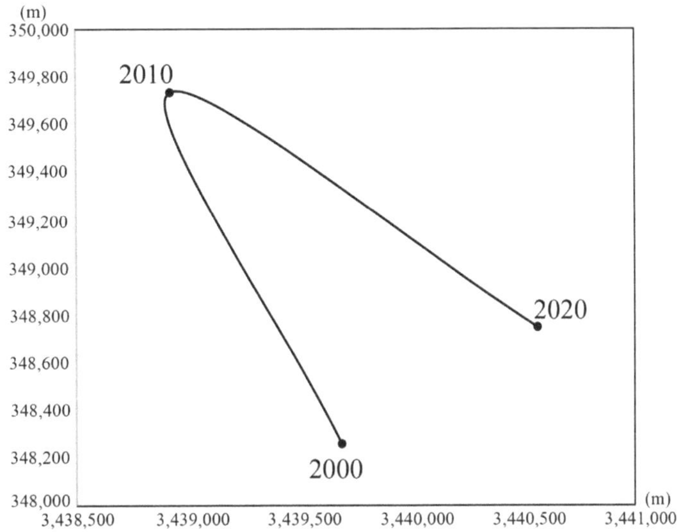

Figure 5. Ellipse centroid movement.

Table 4. Coordinates of the ellipse's center.

Year	CenterX	CenterY	XStdDist	YStdDist	Rotation
2000	348,260.0464	3,439,685.635	31,942.7048	27,870.226	105.205774
2010	349,733.5511	3,438,921.271	33,270.53219	27,802.36061	108.528043
2020	348,749.2127	3,440,560.12	33,410.61056	27,572.50858	97.007432

From 2000 to 2010, the centroid of carbon storage in the mainland urban areas of Shanghai moved northwest, shifting 0.76 km west and 1.64 km north. From 2010 to 2020, the centroid of carbon storage in the mainland urban areas of Shanghai moved southeast, shifting 1.47 km east and 0.98 km south. This suggests that from 2000 to 2010, the northwest region of the mainland urban areas of Shanghai had a higher amount of carbon storage, and from 2010 to 2020, the southeast region of the mainland urban areas of Shanghai experienced an increase in carbon storage. Regarding the degree of spatial dispersion, from 2000 to 2010, the standard deviation of the X axis increased by 1.33 km, while that of the Y axis remained almost unchanged. From 2010 to 2020, the standard deviation of the X axis increased by 0.14 km, and that of the Y axis decreased by 0.23 km (Table 4). This indicates that from 2000 to 2020, the degree of dispersion of carbon storage in the north-south direction in the mainland urban areas of Shanghai continued to increase but at a decreasing rate, while the degree of dispersion in the east-west direction continued to decrease.

3.3. Kernel Density Estimation Results

As shown in Figure 6, the kernel density curves exhibit a single-peak pattern in the years 2000 and 2020, while a double-peak pattern is observed in 2010. From 2000 to 2010, the peak of the kernel density curve shifts to the left, indicating an increase in the number of samples below the mean value. From 2010 to 2020, the curve transitions from a double-peak pattern to a single-peak pattern, still generally showing a leftward movement trend. This suggests that the variability in carbon storage values among samples decreased over the period from 2000 to 2020. Figure 6a,b reveals that from 2000 to 2020, the frequency of samples with moderate carbon storage values continued to increase, while the frequency of samples with higher carbon storage values decreased.

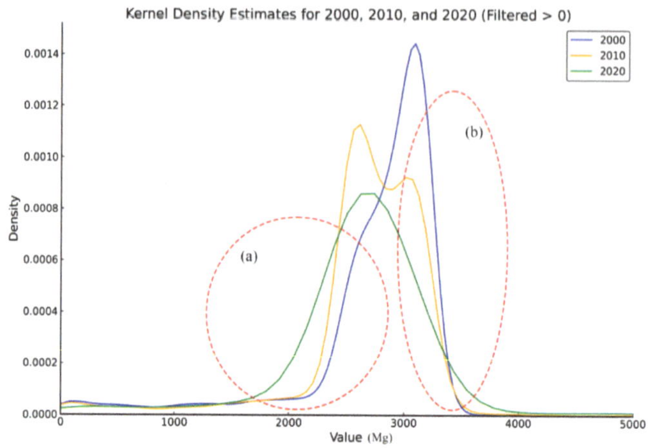

Figure 6. Shanghai's carbon storage kernel density curves from 2000 to 2020. (**a**) moderate carbon storage value region (**b**) higher carbon storage value region.

Figure 7 presents the kernel density curves for the districts of Shanghai (excluding Chongming District) from 2000 to 2020. As can be observed in Figure 6a, the peak values of the kernel density curves in 2000 were concentrated in two regions (Regions I and II). Region I encompassed Huangpu District, Changning District, Xuhui District, Jing'an District, Putuo District, Yangpu District, Hongkou District, and Baoshan District. Region II included Songjiang District, Jiading District, Minhang District, Jinshan District, Fengxian District, Qingpu District, and Pudong District. In 2010, the peak values of the kernel density curves were also concentrated in two regions (Figure 7b), with Region I expanding to include four additional districts and Region II narrowing down to Songjiang District, Jinshan District, and Fengxian District. By 2020, the distribution of carbon storage kernel density peaks across the mainland districts of Shanghai was no longer distinctly divided into two regions. This indicates that from 2000 to 2020, there was a change in the carbon storage distribution patterns among the districts of Shanghai. The carbon storage distribution patterns were clearly divided into two categories in 2000 and 2010, while in 2020, the patterns were broadly similar across all districts.

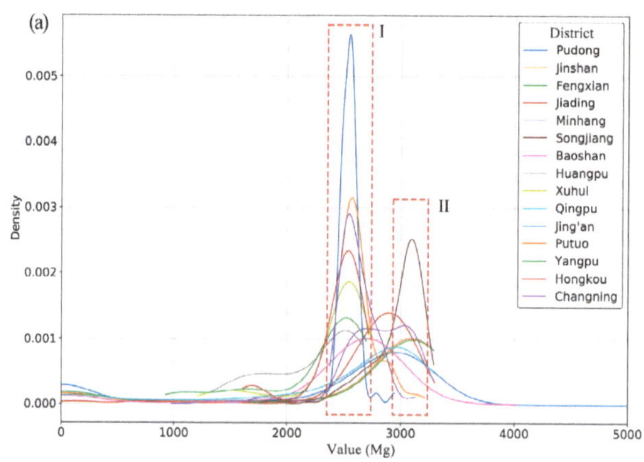

Figure 7. *Cont.*

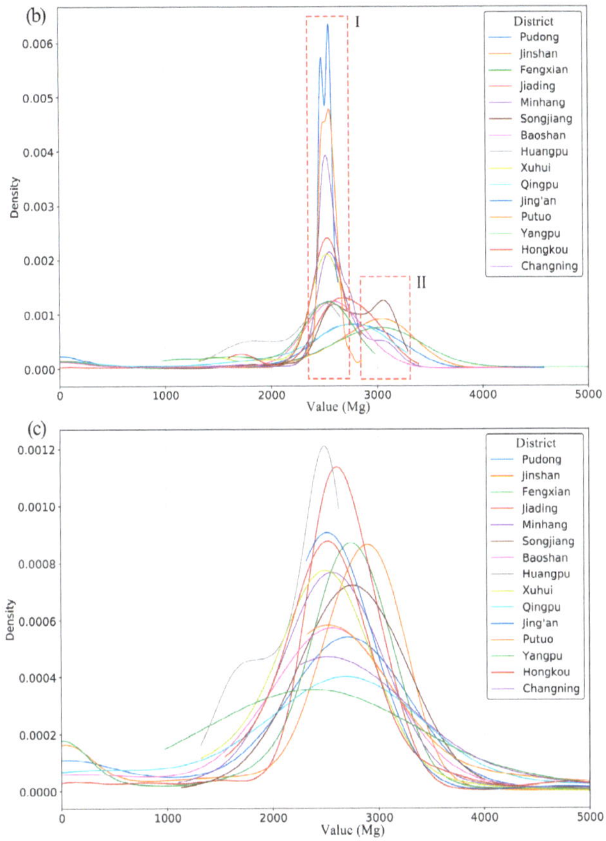

Figure 7. Kernel density curves of carbon storage in different districts of Shanghai from (excluding Chongming District) (**a**) 2000, (**b**) 2010, and (**c**) 2020.

3.4. Impact of Land Use Change on Carbon Storage

From 2000 to 2020, changes in land use in Shanghai led to variations in the city's carbon storage capacity (Table 5). Between 2000 and 2010, Shanghai's carbon sink increased by a total of 0.33 Mt. Land use conversion out contributed to a 1.16 Mt increase in carbon storage, with the transformation of water bodies and the ocean into other land use types accounting for the largest increments, representing 62.71% and 30.52%, respectively. The conversion of artificial surfaces and grasslands into other land use types resulted in an increase in carbon storage, accounting for approximately 6.77%. Additionally, land use conversion out led to a 0.83 Mt decrease in carbon storage, with the conversion of cultivated land into other land use types causing the largest reduction, accounting for about 89.02%, followed by forests and wetlands, which accounted for 5.20% and 5.78%, respectively. In terms of land use conversion in, the increase in land use in 2010 contributed to a 1.16 Mt increase in carbon storage, with the expansion of wetland areas contributing 43.36%, followed by cultivated land, which accounted for 39.36%. The conversion of other land use types into forests resulted in a carbon sink increase, accounting for about 17.29%. Land use conversion in also led to a 0.84 Mt decrease in carbon storage, with artificial surfaces accounting for the largest share of about 61.66%, followed by water bodies, which accounted for about 37.73%. The conversion of grasslands and the ocean contributed to a decrease in carbon storage, totaling 0.26% and 0.34%, respectively. From 2010 to 2020, land use conversion out resulted in a 1.75 Mt increase in carbon storage, with the transformation of artificial surfaces and water bodies into other land use types accounting for the largest increments, representing about 41.55% and 34.48%, respectively. This was followed by cultivated land, the ocean, and grasslands, accounting for 18.82%, 3.47%, and 1.68%, respectively. The reduction in forest and wetland areas caused a 0.27 Mt loss in carbon storage (62.69% and 37.73%, respectively). On the other hand, land use conversion in led to a 0.95 Mt decrease in carbon storage, with other land use types converting into water bodies contributing 54.44%, followed by artificial surfaces, accounting for 41.15%. The conversion of grasslands and the ocean contributed to a decrease in carbon storage, representing 3.91% and 0.50%, respectively. The increase in carbon storage due to land use was approximately 2.43 Mt, with the expansion of forest land contributing the most, accounting for about 77.43%, followed by cultivated land and wetlands, which accounted for 12.35% and 10.18%, respectively. The increase in shrublands contributed a 0.04% increase in carbon storage.

Table 5. The changes in carbon storage resulting from land use changes.

Carbon Storage (t)	2010 Artificial Surfaces	2010 Cultivate Land	2010 Forest	2010 Grassland	2010 Ocean	2010 Shrubland	2010 Water Bodies	2010 Wetland	Total
2020 Artificial Surfaces		−410,396.24	−28,365.70	138.95	1259.18	−2.63	54,219.92	−9447.68	−392,594.19
2020 Cultivate Land	121,182.06		−85,485.79	4689.64	8512.09		268,339.92	−16,708.25	300,529.66
2020 Forest	651,784.98	1,134,112.65		24,782.53		16.08	53,734.62	19,199.54	1,883,630.39
2020 Grassland	−3043.43	−29,642.98	−10,150.20		766.57	−2.73	5227.83	−449.06	−37,294.00
2020 Ocean	−220.87	−28.44	−19.67	−91.36				−4427.82	−4788.16
2020 Shrubland	42.06	772.68					86.19		900.92
2020 Water Bodies	−42,534.36	−371,241.52	−13,964.71	−1704.73		−3.59		−89,899.74	−519,348.64
2020 Wetland	458.84	6066.36	−32,968.07	1562.58	50,211.54		222,293.70		247,624.94
Total	727,669.28	329,642.51	−170,954.14	29,377.60	60,749.38	7.13	603,902.17	−101,733.01	1,478,660.92
2010 Artificial Surfaces		−567,815.2581	−25,148.2531	223.705609	6441.63286	81,425.51823	−11,454.50689	−516,327.1614	
2010 Cultivate Land	24,520.47821		−13,409.20638	4245.792308	11,316.27545	445,034.7865	−12,862.80477	458,845.3213	
2010 Forest	660.880787	106,634.3162		65,889.07864	2438.90641	25,175.80811	763.995628	201,562.9858	
2010 Grassland	−105.70491	−8199.174678	−17.440201		2707.50599	3485.217587	−47.401197	−2176.997409	
2010 Ocean	−507.770989	−1786.031914		−6.6852			−575.279989	−2875.768092	
2010 Shrubland	1.31427	3.736351						5.050621	
2010 Water Bodies	−15,476.76866	−270,913.7389	−4740.13262	−911.415597			−23,929.19955	−315,971.2554	
2010 Wetland	32.103	863.78432			331,489.7938	173,101.1368		505,486.8179	
Total	9124.531701	−741,212.3668	−43,315.0323	69,440.47576	354,394.1145	728,222.4672	−48,105.19677	328,548.9933	

3.5. Impact of Socioeconomic Factors on Carbon Storage in Shanghai

This study endeavored to explore the relationship between socioeconomic factors and carbon storage in Shanghai, selecting six primary socioeconomic indicators through a literature review and interviews with policymakers. Due to the scarcity of relevant statistical data for the year 2000, this study analyzed data from 2010 and 2020 to assess the impact of socioeconomic indicators on Shanghai's carbon storage, divided into three groups: the 2010 data, the 2020 data, and a combined 2010–2020 modeling. The results (Figure 8) demonstrate that from 2010 to 2020, there was a significant positive correlation between per capita green space and per capita industrial output. An increase in the per capita green space area implies an expansion of space available for vegetation growth within cities, which is beneficial for increasing soil carbon storage and biological carbon storage [19,25,35]. Per capita industrial output is also positively correlated with carbon storage, suggesting that a higher per capita industrial output is associated with greater carbon storage in the region. This contrasts with some existing research which suggests that human economic activities often disrupt the natural state of the surface, leading to carbon sink losses [36]. However, there is also research which posits that good economic conditions can provide better financial support for urban green spaces, thereby increasing urban carbon sinks [37]. The discrepancies in these conclusions may be attributed to the scale of analysis. At continental and global scales, economic development often leads to deforestation, wetland loss, and other negative environmental impacts, suggesting that the positive effects of economic growth on the ecological environment are outweighed by the negative effects at larger scales. In contrast, at smaller scales, such as urban and county levels, the enhancement of socioeconomic status prompts urban managers and planners to place greater emphasis on ecological protection and the construction of green infrastructure. Consequently, at this scale, the positive impacts of economic development on the ecological environment exceed the negative impacts.

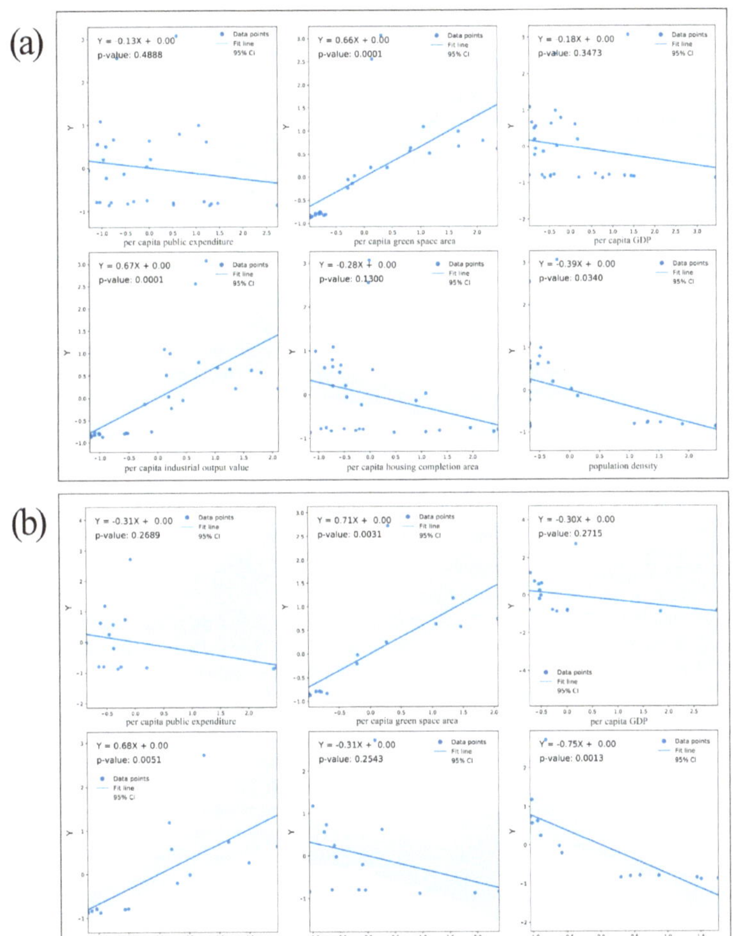

Figure 8. *Cont.*

(c)

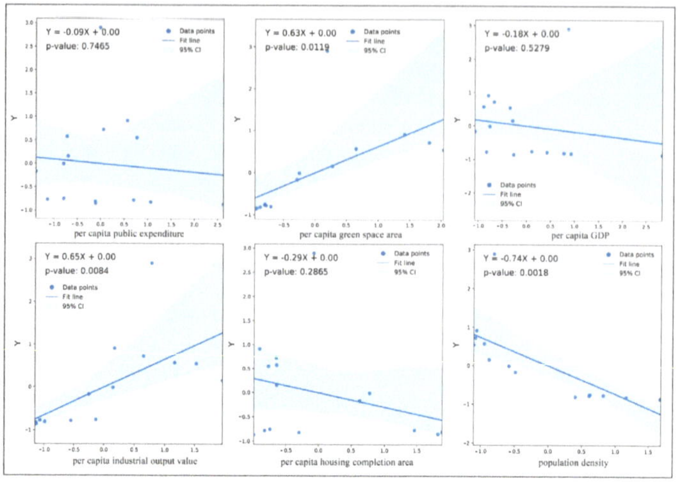

Figure 8. Linear regression results of carbon storage in different districts in Shanghai: (**a**) 2010 and 2020 combined; (**b**) 2010; and (**c**) 2020. CI: Confidence Intervals.

4. Discussion

4.1. Uncertainty of INVEST

The INVEST model has been widely applied in the field of assessing regional carbon storage. This model, which is based on land use, effectively quantifies the impact of spatiotemporal changes in land use and land cover on carbon storage, thereby providing a scientific basis for land management policies [9,38–40]. However, the uncertainties associated with the INVEST model cannot be overlooked. The INVEST model is not a process-based model and does not account for the effects of ecological processes and human activities on carbon storage [41]. For instance, artificial surfaces with different levels of greenery clearly have different carbon densities. Studies have shown that the carbon density of green spaces under different management practices also varies, and the carbon density of vegetation at different growth stages is not equal. Neither of these factors were considered in the model. These factors can introduce significant uncertainty into the assessment of carbon storage. Additionally, there is uncertainty in determining the carbon density parameters for different land use types. At the urban scale, field surveys to measure the aboveground, belowground, and soil carbon densities of different urban land use types require substantial human and material resources, making them nearly impossible to implement. Therefore, studies often combine the literature with field surveys [19,42,43], meaning that sampling times, sampling methods, and measurement techniques all have considerable uncertainty, and the resulting carbon density parameters are also highly uncertain. In this study, Shanghai was selected as the research subject. Located in the Yangtze River Delta's alluvial plain, Shanghai has flat terrain and a relatively stable climate and environment within the region. The carbon density data in this study integrated literature data and field survey data from similar climatic zones and geographical locations, which were then used to analyze urban-scale carbon storage changes. To some extent, this approach can reduce the model's uncertainty and provide scientific guidance for land use management policies within the region.

4.2. Spatiotemporal Variations in Urban Carbon Storage

The findings of this study indicate that from 2000 to 2020, following two decades of rapid urbanization, the carbon storage at the community scale across various urban districts showed a trend toward equilibrium. Over the same period, areas with high and low carbon storage in Shanghai gradually diminished, while regions with carbon storage values close to the mean increased. This trend may be attributed to urban expansion, which has generally encroached upon the area of original forests and green spaces [44]. Concurrently, ecological protection policies in Shanghai have facilitated the transformation of highly urbanized areas from artificial surfaces to high carbon storage land use types, such as urban parks and green spaces. This implies a more dispersed spatial distribution of high carbon storage land use types within the city. The spatial analysis of carbon storage in Shanghai's terrestrial areas reached similar conclusions. Overall, from 2000 to 2010, the center of carbon storage in Shanghai's terrestrial areas shifted northwestward, likely due to the predominance of ecological land in the western regions and the reduction in ecological land in the south and east regions because of urbanization, leading to decreased carbon storage and a subsequent northwestward shift of the center. From 2010 to 2020, the center of carbon storage in Shanghai's terrestrial areas moved southeastward, suggesting a greater increase in carbon storage in the southeastern part. This may be the result of a series of sustainable development policies, including ecological corridors and networks implemented in Shanghai. The kernel density analysis of carbon storage in each district of Shanghai's terrestrial areas revealed that from 2000 to 2020, the distribution of carbon storage in the peripheral urban districts gradually converged with that of the central urban districts. This is primarily due to the higher degree of urbanization in central urban areas, where urban expansion from 2000 to 2020 was relatively limited, and the construction of green spaces and parks resulted in a certain increase in carbon storage. In the peripheral urban areas, rapid urban development has been accompanied by comprehensive planning and construction which includes parks and green spaces, thus maintaining a high level of carbon storage despite the reduction in forested land. A similar result was found in other areas in China.

4.3. The Impact of Socioeconomic Factors on Carbon Storage in Shanghai

The results show that an increase in the per capita industrial output had a more positive impact on carbon storage than an increase in the per capita green space area. The linear models for the three different time periods indicate that the slope of the relationship between the per capita industrial output value and carbon storage was greater than that of the per capita green space for the 2020 group and the combined group, while the impact of the per capita green space was greater for the 2010 group. This may be attributed to policy factors. Shanghai has introduced a series of policies to promote the green and sustainable development of heavy industry, such as the construction of carbon-neutral industrial parks in key industries and green low-carbon demonstration parks in chemical areas (Notice of the General Office of the Shanghai Municipal People's Government on Printing and Distributing the "Three-Year Action Plan for High-Quality Development of Manufacturing in Shanghai (2023–2025)"; https://www.shanghai.gov.cn/nw12344/20230615/328ea17db94546ac8e8709f567bec705.html, accessed on 20 August 2024) and the circular transformation of industrial parks. Shanghai promotes intensive transformation of industrial parks and industrial clusters to achieve shared facilities, cascading energy utilization, and the recycling and reuse of wastewater treatment (Notice of the Shanghai Municipal People's Government on Printing and Distributing the "Implementation Plan for Accelerating the Establishment of a Green, Low-Carbon, and Circular Development Economic System in Shanghai"; https://www.shanghai.gov.cn/nw12344/20211021/bb02574688eb469aaa8a3b2e6a6cc5eb.html, accessed on 20 August 2024).

Shanghai is implementing these measures to achieve the green transformation and sustainable development of heavy industry. These initiatives not only focus on mitigating the negative environmental impacts of industrial activities but also aim to enhance the city's ecological functions and carbon sequestration capabilities. By constructing carbon-neutral industrial parks and green low-carbon demonstration zones, Shanghai encourages the adoption of low-carbon technologies and clean energy, promoting energy efficiency improvements and reductions in greenhouse gas emissions across key industries. Concurrently, the circular transformation of industrial parks facilitates the efficient use of resources and the recycling of waste, which not only reduces industrial waste generation but also decreases environmental pollution. Furthermore, Shanghai enforces stringent environmental regulations and standards to guide industrial enterprises in adhering to higher environmental and energy efficiency requirements, accelerating the green transformation of the industrial sector. Additionally, Shanghai encourages the development of green industrial parks and increases green spaces within industrial areas, which to some extent compensates for the reduction in carbon sequestration capacity due to the expansion of industrial land use. The implementation of these comprehensive measures allows Shanghai to maintain economic vitality while also protecting and improving the ecological environment. This suggests that in highly urbanized areas, urban managers can utilize planning and policy tools to establish a positive relationship between industrial development and carbon storage. Studies have indicated that relevant policies have a significant impact on the relationship between economic development and environmental quality [45–47]. The results also show a significant negative correlation between population density and carbon storage. Similar conclusions have been reached in existing studies. An increase in the urban population implies the need for more housing, commercial, and agricultural land, which encroaches upon urban ecological land. Excessive human activities have a negative impact on carbon storage [37,48].

4.4. Limitations

This study has several limitations which warrant acknowledgment. Firstly, financial and logistical constraints limited the field survey data utilized in this study, resulting in a primary reliance on the literature to establish the relationship between land use and carbon storage. This approach failed to capture the variations in carbon storage among similar land use types, and it did not account for the carbon storage of the numerous small green spaces prevalent in urban areas. Consequently, the outcomes of ongoing urban renewal projects, such as habitat gardens in Shanghai, could not be accurately assessed. In future research, we intend to identify urban green space types based on high-resolution remote sensing imagery and establish a more precise relationship between land use and carbon storage using additional field surveys [35]. Secondly, this study used a 1 km grid to investigate the carbon storage of each district, which may not have aligned with the actual community management divisions. This discrepancy could potentially create difficulties when applying this study's findings in practical settings. Lastly, this study was based on historical data and did not explore future scenarios. Under climate change, Shanghai is experiencing increasing temperatures and humidity [4], which could impact the carbon storage of plants and soil, introducing a degree of uncertainty to urban carbon storage. For instance, increased temperatures and humidity can reduce the carbon sequestration capacity of wetlands [49] while promoting the productivity growth of plants [50]. This uncertainty poses an additional challenge to the management of urban carbon storage.

5. Conclusions

This study employed the INVEST model, complemented by field surveys and literature research, to assess the carbon storage in Shanghai from 2000 to 2020, analyzing the conversion of land use and its impact on carbon storage during this period. Utilizing a 1 km grid for sampling, this study analyzed the spatiotemporal distribution patterns of carbon storage in Shanghai and employed linear regression to discuss the social and economic drivers of carbon storage, arriving at the following conclusions:

(1) The carbon storage in Shanghai is primarily provided by cultivated land and artificial surfaces. From 2000 to 2010, the carbon storage in Shanghai increased from 16.78 Mt to 17.07 Mt, and by 2020, it had risen to 18.40 Mt, with an annual increment of 0.81 Mt.

(2) From 2000 to 2020, the spatial variation in carbon sinks across Shanghai's terrestrial areas remained relatively stable, exhibiting a southeast-northwest distribution, with the degree of dispersion of carbon sinks in the north-south direction continuously increasing, albeit at a decreasing rate, while the degree of dispersion in the east-west direction continuously decreased.

(3) Between 2000–2010 and 2010–2020, the kernel density curves of carbon storage in Shanghai shifted from a bimodal to a unimodal pattern, indicating an overall increase in carbon storage, with a continuous reduction in both high and low carbon storage areas, leading to a more uniform distribution of carbon storage across the city.
(4) There was a significant positive correlation between both the per capita green space area and the per capita industrial output value and carbon storage, which is primarily associated with relevant policies in Shanghai. In contrast, there was a significant negative correlation between the population density and carbon storage.

The results of this study indicate that the effects of urbanization and industrial development on carbon storage are not necessarily detrimental; well-conceived urban ecological planning and management can significantly enhance urban carbon storage. For future research, the interplay and trade-offs between urban ecological protection, economic development, and carbon storage merit attention as important research topics. Additionally, it is intriguing to explore whether these trade-offs and synergies are related to the city's natural geographic conditions and the stage of economic and social development.

Author Contributions: Conceptualization, H.W. and Y.Q.; methodology, H.W., A.L. and C.Y.; software, C.Y.; validation, H.W., A.L. and Y.Q.; formal analysis, A.L.; investigation, H.W. and A.L.; resources, S.C.; data curation, Y.Q.; writing—original draft preparation, H.W. and Y.Q.; writing—review and editing, A.L., D.D., B.I. and C.Y.; visualization, H.W. and Y.Q.; supervision, S.C.; project administration, D.D., B.I. and S.C.; funding acquisition, D.D., B.I., and S.C. All authors have read and agreed to the published version of the manuscript.

Funding: This research was funded by the Shanghai 2022 "Science and Technology Innovation Action Plan" International Science and Technology Cooperation Project (grant number 22230750500) and the Guilin City Scientific Research and Technological Development Program Project (grant number 20230127-3).

Data Availability Statement: The original data presented in the study are openly available in GlobeLand30 (https://www.webmap.cn/commres.do?method=dataDownload, accessed on 6 August 2024), EARTHDATA (https://search.asf.alaska.edu/, accessed on 6 August 2024), Shanghai Statistical Yearbook 2020 (https://tjj.sh.gov.cn/tjnj/20210303/2abf188275224739bd5bce9bf128aca8.html, accessed on 6 August 2024), Shanghai Statistical Yearbook 2010 (https://tjj.sh.gov.cn/tjnj/20210303/2abf188275224739bd5bce9bf128aca8.html, accessed on 15 August 2024), and a dataset of carbon density in Chinese terrestrial ecosystems (Reference [34]).

Acknowledgments: The authors are grateful to the editor and reviewers for their valuable comments and suggestions.

Conflicts of Interest: The authors declare no conflicts of interest.

References

1. IPCC. *Climate Change 2023: Synthesis Report. Contribution of Working Groups I, II and III to the Sixth Assessment Report of the Intergovernmental Panel on Climate Change*; Core Writing Team, Lee, H., Romero, J., Eds.; IPCC: Geneva, Switzerland, 2023; pp. 35–115.
2. Murali, G.; Iwamura, T.; Meiri, S.; Roll, U. Future temperature extremes threaten land vertebrates. *Nature* **2023**, *615*, 461–467. [CrossRef] [PubMed]
3. Jørgensen, L.B.; Ørsted, M.; Malte, H.; Wang, T.; Overgaard, J. Extreme escalation of heat failure rates in ectotherms with global warming. *Nature* **2022**, *611*, 93–98. [CrossRef] [PubMed]
4. Wu, H.; Yang, C.; Xie, C.; Man, Z.; He, S.; Qin, Y.; Che, S. Quantification of contribution of climate change and land use change on urban ecosystem service using multi-scale approach. *Ecol. Indic.* **2024**, *167*, 112619. [CrossRef]
5. Gao, Y.; Gao, X.; Zhang, X. The 2 °C Global Temperature Target and the Evolution of the Long-Term Goal of Addressing Climate Change—From the United Nations Framework Convention on Climate Change to the Paris Agreement. *Engineering* **2017**, *3*, 272–278. [CrossRef]
6. Friedlingstein, P.; O'Sullivan, M.; Jones, M.W.; Andrew, R.M.; Gregor, L.; Hauck, J.; Le Quéré, C.; Luijkx, I.T.; Olsen, A.; Peters, G.P.; et al. Global Carbon Budget 2022. Earth Syst. *Sci. Data* **2022**, *14*, 4811–4900. [CrossRef]
7. He, C.; Zhang, D.; Huang, Q.; Zhao, Y. Assessing the potential impacts of urban expansion on regional carbon storage by linking the LUSD-urban and InVEST models. *Environ. Model. Softw.* **2016**, *75*, 44–58. [CrossRef]

8. Nolan, C.J.; Field, C.B.; Mach, K.J. Constraints and enablers for increasing carbon storage in the terrestrial biosphere. *Nat. Rev. Earth Environ.* **2021**, *2*, 436–446. [CrossRef]
9. Li, L.; Song, Y.; Wei, X.; Dong, J. Exploring the impacts of urban growth on carbon storage under integrated spatial regulation: A case study of Wuhan, China. *Ecol. Indic.* **2020**, *111*, 106064. [CrossRef]
10. Li, C.; Zhao, J.; Thinh, N.X.; Xi, Y. Assessment of the Effects of Urban Expansion on Terrestrial Carbon Storage: A Case Study in Xuzhou City, China. *Sustainability* **2018**, *10*, 647. [CrossRef]
11. Turner, B.L.; Lambin, E.F.; Reenberg, A. The emergence of land change science for global environmental change and sustainability. *Proc. Natl. Acad. Sci. USA* **2007**, *104*, 20666–20671. [CrossRef]
12. Piao, S.; Fang, J.; Ciais, P.; Peylin, P.; Huang, Y.; Sitch, S.; Wang, T. The carbon balance of terrestrial ecosystems in China. *Nature* **2009**, *458*, 1009–1013. [CrossRef] [PubMed]
13. Liu, X.; Wang, S.; Wu, P.; Feng, K.; Hubacek, K.; Li, X.; Sun, L. Impacts of Urban Expansion on Terrestrial Carbon Storage in China. *Environ. Sci. Technol.* **2019**, *53*, 6834–6844. [CrossRef] [PubMed]
14. Cheng, K.; Yang, H.; Tao, S.; Su, Y.; Guan, H.; Ren, Y.; Hu, T.; Li, W.; Xu, G.; Chen, M.; et al. Carbon storage through China's planted forest expansion. *Nat. Commun.* **2024**, *15*, 4106. [CrossRef] [PubMed]
15. Hwang, J.; Choi, Y.; Kim, Y.; Ol, L.N.; Yoo, Y.J.; Cho, H.J.; Sun, Z.; Jeon, S. Analysis of the effect of environmental protected areas on land-use and carbon storage in a megalopolis. *Ecol. Indic.* **2021**, *133*, 108352. [CrossRef]
16. Fryer, J.; Williams, I.D. Regional carbon stock assessment and the potential effects of land cover change. *Sci. Total Environ.* **2021**, *775*, 145815. [CrossRef]
17. Chang, X.; Xing, Y.; Wang, J.; Yang, H.; Gong, W. Effects of land use and cover change (LUCC) on terrestrial carbon stocks in China between 2000 and 2018. *Resour. Conserv. Recycl.* **2022**, *182*, 106333. [CrossRef]
18. Wu, W.; Xu, L.; Zheng, H.; Zhang, X. How much carbon storage will the ecological space leave in a rapid urbanization area? Scenario analysis from Beijing-Tianjin-Hebei Urban Agglomeration. *Resour. Conserv. Recycl.* **2023**, *189*, 106774. [CrossRef]
19. Zhu, L.; Song, R.; Sun, S.; Li, Y.; Hu, K. Land use/land cover change and its impact on ecosystem carbon storage in coastal areas of China from 1980 to 2050. *Ecol. Indic.* **2022**, *142*, 109178. [CrossRef]
20. Chuai, X.; Qi, X.; Zhang, X.; Li, J.; Yuan, Y.; Guo, X.; Huang, X.; Park, S.; Zhao, R.; Xie, X.; et al. Land degradation monitoring using terrestrial ecosystem carbon sinks/sources and their response to climate change in China. *Land Degrad. Dev.* **2018**, *29*, 3489–3502. [CrossRef]
21. Tang, X.; Zhao, X.; Bai, Y.; Tang, Z.; Wang, W.; Zhao, Y.; Wan, H.; Xie, Z.; Shi, X.; Wu, B.; et al. Carbon pools in China's terrestrial ecosystems: New estimates based on an intensive field survey. *Proc. Natl. Acad. Sci. USA* **2018**, *115*, 4021–4026. [CrossRef]
22. Wei, Z.-Q.; Wu, S.-H.; Zhou, S.-L.; Li, J.-T.; Zhao, Q.-G. Soil Organic Carbon Transformation and Related Properties in Urban Soil Under Impervious Surfaces. *Pedosphere* **2014**, *24*, 56–64. [CrossRef]
23. Young, P.J.; Harper, A.B.; Huntingford, C.; Paul, N.D.; Morgenstern, O.; Newman, P.A.; Oman, L.D.; Madronich, S.; Garcia, R.R. The Montreal Protocol protects the terrestrial carbon sink. *Nature* **2021**, *596*, 384–388. [CrossRef] [PubMed]
24. Gampe, D.; Zscheischler, J.; Reichstein, M.; O'Sullivan, M.; Smith, W.K.; Sitch, S.; Buermann, W. Increasing impact of warm droughts on northern ecosystem productivity over recent decades. *Nat. Clim. Chang.* **2021**, *11*, 772–779. [CrossRef]
25. Guo, H.; Du, E.; Terrer, C.; Jackson, R.B. Global distribution of surface soil organic carbon in urban greenspaces. *Nat. Commun.* **2024**, *15*, 806. [CrossRef] [PubMed]
26. Li, H.; Wu, Y.; Liu, S.; Xiao, J.; Zhao, W.; Chen, J.; Alexandrov, G.; Cao, Y. Decipher soil organic carbon dynamics and driving forces across China using machine learning. *Glob. Chang. Biol.* **2022**, *28*, 3394–3410. [CrossRef]
27. Song, Y.; Chen, B.; Ho, H.C.; Kwan, M.-P.; Liu, D.; Wang, F.; Wang, J.; Cai, J.; Li, X.; Xu, Y.; et al. Observed inequality in urban greenspace exposure in China. *Environ. Int.* **2021**, *156*, 106778. [CrossRef]
28. Zhang, W.; Randall, M.; Jensen, M.B.; Brandt, M.; Wang, Q.; Fensholt, R. Socio-economic and climatic changes lead to contrasting global urban vegetation trends. *Glob. Environ. Chang.* **2021**, *71*, 102385. [CrossRef]
29. Jiao, K.; Liu, Z.; Wang, W.; Yu, K.; McGrath, M.J.; Xu, W. Carbon cycle responses to climate change across China's terrestrial ecosystem: Sensitivity and driving process. *Sci. Total Environ.* **2024**, *915*, 170053. [CrossRef]
30. Peng, Y.L.; Cheng, W.Y.; Xu, X.X.; Song, H.F. Analysis and prediction of the spatiotemporal characteristics of land-use ecological risk and carbon storage in Wuhan metropolitan area. *Ecol. Indic.* **2024**, *158*, 111432. [CrossRef]
31. Yang, S.; Li, L.Q.; Zhu, R.H.; Luo, C.; Lu, X.; Sun, M.L.; Xu, B.C. Assessing land-use changes and carbon storage: A case study of the Jialing River Basin, China. *Sci. Rep.* **2024**, *14*, 15984. [CrossRef]
32. Naizheng, X.; Hongying, L.; Feng, W.; Yiping, Z. Urban expanding pattern and soil organic, inorganic carbon distribution in Shanghai, China. *Environ. Earth Sci.* **2012**, *66*, 1233–1238. [CrossRef]
33. Wang, Z.; Cui, X.; Yin, S.; Shen, G.; Han, Y.; Liu, C. Characteristics of carbon storage in Shanghai's urban forest. *Chin. Sci. Bull.* **2013**, *58*, 1130–1138. [CrossRef]
34. Xu, L.; He, N.; Yu, G. A dataset of carbon density in Chinese terrestrial ecosystems (2010s). *China Sci. Data* **2019**, *4*, 2096–2223. [CrossRef]
35. Sun, Y.; Xie, S.; Zhao, S. Valuing urban green spaces in mitigating climate change: A city-wide estimate of aboveground carbon stored in urban green spaces of China's Capital. *Glob. Chang. Biol.* **2019**, *25*, 1717–1732. [CrossRef]
36. Cao, M.; Tian, Y.; Wu, K.; Chen, M.; Chen, Y.; Hu, X.; Sun, Z.; Zuo, L.; Lin, J.; Luo, L.; et al. Future land-use change and its impact on terrestrial ecosystem carbon pool evolution along the Silk Road under SDG scenarios. *Sci. Bull.* **2023**, *68*, 740–749. [CrossRef]

37. Wu, B.W.; Zhang, Y.Y.; Wang, Y.; Lin, X.B.; Wu, Y.F.; Wang, J.W.; Wu, S.D.; He, Y.M. Urbanization promotes carbon storage or not? The evidence during the rapid process of China. *J. Environ. Manag.* **2024**, *359*, 121061. [CrossRef]
38. Rachid, L.; Elmostafa, A.; Mehdi, M.; Hassan, R. Assessing carbon storage and sequestration benefits of urban greening in Nador City, Morocco, utilizing GIS and the InVEST model. *Sustain. Futures* **2024**, *7*, 100171. [CrossRef]
39. Chen, D.; Liu, R.R.; Zhou, M.X. Delineation of Urban Growth Boundary Based on Habitat Quality and Carbon Storage: A Case Study of Weiyuan County in Gansu, China. *Land* **2023**, *12*, 1006. [CrossRef]
40. Tang, L.P.; Ke, X.L.; Zhou, T.; Zheng, W.W.; Wang, L.Y. Impacts of cropland expansion on carbon storage: A case study in Hubei, China. *J. Environ. Manag.* **2020**, *265*, 110515. [CrossRef]
41. Wang, R.Y.; Mo, X.Y.; Ji, H.; Zhu, Z.; Wang, Y.S.; Bao, Z.L.; Li, T.H. Comparison of the CASA and InVEST models' effects for estimating spatiotemporal differences in carbon storage of green spaces in megacities. *Sci. Rep.* **2024**, *14*, 5456. [CrossRef]
42. Nie, X.; Lu, B.; Chen, Z.P.; Yang, Y.W.; Chen, S.; Chen, Z.H.; Wang, H. Increase or decrease? Integrating the CLUMondo and InVEST models to assess the impact of the implementation of the Major Function Oriented Zone planning on carbon storage. *Ecol. Indic.* **2020**, *118*, 106708. [CrossRef]
43. Liang, Y.J.; Hashimoto, S.; Liu, L.J. Integrated assessment of land-use/land-cover dynamics on carbon storage services in the Loess Plateau of China from 1995 to 2050. *Ecol. Indic.* **2021**, *120*, 106939. [CrossRef]
44. Wang, C.; Li, M.; Wang, X.; Deng, M.; Wu, Y.; Hong, W. Spatio-Temporal Dynamics of Carbon Storage in Rapidly Urbanizing Shenzhen, China: Insights and Predictions. *Land* **2024**, *13*, 1566. [CrossRef]
45. He, J.; Wang, H. Economic structure, development policy and environmental quality: An empirical analysis of environmental Kuznets curves with Chinese municipal data. *Ecol. Econ.* **2012**, *76*, 49–59. [CrossRef]
46. Long, X.; Ji, X. Economic Growth Quality, Environmental Sustainability, and Social Welfare in China—Provincial Assessment Based on Genuine Progress Indicator (GPI). *Ecol. Econ.* **2019**, *159*, 157–176. [CrossRef]
47. Feng, S.; Mohd Shafiei, M.W.; Ng, T.F.; Ren, J.; Jiang, Y. The intersection of economic growth and environmental sustainability in China: Pathways to achieving SDG. *Energy Strategy Rev.* **2024**, *55*, 101530. [CrossRef]
48. Grimm, N.B.; Faeth, S.H.; Golubiewski, N.E.; Redman, C.L.; Wu, J.; Bai, X.; Briggs, J.M. Global Change and the Ecology of Cities. *Science* **2008**, *319*, 756–760. [CrossRef]
49. Bao, T.; Jia, G.; Xu, X. Weakening greenhouse gas sink of pristine wetlands under warming. *Nat. Clim. Chang.* **2023**, *13*, 462–469. [CrossRef]
50. Zhang, Y.; Piao, S.; Sun, Y.; Rogers, B.M.; Li, X.; Lian, X.; Liu, Z.; Chen, A.; Peñuelas, J. Future reversal of warming-enhanced vegetation productivity in the Northern Hemisphere. *Nat. Clim. Chang.* **2022**, *12*, 581–586. [CrossRef]

land

MDPI

Article

From Fertile Grounds to Sealed Fields: Assessing and Mapping Soil Ecosystem Services in Forlì's Urban Landscape (NE Italy)

Fabrizio Ungaro [1,*], Paola Tarocco [2], Alessandra Aprea [2], Stefano Bazzocchi [3] and Costanza Calzolari [1]

[1] National Research Council, Institute of BioEconomy CNR IBE, Via Madonna del Piano 10, 50019 Sesto Fiorentino, Italy; mariacostanza.calzolari@ibe.cnr.it
[2] Geology, Soil and Seismic Risk Area, Regione Emilia-Romagna, Viale A. Moro 52, 40127 Bologna, Italy; paola.tarocco@regione.emilia-romagna.it (P.T.); alessandra.aprea@regione.emilia-romagna.it (A.A.)
[3] Servizio Ambiente e Urbanistica, Comune di Forlì, Corso Diaz 21, 47121 Forlì, Italy; stefano.bazzocchi@comune.forli.fc.it
* Correspondence: fabrizio.ungaro@ibe.cnr.it

Abstract: Between 2022 and 2023, the urban soils of Forlì (NE Italy) were surveyed, sampled, analyzed, and mapped over an area of ca. 5700 ha, of which 2820 were sealed. The outcomes of the survey allowed the integration of the existing knowledge about soil and land use with the urban plan and provided the basis to produce a 1:10,000 map of urban soils along with their land capability and an updated 1:50,000 soil map of the municipality. Soil data (textural fractions, pH, organic carbon content) were interpolated over the entire case study area, providing the inputs for locally calibrated pedotransfer functions whose outputs were used to assess a set of seven indicators for the potential supply of soil ecosystem services (SESs): soil biodiversity, buffer capacity, carbon storage, agricultural production, biomass production, water regulation, and water storage. Maps of the seven ecosystem services on a hybrid resolution grid of 25 and 100 m were complemented with an overall urban soil quality map based on the combinations of four different SES indicators. Results show that for several services, hotspots occur not only in the peri-urban agricultural areas but also in unsealed soils within the urban fabric, and that different soils provide high-quality services in diverse constellations depending on the soil characteristics, age and extent of disturbance and degree of sealing.

Keywords: urban soils; ecosystem services; soil indicators; soil sealing; urban planning; pedotransfer functions; Emilia–Romagna

Academic Editors: Guangju Zhao and Hanoch Lavee

Received: 11 February 2025
Revised: 11 March 2025
Accepted: 27 March 2025
Published: 27 March 2025

Citation: Ungaro, F.; Tarocco, P.; Aprea, A.; Bazzocchi, S.; Calzolari, C. From Fertile Grounds to Sealed Fields: Assessing and Mapping Soil Ecosystem Services in Forlì's Urban Landscape (NE Italy). *Land* **2025**, *14*, 719. https://doi.org/10.3390/land14040719

1. Introduction

"La terra che fé già la lunga prova
e di Franceschi sanguinoso mucchio,
sotto le branche verdi si ritrova"
Dante Alighieri, Inferno, Canto XXVII

Despite a growing body of scientific literature in the last decade focusing on the role of urban soils as providers of ecosystem services, there is still a generalized lack of implementation of urban soils ecosystem services in the practice of urban planning [1]. If on one side the hyperbolic growth of urban population is acknowledged as one of the signatures of the Anthropocene [2,3], leading planners and policy makers to reframe the role of cities in terms of environmental quality, social justice and sustainable development [4,5], on the other, the potential of urban soils to contribute to urban resilience in facing the challenges posed by the ongoing climate crisis is far from being fully understood and remains untapped [6–8]. The assessment of soil-based ecosystem services (SESs) has recently achieved more importance, mostly due to the recognition of the relevance of the regulation services they provide, primarily their capacity to trap greenhouse gases from the atmosphere [9,10]. Nonetheless, as for ecosystem services in the built-up environment, Blanchart et al. [11], in considering the role of soil in urban planning documents, reported that "urban soils are predominantly seen as surface areas to be converted or as a potential threat due to their level of contamination or geotechnical properties", and in a recent meta-analysis of published literature on mapping urban and peri-urban ecosystem services from more than 200 research papers extracted following the Preferred Reporting Items for Systematic Reviews and Meta-alpha Methods, not a single reference to SESs is mentioned [12]. This is likely to be due to several factors: the inherent complexity and heterogeneity of urban soils [13,14], which requires ad hoc surveys and dedicated human and financial resources, the widespread gaps in soil literacy among city planners and policy makers at the different administrative levels leading to ineffective soil governance [15,16], and the lack of standardized and widely tested approaches in assessing and mapping the ecosystem services of urban soils [17–19]. Hyun et al. [20]. proposed an urban soil quality index to evaluate the soil status in various spatial types of urban greenery considering comprehensively ecosystem services and functions of urban soil using ten quantitative soil indicators. More recently, a pedon-based approach called DESTISOL was proposed by Séré et al. [21] and tested in 37 urban soils under various situations and pedoclimates. The architecture of the model is based on 20 physico-chemical–biological soil indicators used to score 15 soil functions based on a detailed set of expert-based decision rules. Among the major research gaps in the assessment of urban soil ecosystem services, there is a lack of universally accepted metrics and indicators for assessing urban soil ecosystem services, making it difficult to compare studies and integrate findings across different regions and contexts. Furthermore, the gap in understanding how to effectively integrate urban soil ecosystem services into urban planning and policy frameworks has limited the adoption of sustainable management practices of urban soils. Addressing such issues in the EU-funded demonstrative LIFE project SOS4Life (Save Our Soils for Life, https://www.sos4life.it/en/project/, accessed 26 March 2025), Calzolari et al. [22] surveyed, assessed and mapped six ecosystem services of urban soils in the city of Carpi (NE Italy), tailoring to the urban context a methodology originally developed for agricultural soils [23]. Among the aims of the project was in fact the definition of a methodology for the detection, evaluation and mapping of ecosystem services provided by urban soils, which was aimed at quantifying ecosystem services and planning actions for their maintenance and improvement, and to identify areas suitable for urban regeneration to compensate the loss of ecosystem services due to land take following urban expansion. Among the activities of the after-LIFE plan, the Municipality of Forlì (NE Italy), a coordinating beneficiary of the SOS4Life project, requested the implementation of the same SOS4Life methodology to integrate the knowledge about urban soils and their ecosystem services within the new General Urban Plan (PUG), as foreseen by the Emilia–Romagna regional law 24/2017 on land planning to reduce and mitigate land take toward the EU target of zero net land take by 2050 [24]. The present work illustrates the outcomes of the urban soils survey carried out in the Forlì urban area between 2022 and 2023, resulting in (i) a map of urban soils at the scale of 1:10,000, (ii) the assessment of seven SESs including habitat for organisms, and (iii) the assessment of the impact of urbanization on SESs supply. Differently from other SES assessments and mapping approaches, the methodological steps followed in this work can be implemented in any urban context if the basic soil properties of urban soils are assessed via an ad hoc survey of urban soils. Furthermore, resorting to standardized indicators for SESs would easily allow a comparison between different urban contexts, and the interpretation of results would be straightforward also for non-soil practitioners. Our results contribute to raising awareness about the role of soil in the built-up environment and provide tools to assess and map urban ecosystem services, highlighting the possibility of integrating soil knowledge into urban planning.

2. Materials and Methods

2.1. Study Area

The study was conducted in the urban and peri-urban area of the municipality of Forlì (228.2 km^2, 117.430 inhabitants). Forlì (Figure 1; 44°13′21″ N 12°02′27″ E) is located at an average height of 36 m above mean sea level in the plain area of southeastern Emilia–Romagna south of the Po River, a few km away from the foothills of the Tuscan-Romagnolo Apennines to the south, and about 25 km from the Adriatic sea to the east. The urban area, with a history dating back more than 22 centuries, developed over Holocene sediments of the alluvial fans and terraced deposits of three Apennine rivers flowing SW to NE, namely the Montone and the Rabbi to the west and the Ronco to the east. The climate according to the Köppen–Geiger classification is warm temperate and constantly humid with a hot summer (Cfa). For the reference period 1991–2020, an average annual cumulative precipitation of 769 mm was recorded with a minimum average temperature of 3.9 °C (January) and a maximum average temperature of 24.6 °C (July) [25].

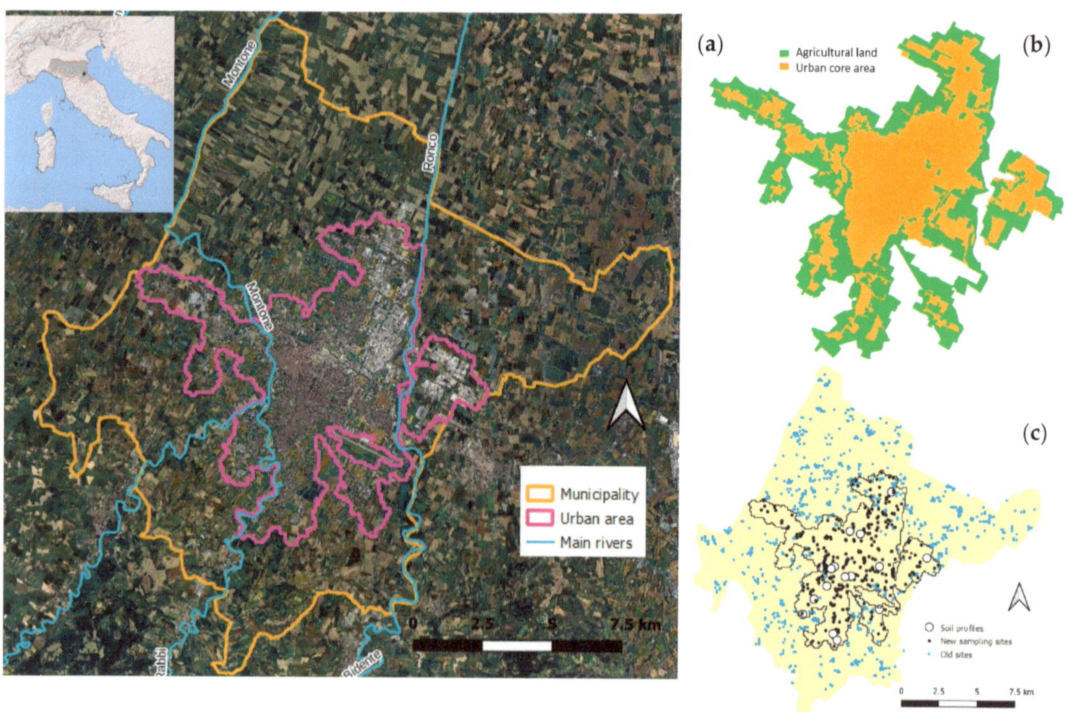

Figure 1. Study area: (**a**) municipality of Forlì in southeastern Emilia–Romagna (NE Italy) and delineation of the urban area; (**b**) urban core area and peri-urban agricultural land; (**c**) location of soil sampling points, showing old (blue dots) and new (black dots) sampling sites.

In May 2023, two separate extreme rainfall events occurred in the area with a cumulative rainfall exceeding 400 mm and locally 600 mm [26]. The local return period of the total accumulated rainfall recorded at the ground according to WMO standards [27] was more than 500 years, and the precipitation resulted in severe flooding in the whole Romagna basins with severe and widespread damage to buildings and infrastructures, which was amplified by the high rate of land take. In 2023, the sealing of the soil surface affected a total of ca. 3832 ha, corresponding to 16.8% of the area of the Forlì municipality (Figure 2), with an increase of 167 ha in the last decade. This share is almost double the regional average, which is equal to 8.9% [28].

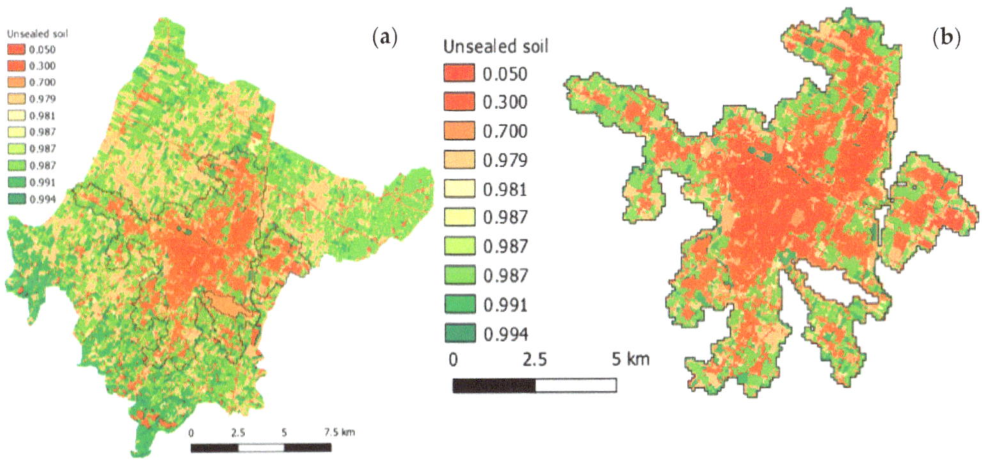

Figure 2. Share of unsealed soil surface in the municipality of Forlì (**a**) and in the surveyed urban core area (**b**). Source: Regione Emilia–Romagna, modified [27].

The study area corresponds to the continuous built-up area, with an extension of 5691 ha, nearly half of which has urbanized (2645 ha) but with a relevant share of undisturbed agricultural soil, which is mostly in the peri-urban buffer (2771 ha). Urban green areas make up about 420 ha, comprising mainly public and private parks, about 178 ha, followed by urban wastelands, about 144 ha, sport and leisure green, about 66 ha, and roadside and railway side green, about 37 ha [29].

2.2. Available Soil Data and Urban Soil Survey

As most SESs in urban areas are provided by green areas, the semi-quantitative assessment of soil functions and related services focused on gardens and parks, enclosed or peri-urban agricultural areas, sports green areas and flower beds generally larger than 0.1 ha. Private gardens and appurtenances were not considered, having verified their high degree of sealing (paths, parking lots, access ramps to garages, etc.). However, even these soils, although strongly altered, perform some functions to some extent. Differently from most of the cities of the Emilia–Romagna plain, whose urban areas are rather compact, and the peri-urban area is effectively such, in Forlì, enclosed agricultural lands are rather widespread, so the peri-urban area is quite fragmented.

A set of preliminary "urban pedolandscape units" were defined considering the distribution of "natural soils" underlying the town as derived from the soil map of Emilia-Romagna plain at a 1:50,000 scale [30], and the urbanistic structure was defined in terms of (i) age of the different buildings and parts of the town, (ii) typology of the various neighborhoods, and (iii) occurrence of public and private green areas. With the support of these maps, and resorting to multitemporal aerial images, eighty-three urban pedolandscape units were defined. Accordingly, 240 sites were in the urban (N = 160) and peri-urban (N = 80) area (Figure 1b) with the aim of covering the different pedolandscape units identified. Each unit was also characterized in terms of the degree of disturbance the soil was subject to, resorting to a multitemporal aerial photo examination (1954, 1978, 1989, 1994, 1995, 1997, 2003, 2006, 2008, 2011, 2014, 2017, 2018, and 2020) to assess the age and extent of soil disturbance. This was accomplished via an aerial photograph interpretation of stereoscopic images until 2014 and then via assessment of the differences in the Normalized Difference Vegetation Index (NDVI) derived from Sentinel-2 images via the Google Earth Engine as described in [28]. Additionally, 8 soil profiles were dug, described, sampled and analyzed at the end of summer 2023. In total, there were 762 soil observations in the municipal area (3/km^2) and 346 in the study area (6/km^2), 248 of which were new observations from the ad hoc survey of urban and peri-urban soils and 98 previous ones. The collected samples for the reference depth 0–30 cm were analyzed for texture (sand, silt and clay) limestone content %, pH, organic carbon %, and cation exchange capacity; in total, 592 samples from 172 sites were analyzed. Soil textural fractions (USDA limits) [31] were determined with the pipette method (DM 13/09/1999, Method II.5, II.6) in the case of clay and silt and with the sieve method in the case of sand (DM 13/09/1999, Method II.5); pH was determined in H_2O with a 1:2.5 soil to water ratio (DM 13/09/1999, Method III.1), the percentage of total $CaCO_3$ was measured with the gas volumetric method (DM 13/09/1999, Method V.1), and organic C % was assessed with an elemental analyzer (DM 13/09/1999, Method VII.1) [32].

Through the integration of the multitemporal aerial photos' series with the map of the sealed areas available for the whole of Italy at a 10 m resolution [28], the LIDAR digital model with a 2 m resolution, and the soil observations and analyses, the preliminary map of urban pedolandscapes (1:10,000) and the soil map of the entire municipality (1:50,000) were updated and finalized.

2.3. Soil Functions and SESs Assessment

The assessment of the soil contribution to seven urban SESs resorted to indicators based on measured or estimated soil properties assumed as a proxy of soil functions according to the methodology described in Calzolari et al. [23]. The method considers a set of indicators of soil functionality linked to the provision of ecosystem services. The ecosystem services, the underpinning soil functions and the soil data required as input for their calculation are presented in Table 1.

Table 1. Ecosystem services (ESs), underpinning soil functions, indicators and input data.

Ecosystem Service [a]	CICES Code 5.1 [b]	Soil Contribution to ES [c]	Soil Function	Indicator	Input Data for Calculation	Code
Regulating	2.2.1.1 2.3.3.2	Buffering capacity for nutrients and pollutants: natural attenuation (potential)	Storing filtering and transforming nutrients, substances and water	Cation exchange capacity (CEC) Soil reaction Rooting depth	C org. % Clay % pH Coarse fraction %	BUF
Regulating	2.1.1.2 2.3.3.2	Carbon sequestration (potential)	Carbon pool	Carbon sequestration actual	C org % Bulk density	CST

Table 1. *Cont.*

Ecosystem Service [a]	CICES Code 5.1 [b]	Soil Contribution to ES [c]	Soil Function	Indicator	Input Data for Calculation	Code
Provisioning	1.1.1.1	Food provision (potential)	Biomass production	Land capability (LC) map	LCC and integrades	PRO
Provisioning	1.1.1.x 1.1.5.x	Biomass supply (potential)	Biomass production	NDVI average 2015–2020	NDVI (LANDSAT8)	BIOMASS
Regulating	2.2.1.3	Water regulation: Runoff-flood control (potential)	Storing filtering and transforming nutrients, substances and water	Infiltration capacity	Ksat (mm/h) Psi_e (cm)	WAR
Regulating (Provisioning)	2.2.1.3 (4.2.2.2)	Water regulation: Water storage (potential)	Storing filtering and transforming nutrients, substances and water	Water content at field capacity Presence of water table	Field capacity (−33 kPa)	WAS
Supporting	2.2.2.3	Habitat for soil organisms	Biodiversity pool	Potential habitat for soil organisms	Index QBS-ar Covariates DSM	BIO

[a] MEA, 2005 [33]; [b] CICES Haines-Young, R., Potschin, M.B., 2018 [34]; [c] Dominati et al., 2010 [35].

Table 2 reports the calculations applied to derive each indicator, as specified in Calzolari et al. [23], for all of the indicators reported in Table 1, except for BIO and BIOMASS. The former was derived from applying the digital soil mapping (DSM) technique based on machine learning algorithms [36] to the set of Soil Biological Quality index values [37] available at a regional scale, while the latter is based on the value of the Normalized Difference Vegetation Index (NDVI) [38] derived from Landsat8 via Google Earth Engine [39].

Table 2. Input data and calculations of soil ecosystem service (SESs) indicators.

SES Code	Input Data	Calculation
BUF	CEC (cmolc/kg) depending on OC (%) and clay (%) CEC = 6.332 + 0.404 clay + 1.690 OC (R^2 = 0.75) pH Coarse fragments content, sk (%) Average depth of shallow water table, WT (cm)	BUF0-1 = Log CEC (pH; sk)0-1 with pH < 6.5 reduction by 0.25 or 0.5 depending on CEC and by 0.25 for sk > 30% Water Table (WT) depth < 30 cm BUF0-1 = Log CSC (pH; sk)0-1 × WT/30
CST	Organic carbon, OC (%) Bulk density, BD (Mg m^{-3})	$CST_{0\text{-}1} = \log [OC \times BD \times (1\text{-}sk)]_{0\text{-}1}$
PRO	Land capability (LC) classes and intergrades [39]	LCC reclassification (0-1)
BIOMASS	NDVI (Normalized Difference Vegetation Index)	Standardization (0-1) NDVI (average median values 2015–2020)
WAR	Saturated hydraulic conductivity, Ksat (mmh^{-1}) Air entry potential, PSIe (cm)	$WAR_{0\text{-}1} = \log Ksat_{0\text{-}1} - PSIe_{0\text{-}1}$

Table 2. *Cont.*

SES Code	Input Data	Calculation
WAS	Water content at field capacity (−33 kPa), WCFC (vol/vol) Average depth of water table, WT (cm) sk, coarse fragments (Ø > 2 mm, vol/vol)	$WAS_{0\text{-}1}$ = $(WC_{FC} \times 1\text{-sk})_{0\text{-}1}$ if WT > 100 cm, and $WAS_{0\text{-}1}$ = $(WC_{FC} \times 1\text{-sk}) \times WT/100$ if WT < 100 cm
BIO	Soil Biological Quality index, QBS_{ar} [40] Covariate per digital soil mapping	Spatialization of QBS_{ar} point data values via DSM (Quantile Random Forest)

The Land Capability Classes (LCCs), being originally assessed for agricultural and forest lands [40,41], were also defined for the urban area based on the urban soil units of the new map on scale 1:10,000. The hydraulic soil properties listed in Table 2, BD (bulk density, Mg/m^3), Ksat (saturated hydraulic conductivity, mmh^{-1}), PSIe (air entry potential, cm), WCFC (water content at field capacity, vol/vol) were calculated, resorting to a set of locally calibrated pedotransfer functions (PTFs) [42,43]. The average shallow groundwater depth WT (cm) was derived from Calzolari and Ungaro (2012) [44].

The values of each SES indicator are given as numbers in the range from 0 to 1, resorting to an interval normalization transformation [45]:

$$Xi_{0\text{-}1} = (X_i - X_{min})/(X_{max} - X_{min}) \tag{1}$$

where $Xi_{0\text{-}1}$ is the standardized value [0, 1], Xi is the current value, and X_{min} and X_{max} are, respectively, the maximum and minimum of the indicator observed in the considered territory. The maximum observed value is set equal to 1, and the value 0 indicates the relative minimum in the area considered. The results are then strongly influenced by the degree of variability of the measured and estimated soil properties observed at the scale of investigation. By considering specific portions of territory of administrative, planning or management relevance (e.g., province, union of municipalities, municipality), the indicators can be normalized over the area of interest at the scale of investigation. In this case, the indicators have been normalized in the range 0–1 over the entire municipal territory. A synthetic index of soil quality (IQ4) based on the potential supply of ecosystem regulating and provisioning services was eventually calculated as the sum of four indicators: BUF, CST, PRO and WAR. The IQ4 values were then classed 1 to 5 based on the breakdown of the observed distribution, with 1 and 5 indicating the highest and the lowest quality, respectively.

2.4. Mapping Urban Soils Properties and Ecosystem Services

Through the geostatistical processing of point data, followed by the use of locally calibrated PTFs, the maps of soil properties at the basis of the estimation of the SESs indicators were produced for the entire municipal territory. Seven maps were therefore derived, one for each SESs, in raster format with a hybrid resolution, i.e., 25 m for the core urban area and 100 m for the rest of the municipal territory.

The spatialization of the point data over the entire municipal area was carried out using sequential Gaussian simulations (SGSs) [46]. Being based on multiple realizations, SGSs allow for the assessment of uncertainty in predictions, which is crucial for risk-based decision making. Furthermore, SGS realizations maintain the spatial autocorrelation and statistical properties of the data, ensuring realistic representations of soil variability, avoiding the shortcomings of deterministic interpolation methods and the inability of machine learning algorithms to deal with spatial autocorrelation. Although computationally demanding, SGS can easily integrate auxiliary data in the process of simulation: in the case presented in this work, the outcomes were conditioned on the mean values of the urban soil map delineation.

The implementation of the SGSs (N = 25) required data transformation into normal scores, i.e., the normalized residues of the average values calculated for the individual properties (sand, silt, clay, pH and C org) at the level of the soil map delineations at a scale of 1:50,000 for the entire municipal territory and at a scale of 1:10,000 for the urbanized core area. Once each point was associated with the mean value of the delineation in which it falls, the residual value (deviation from the mean value) was calculated by difference. Once normalized via a normal score transform, this was used to estimate and model the experimental semivariograms needed to implement the geostatistical simulations through ordinary kriging on a regular grid of 25 m in the urban area and 100 m in the remaining municipal territory. The mean value of the 25 simulations was then subjected to an inverse transformation to obtain the residual value to which the mean value of the delineation was finally added to obtain the estimate value of each variable. An exception to this procedure is represented by the percentage of coarse fragments, whose distribution is mainly represented by values equal to zero, a situation in which a geostatistical approach is not applicable. For this reason, in the case of the coarse fragments content, the mean value attributed to the delineations of the soil map was used based on the available data (N = 1121). All the geostatistical analyses presented in this paper were carried out with the software Wingslib1.3.1 [47], which works in conjunction with the GSLIB90 executables [48]. All GIS operations and mapping were performed using QGIS v3.22.11 [49].

3. Results

3.1. Urban Soil Map 1:10,000 and Degree of Soil Disturbance

Evidence from the survey allowed in the first place to assess the degree of soil disturbance at each sampling site (Figure 3a), which was summarized in seven different classes as follows:

1. Undisturbed "natural" soils found in agricultural areas but also in public parks and uncultivated land;
2. Disturbed soils with superficial fill of allochthonous soil material in the topsoil;
3. Deep disturbance in place: like 2 but the disturbance is deeper with overturned horizons and buried soil;
4. All fill: often it is made of building material mixed with soil with high percentages of coarse elements so that augering and sampling are impeded;
5. Disturbed after sampling;
6. Urbanized and sealed;
7. Natural topsoils reworked: no fill, mixing of Ap with bricks and artifacts. The bricks can be from the Roman age. They occur quite frequently also in agricultural areas.

Furthermore, urban soils are almost always characterized by thin surface layers enriched in organic matter, as they are almost always found under permanent herbaceous cover. The seven classes of soil disturbance were reduced to four (Figure 3b) when applied to the 79 mapping units identified in the urban core area (Figure 4): (i) undisturbed "natural" soils (merging classes 1 and 7); (ii) disturbed soils (merging classes 2 to 5); (iii) urbanized soils at different degree of sealing (class 6); and (iv) urbanized/disturbed mixed class.

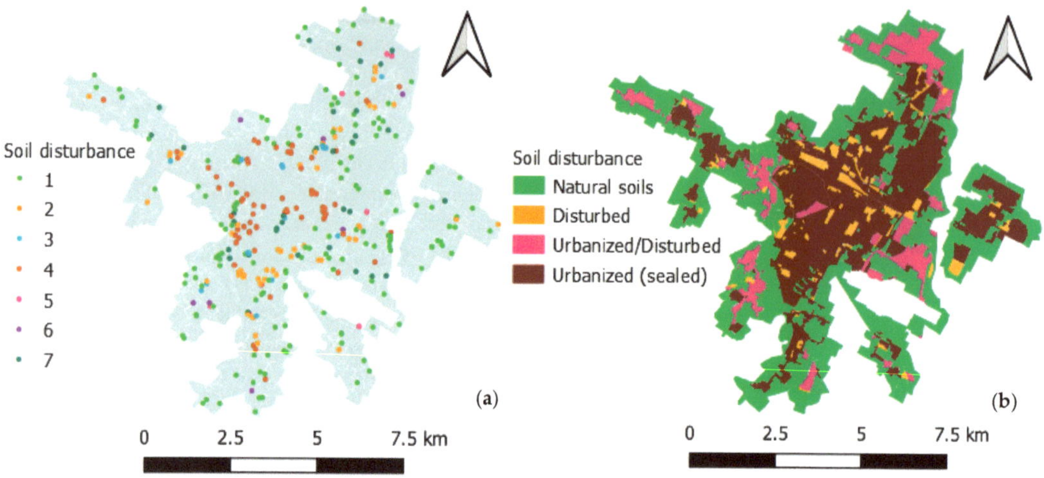

Figure 3. Degree of soil disturbance (**a**) at sampling points and (**b**) over the surveyed urban core area.

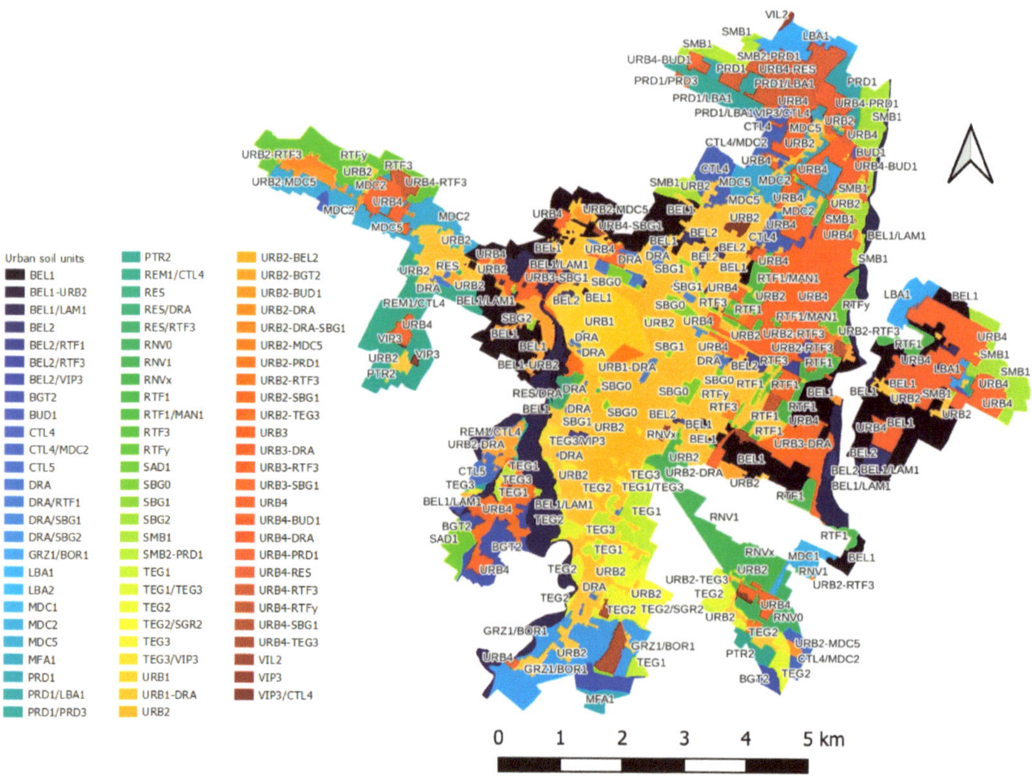

Figure 4. Soil map (scale 1:10,000) of the urban core area. A description of urban soils mapping unit is provided in the Supplementary Materials (Table S1).

Undisturbed soils are the majority (48.4% of the area, 2771.43 ha) and are mostly found in peri-urban and intra-urban agriculture, the latter being quite widespread in Forlì. A small percentage of these soils is also found in urban green areas or in areas included in industrial zones. Disturbed soil was mostly observed within urban and industrial areas; in small percentages, they were also found in the agricultural land (mostly abandoned; the most frequent use is lawn). The percentage of mapped occurrence is about 5% (427 ha), but it is believed that it is probably higher if we consider small gardens, urban flowerbeds, roundabouts, etc. that were not mapped due to their small size.

To identify the extent of the disturbance, we relied on the survey data and the examination of aerial photos (from 1936 to 2020). With a few exceptions, the disturbed urban soils are derived from overturning and carrying over of soils present on site or from nearby locations with a very low occurrence of allochthonous soil material. Urban soils encompass all soils that had artifacts such as bricks, ceramic, cement, plastic, and building materials within the profile. Four soil typologies (Figure 5) were created for the urbanized area (36.02%, 2051.25 ha) based on the degree of soil sealing and risk of flooding: (i) URB1 (134.26 ha, 2.36%): historical center, where the occurrence of unsealed soil is variable but overall quite low; (ii) URB2 (1177.15 ha, 20.67%): residential areas generally characterized by single houses or buildings with small gardens and green areas characterized by the presence of disturbed soils; (iii) URB3 (11.45 ha, 0.20%): urbanized area comparable to URB2 in terms of soil sealing but located in areas at higher risk of flooding (floodplain areas); (iv) URB4 (728.39 ha, 12.79%): industrial areas where the percentage of soil sealing is very high—higher than the URB2 and URB3 units. The unsealed areas were mapped in most cases; they are private green areas such as lawns or small gardens.

Figure 5. Examples of the four urbanized mapping units: (**a**) URB1; (**b**) URB2; (**c**) URB3 and (**d**) URB4. Source: Google Earth, AIRBUS image 28 May 2023.

Very often, these urbanized areas, even if contiguous, were divided into separate mapping units based on the underlying original soils as derived from the 1:50,000 soil map. Finally, some urbanized areas have larger unsealed surfaces for which some cartographic units have been identified, which are associations of urbanized and interlocked disturbed soils (612.93 ha, 10.78%).

3.2. Maps of Soil Properties

Table 3 summarizes the descriptive statistics of the soil properties requested for the assessment of the indicators of potential SESs supply.

Table 3. Descriptive statistics of selected soil properties (0–30 cm) for the available soil data used for SESs assessment in the municipality of Forlì.

Variable		Agricultural Soils	Undisturbed Soils	Disturbed Soils	Urbanized Soils	Urbanized/ Disturbed Soils	Urban Core Area	Forlì Whole Area
	Num. Obs.	792	195	71	37	26	355	1121
Sand %	Mean	20.64	19.64	20.11	19.16	19.45	19.62	20.35
	Dev. Std.	7.09	9.16	7.47	8.29	10.41	8.66	7.64
	Min	3.87	3.38	6.22	6.22	3.87	3.38	3.38
	Max	46.97	43.38	27.20	35.43	43.38	43.38	46.97
Silt %	Mean	50.58	51.12	52.41	50.36	52.04	51.33	50.82
	Dev. Std.	4.88	3.78	2.71	3.57	5.79	3.81	4.60
	Min	33.65	36.05	47.19	43.76	36.05	36.05	33.65
	Max	62.94	60.36	56.74	56.74	60.36	60.36	62.94
Clay %	Mean	28.78	29.24	27.48	30.48	28.51	29.04	28.83
	Dev. Std.	6.36	8.35	7.64	7.51	8.12	7.97	6.91
	Min	16.41	16.41	21.42	20.24	16.41	16.41	16.41
	Max	43.96	42.78	42.78	42.78	42.78	42.78	43.96
C org. %	Mean	1.03	1.33	1.77	1.50	1.31	1.43	1.15
	Dev. Std.	0.29	0.46	0.45	0.44	0.47	0.48	0.41
	Min	0.46	0.76	1.08	0.81	0.76	0.76	0.46
	Max	2.34	2.21	2.68	2.21	2.21	2.68	2.68
pH	Mean	7.81	7.82	7.71	7.82	7.87	7.80	7.80
	Dev. Std.	0.27	0.18	0.16	0.17	0.18	0.19	0.24
	Min	6.56	7.44	7.48	7.57	7.56	6.99	6.56
	Max	8.29	8.13	8.05	8.05	8.13	8.13	8.29
Skel %	Mean	0.09	0.98	1.59	1.33	0.70	1.08	0.40
	Dev. Std.	0.51	1.58	1.63	1.93	1.12	1.62	1.09
	Min	0.00	0.00	0.00	0.00	0.00	0.00	0.00
	Max	6.97	6.97	6.97	6.97	3.25	6.97	6.97

From the values reported in Table 3, it is concluded that compared to agricultural soils (N = 792), the soils of the urban area (N = 355) are characterized by a significantly lower sand content ($p = 0.1$) as well as significantly higher silt, skeleton and organic C contents ($p = 0.05$). Likewise, although not statistically significant, higher pH values are found in highly urbanized areas. Analyzing the data in terms of type of disturbance within the urban area, it can be observed that the parameters considered are generally characterized by greater variability (higher standard deviation values), especially in the case of urbanized and urbanized/disturbed soils, as can be seen in the box and whisker plots shown in Figure 6.

The description of eight soil profiles in the urban green areas (Table 4) was relevant for urban soils mapping purposes, as their field evidence and analyses provided the information necessary to highlight the relationships between in situ soil materials and exogenous material, either soil or of anthropic nature. In this way, it became possible to assess the actual nature and extent of anthropic soil disturbance as well as its effects on the provision of ecosystem services. Analytical data from the soil profiles are provided in the Supplementary Materials (Table S2).

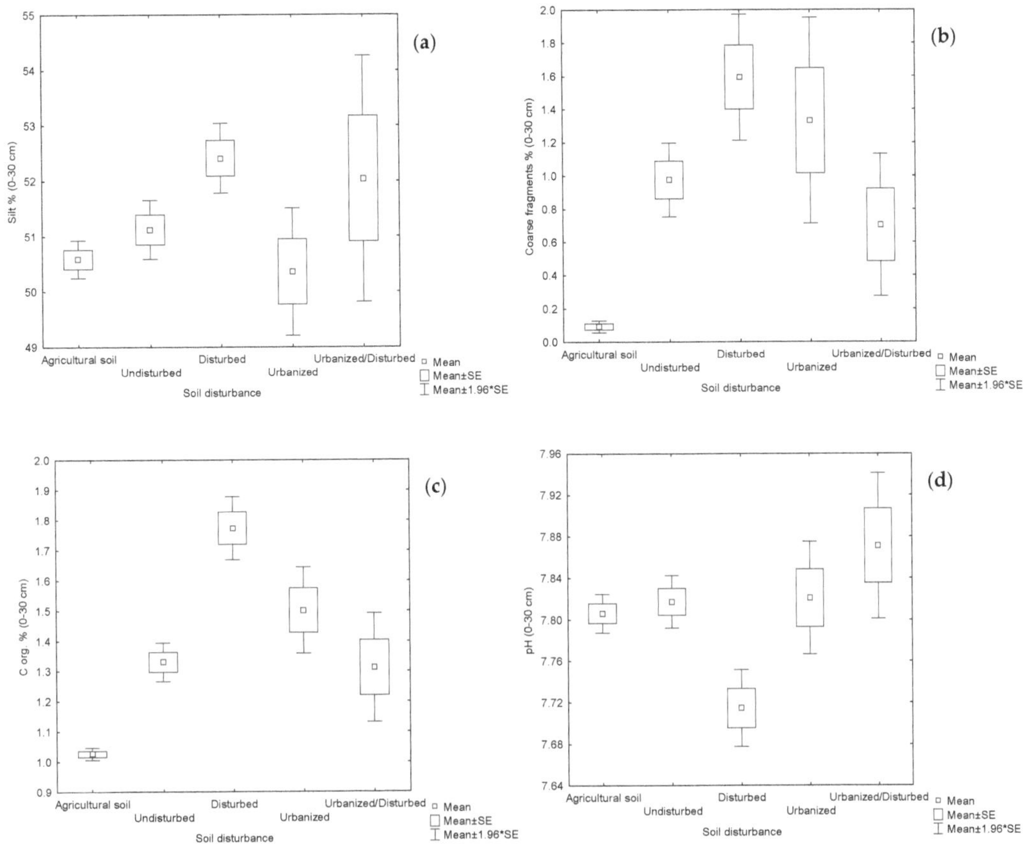

Figure 6. Box and whiskers plots for selected soil properties (0–30 cm) grouped in terms of soil disturbance: (**a**) silt %; (**b**) coarse fragments %; (**c**) organic C % and (**d**) pH.

The procedure for estimating the indicators of ecosystem services was preceded by the estimation of the necessary chemical–physical parameters. To this end, the point values measured in the sampling sites were interpolated via geostatistical conditional simulations over the study area to reconstruct the continuous spatial distribution of the parameters reported in Table 3, which was followed by the application of a set of PTFs calibrated on regional data sets from Emilia–Romagna.

The parameters of the semivariograms models (Figure S1), which were used for the spatial interpolation of the normalized residuals of the three textural soil fractions, pH and organic C content, are shown in Table 5. For all variables, a double spherical model with nuggets provided the best interpolation of the experimental semivariograms. The normalized residuals of the considered soil properties have a share of spatially uncorrelated variance (nugget, C_0) ranging between 24 and 42%. The two spatially structured components, C_1 and C_2, accounted for similar shares of the residual variance only in the case of pH, while for the textural fractions and organic C content, the short-range component explained most of the spatially structure variance. The resulting maps are shown in Figure 7a–e; in the case of coarse fragments content, Figure 7f reports the average values of the typological soil units.

Table 4. Site description and classification of eight urban soil profiles in Forlì.

Site	Site Description	Soil	Classification WRB 2014
Public park	Permanent meadows, not irrigated Natural soil with surface fill	BEL2	Hypereutric Cambisols (Transportic)
Public park	Permanent meadows, not irrigated Surface fill down to 120 cm	DRA	Calcaric Regosols (Prototechnic, Transportic)
Public park	Permanent meadows, not irrigated Surface fill 0–60 cm over in situ alluvial deposits	RTFy	Hypereutric Cambisols (Transportic)
Public park	Permanent meadows, not irrigated Surface fill down to 120 cm	RES	Calcaric Regosols (Prototechnic, Transportic)
Public park	Meadows, old trees, irrigated Surface fill 0–55 cm over in situ alluvial deposits	TEG3	Hypereutric Cambisols (Transportic)
Public park	Sealed with trees is small allotments Planned de-sealing area	TEG3	Hypereutric Cambisols (Transportic)
Public park	Permanent meadows, not irrigated Surface fill down to 150 cm	DRA	Calcaric Regosols (Prototechnic, Transportic)
Public park	Permanent meadows, not irrigated Surface fill down to 190 cm	SBG1	Urbic Ekranic Technosols (Calcaric Epitechnoskeletic)

E7809P0001 E7809P0002 E7809P0003 E7809P0004 E7809P0005 E7809P0006 E7809P0007 E7809P0008

Table 5. Semivariogram model parameters for selected soil properties. For all properties, the parameters refer to nested spherical models. The spherical semivariogram model can be written as $\gamma(h) = C_0 + \Sigma_{ni=1} C_i\ (1.5\,h/r_i - 0.5h^3/ri^3)$, for $h \leq r_i$, where h is the distance (m), C_0 is the nugget, C_i (i = 1, . . ., n) is the sill of the i nested structure, and r_i is its spatial range (m).

Variable	Nugget C_0	Model	Sill C_1	Range a_1 (m)	Sill C_2	Range a_2 (m)
Sand %	0.30	Sph. + Sph.	0.60	600	0.10	10,000
Silt %	0.25	Sph. + Sph.	0.63	700	0.14	12,000
Clay %	0.24	Sph. + Sph.	0.57	450	0.18	6700
Organic C %	0.42	Sph. + Sph.	0.46	1058	0.19	5353
pH	0.37	Sph. + Sph.	0.40	996	0.37	2872

At the nodes of a 25 m regular grid over the urban core area, and of a 100 m regular grid for the rest of the municipality, a set of locally calibrated pedotransfer functions estimated the values of the soil properties requested to calculate the indicators for the selected soil ecosystem services. These PTF estimated properties, depicted in Figure 8, were bulk density (BD, Mg m^{-3}), which provided the necessary input to estimate the soil C stock (Mg/ha, for the reference 0–30 cm depth), saturated water conductivity (Ksat, mm h^{-1}), air entry point (PSIe, cm), water content at field capacity (WCFC, vol./vol.), and cation exchange capacity (cmol/kg).

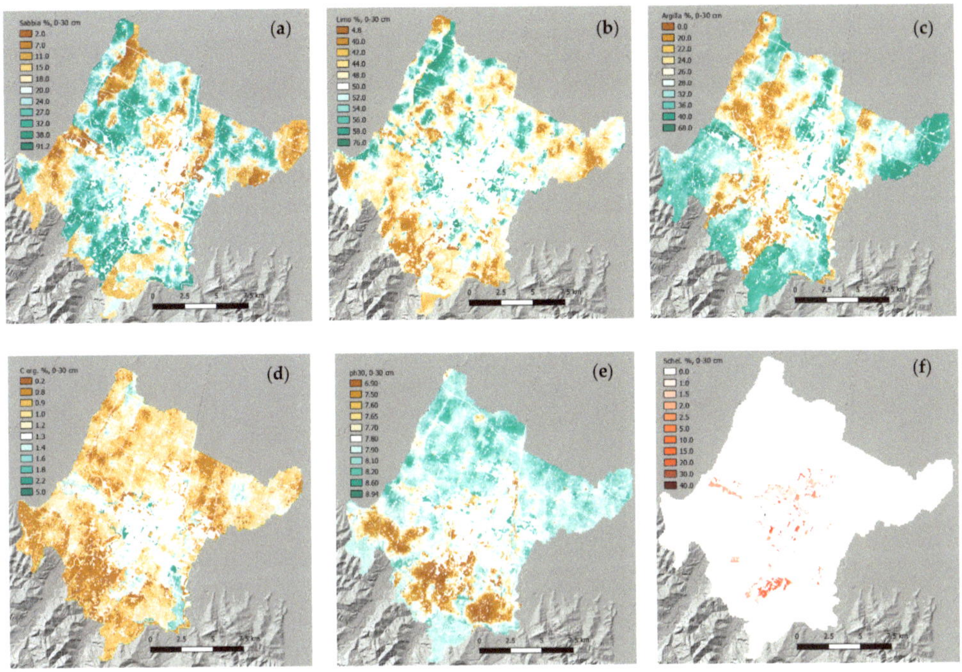

Figure 7. Raster maps of basic soil properties (0–30 cm): (**a**) sand %, (**b**) silt %, (**c**) clay %, (**d**) organic C %, (**e**) pH, (**f**) coarse fragments %.

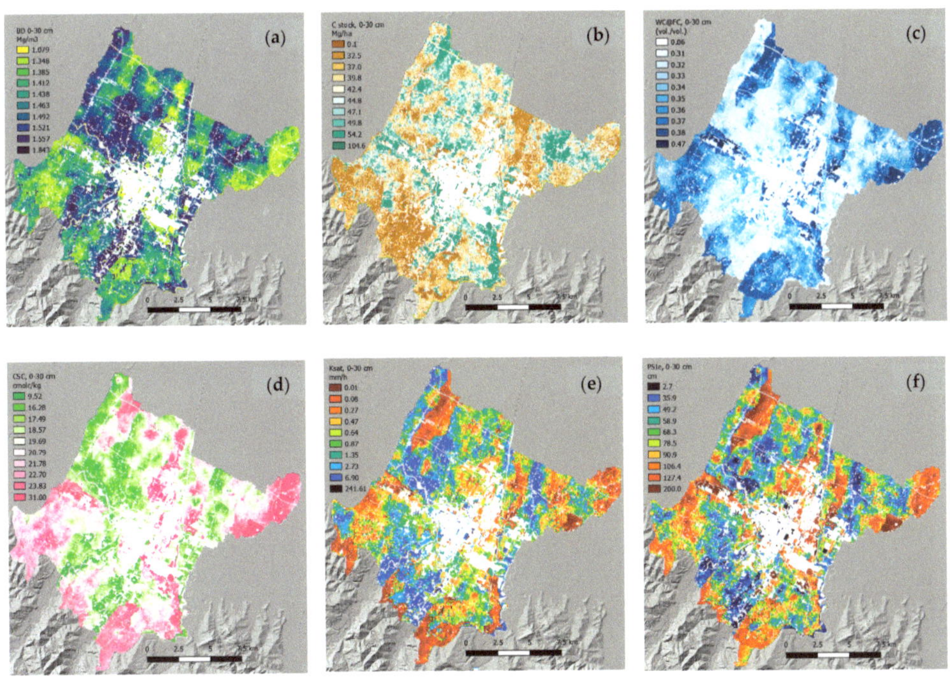

Figure 8. Raster maps of PTFs-derived soil properties (0–30 cm): (**a**) bulk density (Mg/m^3), (**b**) organic C stock (Mg/ha), (**c**) water content at field capacity (vol./vol.), (**d**) cation exchange capacity (cmol/kg), (**e**) saturated hydraulic conductivity (mm/h), and (**f**) air entry tension (cm).

3.3. *Urban Soils Ecosystem Services: Assessment and Mapping*

The maps of the seven indicators of soil ecosystem services considered are shown in Figure 9a–g. All indicators were rescaled on local variability (minimum and maximum values) so to have values between zero and one. The raster maps have a hybrid resolution, which is equal to 25 m in the urban area specifically surveyed and 100 m in the predominantly agricultural areas in the rest of the municipal territory. In the case of PRO, the indicator was based on the 0–1 standardization of the LC classes detected on the new LC map; both map and LC standardized values are provided in the Supplementary Materials in Figure S1 and Table S3, respectively.

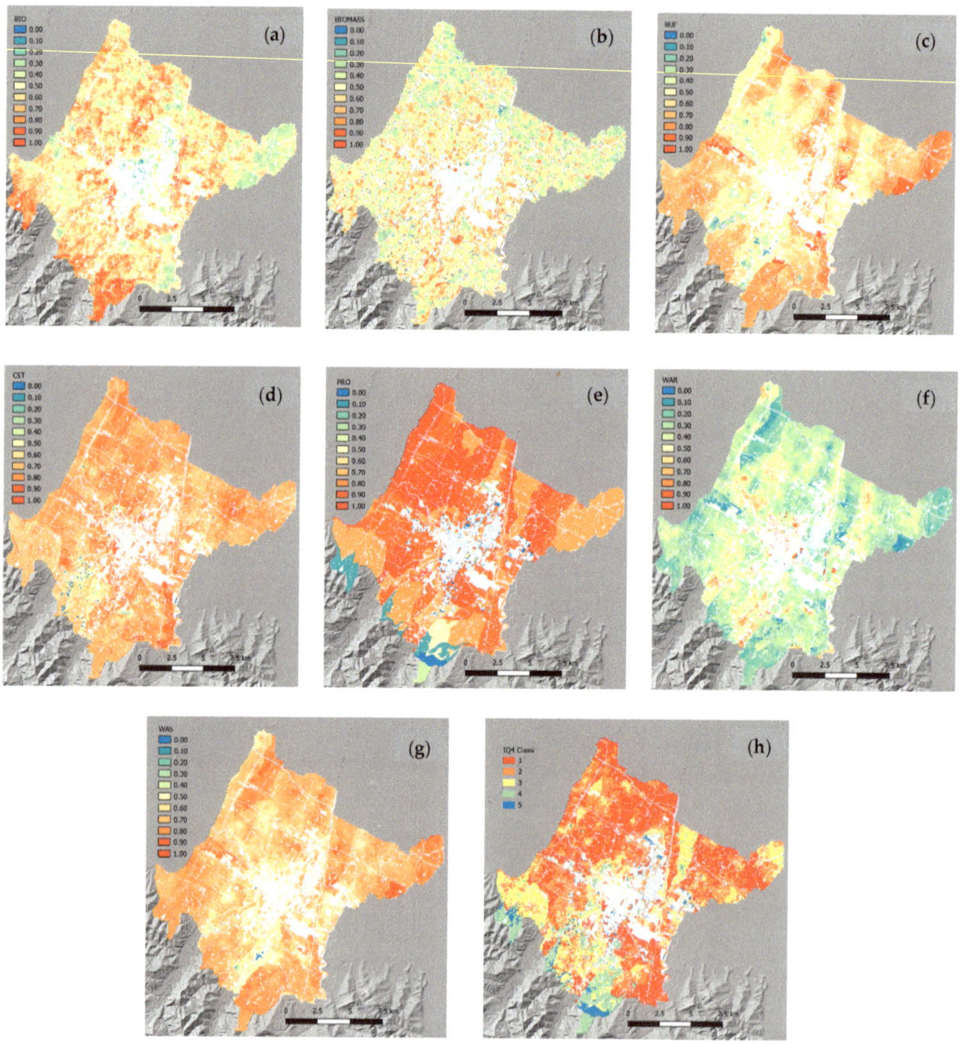

Figure 9. Raster maps of soil ecosystem services (SESs) in the municipality of Forlì (0–30 cm): (**a**) BIO (habitat for soil organisms), (**b**) BIOMASS (biomass supply), (**c**) BUF (buffering capacity), (**d**) CST (carbon sequestration), (**e**) PRO (food provision), (**f**) WAR (water regulation), (**g**) WAS (water storage), (**h**) SESs-based quality indicator (IQ4). The value of IQ4 was classed based on the breakdown of the observed distribution of the values obtained by summing the scores of the indicators BUF, CST, PRO and WAR.

The values of the indicators displayed in the maps in Figure 9 were eventually used to characterize the potential ecosystem services provision by soils considering the typological units of the urban soils map (N = 79, Figure 4) and as a function of the degree of disturbance to which the soils of the urbanized area surveyed have been subjected (N = 4, Figure 3b). The descriptive statistics of each indicator for the different cartographic units of the urban soil map can be used to verify which ecosystem services are in synergy with each other and which are antagonistic. Limiting ourselves to considering the first 30 units by territorial extension, which cover a total of 92% of the urbanized area (2546 ha) and considering only the soil surfaces free from infrastructures and buildings, in terms of average values of each indicator, we obtained the figures shown in Table 6 and graphically summarized by the radar chart in Figure 10.

Table 6. Average values of ecosystem services indicators and soil quality index IQ4 for the mapping units of the urban soil map.

Mapping Units	Soil Disturbance	Area, ha	Area Share	BIO	BIOMASS	BUF	CST	PRO	WAR	WAS	IQ4
BEL1	undisturbed	468.8	0.17	0.60	0.60	0.52	0.70	0.86	0.35	0.59	2.424
BEL1/LAM1	undisturbed	274.9	0.10	0.69	0.68	0.52	0.75	0.71	0.40	0.58	2.381
MDC2	undisturbed	156.1	0.06	0.54	0.60	0.70	0.73	0.86	0.27	0.67	2.561
URB2	urbanized/sealed	130.1	0.05	0.34	0.36	0.59	0.78	0.00	0.50	0.61	1.862
SMB1	undisturbed	128.6	0.05	0.59	0.56	0.53	0.66	1.00	0.41	0.58	2.598
GRZ1/BOR1	undisturbed	116.6	0.04	0.46	0.59	0.44	0.77	0.65	0.41	0.44	2.273
RTFy	undisturbed	116.2	0.04	0.61	0.57	0.87	0.75	0.86	0.11	0.81	2.588
REM1/CTL4	undisturbed	94.6	0.03	0.68	0.62	0.61	0.65	0.86	0.41	0.61	2.532
TEG2	undisturbed	93.3	0.03	0.64	0.64	0.53	0.73	0.86	0.33	0.60	2.438
TEG1	undisturbed	86.3	0.03	0.62	0.61	0.46	0.72	1.00	0.41	0.55	2.587
BGT2	undisturbed	84.8	0.03	0.61	0.58	0.68	0.54	0.86	0.31	0.63	2.391
CTL4	undisturbed	81.9	0.03	0.51	0.59	0.67	0.70	0.86	0.31	0.65	2.532
RNV1	undisturbed	79.1	0.03	0.61	0.63	0.72	0.70	0.86	0.30	0.67	2.579
RTF1	undisturbed	71.4	0.03	0.56	0.60	0.73	0.77	0.86	0.23	0.71	2.591
URB4	urbanized/sealed	70.5	0.03	0.39	0.38	0.62	0.70	0.00	0.38	0.62	1.689
LBA1	undisturbed	68.7	0.02	0.54	0.56	0.81	0.70	0.71	0.18	0.75	2.395
PTR2	undisturbed	59.9	0.02	0.59	0.63	0.37	0.60	1.00	0.60	0.50	2.569
TEG2/SGR2	undisturbed	40.3	0.01	0.73	0.78	0.60	0.79	0.92	0.31	0.62	2.622
PRD1	undisturbed	32.2	0.01	0.51	0.52	0.58	0.69	0.86	0.34	0.62	2.471
BEL2	disturbed	31.6	0.01	0.49	0.57	0.49	0.72	0.86	0.68	0.56	2.746
URB4-RES	urb./disturbed	31.5	0.01	0.38	0.44	0.65	0.62	0.39	0.20	0.65	1.865
RNV0	undisturbed	30.8	0.01	0.49	0.61	0.88	0.76	0.79	0.17	0.77	2.604
DRA	disturbed	28.8	0.01	0.35	0.65	0.56	0.71	0.86	0.50	0.60	2.621
MDC1	undisturbed	26.9	0.01	0.70	0.68	0.89	0.90	0.71	0.32	0.76	2.816
PRD1/LBA1	undisturbed	25.3	0.01	0.55	0.61	0.71	0.66	0.79	0.31	0.66	2.477
RTF1/MAN1	undisturbed	25.1	0.01	0.39	0.62	0.72	0.74	0.79	0.25	0.69	2.501
SAD1	undisturbed	24.1	0.01	0.58	0.63	0.61	0.45	0.71	0.29	0.63	2.059
CTL4/MDC2	undisturbed	23.8	0.01	0.50	0.56	0.64	0.77	0.86	0.32	0.61	2.582
SBG0	disturbed	22.2	0.01	0.25	0.54	0.47	0.65	0.57	0.77	0.47	2.458
PRD1/PRD3	undisturbed	21.6	0.01	0.81	0.63	0.62	0.74	0.86	0.27	0.65	2.487

From the values in Table 6 and the trend of the values in Figure 10, we can appreciate the severe drop in the average values of the PRO, BIO, and BIOMASS indicators in urbanized and urbanized/disturbed soils. The decrease was on average more contained for WAS and BUF, while in the case of WAR in the URB1 unit, an average value higher than 0.8 was observed. On the other hand, the average of CST remained high with values > 0.7, which was close to 0.8 in the URB2 and URB1 units.

Table 7 reports the correlation coefficients between the average values of the ecosystem services indicators of the mapping units of the urban soil map. In most cases, the services were in synergy with each other, except for WAR and WAS, and WAR and BUF.

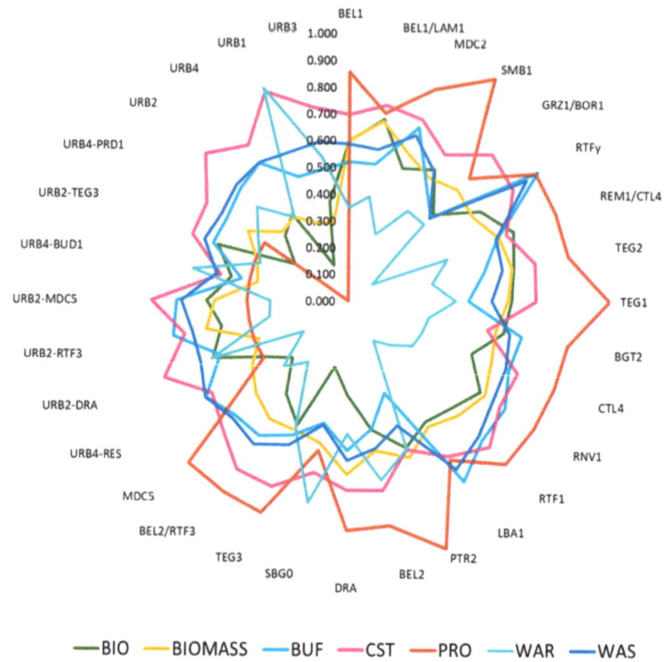

Figure 10. Average values of seven SESs indicators in the most widespread mapping units of the urban soil map.

Table 7. Pearson correlation coefficients between the mean values of the ecosystem services indicators of the mapping units of the urban soil map. Statistically significant correlations are in bold ($p = 0.05$).

Indicator	Mean	Std. Dev.	BIO	BIOMASS	BUF	CST	PRO	WAR	WAS
BIO	0.484	0.180	1.000	**0.506**	**0.427**	**0.373**	**0.437**	−0.102	**0.421**
BIOMASS	0.519	0.156	**0.506**	1.000	**0.328**	**0.351**	**0.762**	0.218	**0.330**
BUF	0.580	0.152	**0.427**	0.328	1.000	**0.711**	**0.277**	**−0.342**	**0.890**
CST	0.680	0.156	**0.373**	0.351	0.711	1.000	**0.258**	0.089	**0.838**
PRO	0.649	0.281	**0.437**	0.762	0.277	0.258	1.000	−0.001	**0.305**
WAR	0.395	0.183	−0.102	0.218	−0.342	0.089	−0.001	1.000	−0.217
WAS	0.587	0.137	**0.421**	**0.330**	**0.890**	**0.838**	**0.305**	−0.217	1.000

What was highlighted at the mapping units level became even more evident when observing the trend of the weighted average values as a function of the type of disturbance that characterizes each mapping unit (Figure 11). The BIO indicator was on average significantly higher in the "undisturbed" soils of peri-urban agricultural areas with a decreasing trend as the degree of disturbance increased. In the case of the BIOMASS indicator, the average value of disturbed soils was close to that of undisturbed soils and was significantly higher than that of urbanized and urbanized/disturbed soils. As for PRO, the average values of the mapping units characterized by undisturbed agricultural soils and disturbed soils were both higher than 0.8, and they were significantly higher than the average values characterizing the units with intermediate typologies urbanized/disturbed (0.377 ± 0.054) and obviously of the completely urbanized ones where the PRO value was null. The average values of the BUF, CST and WAS indicators were not significantly affected by the degree of disturbance and turned out to be very close to each other. It is interesting to note that in the case of CST, the soils of the green areas of the urbanized fabric had the highest average value (0.708 ± 0.098), which was slightly higher than that estimated for the agricultural soils of the peri-urban area (0.700 ± 0.031). The BUF and WAS indicators were on average higher in undisturbed agricultural soils and in urbanized soils. Finally, in the case of WAR, the indicator showed an average and significantly higher value in the mapping units characterized by the presence of disturbed soils (0.531 ± 0.088) and urbanized soils (0.441 ± 0.329). In this last case, however, a high variability was observed.

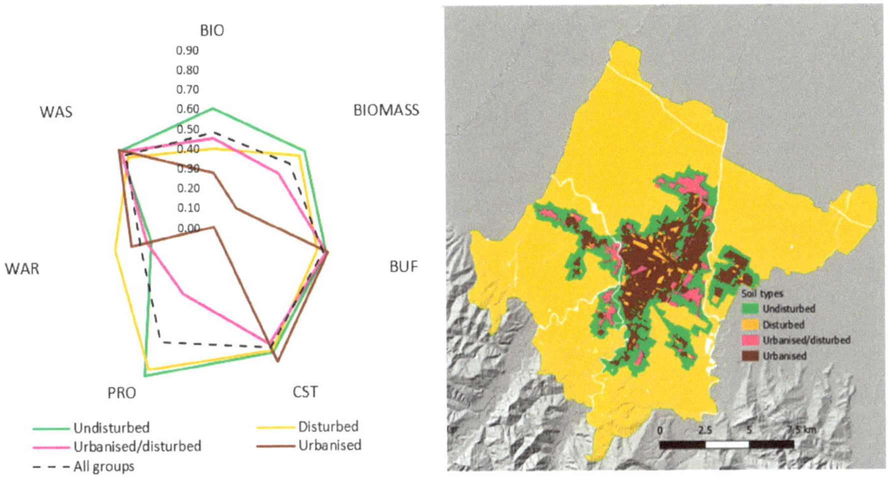

Figure 11. Average values of seven SESs indicators as a function of the degree of disturbance of urban soils. The yellow color identify the municipality of Forlì.

4. Discussion

The findings of this work need to be framed in the context of the urbanization dynamics which characterized the territory of Emilia–Romagna since the last two decades with a significant impact on the provision of soil-based ecosystem services. In 2023, the general figure in terms of soil sealing for the whole region reports an average value of 8.91%, which is equivalent to 2005.5 km^2 [28]; the same figures in 2006 were 8.36% and 1880.7 km^2 with an average loss of soil equal to 734 ha per year. This is indeed quite high a figure, as Emilia–Romagna is the third Italian region for land take compared to the national average (7.16%). However, this does not give a realistic idea of the level of urbanization of part of the plain areas of the region, as it obviously also accounts for all the hilly and mountainous areas, which sum up to nearly half of the region and are very sparsely inhabited. When considering the data at the municipal level, in fact, a much more worrying image emerges for the inhabited centers of the plains, especially for those located along the major NW–SE road axes with values often above 15%, as in the case of Forlì (16.5%, 3832 ha) [28]. Such a continuous loss of agricultural soil was not coupled with a positive demographic trend or with significant emigration from rural areas toward the city; rather, it is associated with changes associated with recent needs for new industrial settlements, large storage spaces for goods and increasingly widespread infrastructures. Considering the whole region, between 2003 and 2020, sealed areas grew constantly and significantly: urbanized areas, which include residences, commercial and industrial infrastructures and everything classified by the Corine Land Cover Classification (CLC) with code 1 increased by 10.4% (+25,984.2 ha). It must be noted that the most significant increases were recorded in some specific classes such as goods sorting plants (+126.4%), landfills and deposits of quarries, mines and industries (+69.6%), networks for the distribution and production of energy (+44.9%) and railway networks (+25.7%) [50]. In this scenario of constantly increasing soil sealing, the implementation of the Regional Law n. 24/2017 "Regional discipline on the protection and use of the land", aiming to zero net land take by 2050, is still limited and so far, it has missed its goals. This on the one hand depended on the possibility for the municipalities to postpone for three years the approval of their new General Urban Plans (PUGs), which would have introduced a limit of 3% on additional land take by 2050. On the other hand, it was determined by the exclusion from the 3% ceiling, established by the law, of public or public interest interventions, of the expansion of already established business activities and of new production facilities of significant regional and/or state interest. Nevertheless, even if the land take did not reverse its growing trend, the application of the Regional Law by the municipalities which implemented it through the approval of their PUGs spared from land take 15.274 ha of agricultural soil out of 21.922 which were planned for being built compliant with the old urban plans.

Similarly, Forlì increasingly lost an additional 127.7 ha of agricultural soil between 2017 and 2023 with a peak of 35 ha in 2023, further reducing the stock of natural capital represented by soils and impacting the potential supply of ecosystem services. It was estimated that between 1997 and 2016, Forlì lost due to land take approximately 570 ha of agricultural soils, corresponding to ca. EUR 194 million of average agricultural value and the capacity to produce approximately 370,000 tons of wheat per year. Of these soils, 78% were high-quality soils: deep alluvial soils with excellent chemical–physical fertility characteristics. In terms of land capability class (LCC), approximately 21% of the LCC I soils were lost, and more than 34% of LCC II soils were lost, suggesting the best soils are those consumed the most. In the same time span, the capacity of storing approximately 3.8 million m^3 of water was lost along with 319,000 tons of soil organic carbon, which was equivalent to 1,171,730 tons of CO_2 being released back into the atmosphere [51]. Furthermore, the recent and repeated severe flood events of 2023 and 2024 following unprecedented rainfall events highlighted the inadequacy of the prevention and mitigation strategies put in place so far and the absolute necessity to prevent further land take. To this goal, the municipality of Forlì in 2024 adopted the SES maps presented in this work as an integral part of its PUG in the attempt to reach two results: preserving the best-quality soils as identified by the IQ4 index from additional sealing and using the index to guide compensation measurements in all circumstances. Sealing is not avoidable through the identification of candidate areas for de-sealing whose quality and extent could compensate the further soil loss.

The methodology presented in this work, initially tested in the municipality of Carpi [22], was further refined for its application to the municipality of Forlì by adopting a finer estimation grid with a hybrid resolution to tackle the major inherent complexity of the Forlì urban soils and the higher spatial variation in the natural soils underlying the urban fabric. The SES assessment approach, although providing outputs in terms of standardized indicators scores and maps, is based on quantitatively assessed soil functions based on measured or PTF derived soil data. For example, in the case of the regulating ES indicators WAS and CST, it was calculated that in the 420 ha of urban green areas, the first 30 cm of soil in the green areas of the municipality of Forlì have the potential to store 17,707 m^3 of water and have sequestered 17,409 t of soil carbon, which was equivalent to 63,890 t CO_2 removed from the atmosphere. The approach, making it possible to post-process the results with reference to different spatial domains, allowed also verifying that the SESs potential supply in green urban areas depends on soil functional recovery after anthropic disturbance, highlighting that urban green spaces may possess very different functionality levels in terms of the potential supply of specific ecosystem services, as already pointed out in other studies in different urban contexts [18,52–56], and that soil quality assessment should be one of the core elements of designing urban green infrastructure [57–59].

To characterize and rank the areas of potential urban development in terms of loss of potential ecosystem service supply in case of future urbanization, a soil quality index IQ4 was calculated considering four ecosystem services selected with the local stakeholders, namely BUF, CST, PRO and WAR. The definition of indices of urban soil quality demands a balance between comprehensiveness and feasibility: the selection should from one side encompass a sufficient range of soil processes relevant to the environmental quality of the urban area and on the other be realistic in terms of data inputs required for their assessment. According to Lehmann and Stahr [60,61], a multi-level approach based on four different dimensions of soil multifunctionality, i.e., hazard protection, production, environmental quality and cultural heritage, should be integrated in the planning process. Quite surprisingly though, the required information would be derived from available data, i.e., geological maps, hydrological maps, historical information, and information on building ground, as soil mapping was considered only necessary for calibration. Tresch et al. [62] pointed out that understanding soil quality in urban ecosystems would need multi-indicator frameworks to capture the complexity of soil characteristics and the influencing factors in space and time. In their study on the urban soil quality in Zurich, Switzerland, they resorted to a multivariate soil quality assessment considering a set of 44 soil quality indicators (17 biological, 9 chemical, 10 physical and 8 addressing SOM) measured at 170 sites in 85 city gardens along a gradient of urban density. Although providing a deep insight into the processes determining the quality and the function of garden soils in the urban ecosystems, such a complex multivariate framework could hardly become part of the planning process. A quantitative assessment of soil quality was implemented by Mamehpour et al. [63] comparing two models under different urbanization scenarios in West Azerbaijan Province, northwestern Iran, based on a set of twenty-four soil attributes. Those included a combination of fertility, salinity, and sodicity attributes and heavy metals (0–50 cm) and were tested for 14 soil profiles of urban and non-urban fields, for which the two approaches returned equal rankings. More recently, Séré et al. [64] selected and tested soil quality indicators for 109 different urban soils located in seven cities of western Europe and under various land uses. Nine soil functions and sub-functions (SOM cycle, nutrient storage, nutrient retention capacity, nutrient recycling, vegetation support, water retention, water infiltration and biodiversity) were ranked in four classes from 0 to 3, resorting to 29 indicators derived from laboratory measurements of soil properties, profile descriptions and site observations, with the lowest scores for sealed soils and soils located in urban brownfields, whereas the highest were found for soils located in city parks or urban agriculture. Although based on an extensive set of soil data and encompassing a wider range of soil functions, the implementation of the scoring approach requires a solid soil science expertise, time and substantial budgets, making its integration in urban planning difficult.

In the case of Forlì, we combined an indicator-based approach with a spatially explicit geostatistical approach which allowed for incorporating spatial variability to analyze and visualize an overall soil quality index (IQ4) across urban landscapes and trying to address urban-planning specific challenges. In doing so, intermediate outcomes maps of basic (textural fractions, pH, organic C) and PTFs-derived soil properties were produced for the 0–30 cm reference layer. The selected SES indicators cover specific dimensions of soil multifunctionality which are deemed strategic by the Forlì municipality for urban and peri-urban environmental quality: hazard mitigation (WAR), food production (PRO), climate regulation (CST) and pollution attenuation (BUF). In agreement with the Regional Soil Survey staff, the selection of the indicators to be included in the quality index was also guided by the criteria of lower correlation between the four indicators in order to reduce the effect of existing synergies and trade-offs between services, as highlighted in Table 7, so as not to overemphasize the effect of existing service bundles [65]. Given the data requirement for assessing the four single indicators, the final IQ4 score relies on a total of nineteen different soil properties. The average soil quality score for the whole urban area is equal to 2.42 (±0.12): 44% of the urban soil units have an average score lower than the average, encompassing 32% of the urban core area (908.3 ha). The soil units with the lowest quality score are those characterized by the presence of urbanized (IQ4 avg. 1.88, std. dev. 0.21) and highly disturbed soils (IQ4 avg. 2.06, std. dev. 0.10), whose score are in all cases the lowest for PRO and in some units also for WAR. Undisturbed and disturbed soils have very similar IQ4 scores, equal, respectively to 2.53 and 2.51, but in the case of the disturbed soils, the observed variability, expressed in terms of standard deviation is almost double, i.e., 0.15 vs. 0.08, respectively.

In our approach, as suggested by Suleymanov et al. [66], we supported the geostatistical interpolation of soil properties using the information provided by the urban soil map whose units account for the degree of disturbance and differentiate into industrial, residential, recreational and other areas within the urban fabric. In this way, the inherent heterogeneity of the urban soils and of their properties was explicitly accounted for in the evaluation of the ecosystem services provided by urban soils and in the assessment of urban soil quality.

5. Conclusions

This work presented an approach to foster the integration of urban soils and their potential ecosystem services supply into the practice of urban planning, allowing simultaneously the quantification of soil-based ecosystem services loss due to urbanization. Spatially explicit knowledge of the properties of urban and peri-urban soil allowed the implementation of an ecosystem service assessment framework originally designed for agricultural soils. This required on one hand the identification of urban soil typological units characterized by different degrees of anthropic disturbance and on the other the spatial estimation and mapping of soil properties measured at sampling points. Both pieces of information were provided by an ad hoc survey of urban and peri-urban soils which complemented the existing soil information for the agricultural soils surrounding the urban area. Once integrated into the general urban plan, the classed soil quality index allowed the identification of high-quality soil areas whose urbanization should be avoided or limited, providing at the same time the basis for compensating for further soil loss by de-sealing urbanized areas comparable in terms of overall soil quality or in the provision of targeted ecosystem services. Results showed that notwithstanding a substantial degree of anthropic disturbance, the supply of regulating services can still be of good quality also in urbanized areas and then relevant if not crucial in terms of extreme events mitigation.

To date, 166 municipalities out of 330 in Emilia–Romagna requested from the regional soil services the information on the ecosystem services provided by agricultural and forest soil in the areas of their administrative competence to integrate them into the general urban plans. Such information resulted from the implementation of the SES assessment framework presented in this work to the whole region. Further research, though, is required soon to assess the effectiveness of the integration of such information in the PUGs to verify if and to which extent it provided stakeholders and planners with useful knowledge to preserve, or eventually restore, the multifunctionality of urban soils in terms of ecosystem service supply and possibly trigger further ad hoc soil surveys in the main urban areas of the region.

Supplementary Materials: The following supporting information can be downloaded at https://www.mdpi.com/article/10.3390/land14040719/s1, Figure S1: Experimental (dots) and model (continuous line) semivariograms for soil properties (0–30 cm); Table S1: Description of the urban soils mapping units; Table S2: Soil profiles analyses; Figure S2: Land Capability Class (LCC) map 1:50,000 of the Forlì municipality and LCC map 1:100,000 of the urban soils; Table S3: Score of the indicator for agricultural productivity (PRO) according to the land capability class (LCC).

Author Contributions: Conceptualization, C.C., P.T. and F.U.; methodology, C.C., F.U. and P.T.; software, F.U. and P.T.; validation, P.T., A.A. and S.B.; formal analysis, F.U. and C.C.; investigation, F.U., C.C., P.T. and A.A.; resources, S.B.; data curation, F.U. and P.T.; writing—original draft preparation, F.U.; writing—review and editing, P.T., A.A., S.B. and C.C.; visualization, F.U., P.T. and C.C.; supervision, C.C. and P.T.; project administration, S.B.; funding acquisition, S.B. All authors have read and agreed to the published version of the manuscript.

Funding: This research was conducted within the framework of the project "Climate vulnerability and ecosystem services of urban soils functional to the drafting of the general urban plan" committed by the Municipality of Forlì, grant number R7-1-2022/26.01.2021, with funds of the Ministry of Ecological Transition relating to the 2021 "Experimental program of interventions for adaptation to climate change in urban areas" (D.D. n. 117 15 April 2021).

Data Availability Statement: The 1:50,000 soil map of Emilia–Romagna (Ed. 2021) is available at the following link: https://ambiente.regione.emilia-romagna.it/it/geologia/suoli/conoscere-suolo/siti-web-sul-suolo-in-emilia-romagna/cartografia-dei-suoli-dellemilia-romagna (accessed on 26 March 2025). The soil map is also available for download at the following link: URL: https://mappegis.regione.emilia-romagna.it/moka/ckan/suolo/Carta_Suoli_50k.zip (accessed on 26 March 2025). The maps of soil based ecosystem services and of the IQ4 soil quality index for the whole territory of Emilia–Romagna at 100 m resolution in raster GEO TIF format are available at the following link: https://mappegis.regione.emilia-romagna.it/moka/ckan/suolo/Servizi_ecosistemici_rst.zip (accessed on 26 March 2025). SES indicator maps rescaled at the level of provinces, unions of municipalities and municipalities, supplemented with a knowledge framework on local soils, are available for the drafting of urban plans (PUGs) by sending a request via certified electronic mail to segrgeol@postacert.regione.emilia-romagna.it. Point soil analytical data for the municipality of Forlì are available upon request to the authors.

Acknowledgments: The authors wish to thank Filippo Sarti for his valuable work in surveying the urban soils and the administrative and technical staff of the Forlì municipality for their precious support during the survey of urban soils and all the related field activities.

Conflicts of Interest: The authors declare no conflicts of interest.

Abbreviations

The following abbreviations are used in this manuscript:

BIO	habitat for biodiversity
BIOMASS	biomass production
BUF	buffering capacity
CLC	CORINE Land Cover
CST	carbon sequestration
DSM	digital soil mapping
IQ4	soil quality index
LCC	Land Capability Class
NDVI	Normalized Difference Vegetation Index
PRO	food production
PTF	pedotransfer function
PUG	General Urban Plan
QBSar	soil biological quality
SES	soil-based ecosystem service
SGS	sequential Gaussian simulation
SOM	soil organic matter
WAR	water regulation
WAS	water storage

References

1. O'Riordan, R.; Davies, J.; Stevens, C.; Quinton, J.N.; Boyko, C. The ecosystem services of urban soils: A review. *Geoderma* **2021**, *395*, 115076. [CrossRef]
2. Nielsen, R.W. Mathematical Analysis of Anthropogenic Signatures: The Great Deceleration. Open-Access E-Print Archive. 2018, pp. 1–41. Available online: https://arxiv.org/pdf/1803.06935 (accessed on 14 January 2025).
3. Head, M.J.; Steffen, W.; Fagerlind, D.; Waters, C.N.; Poirier, C.; Syvitski, J.; Zalasiewicz, J.A.; Barnosky, A.D.; Cearreta, A.; Jeandel, C.; et al. The Great Acceleration is real and provides a quantitative basis for the proposed Anthropocene Series/Epoch. *Episodes* **2022**, *45*, 359–376. [CrossRef]
4. Yigitcanlar, T.; Kamruzzaman, M. Planning, Development and Management of Sustainable Cities: A Commentary from the Guest Editors. *Sustainability* **2015**, *7*, 14677–14688. [CrossRef]
5. Haque, M.N.; Sharifi, A. Justice in access to urban ecosystem services: A critical review of the literature. *Ecosyst. Serv.* **2024**, *67*, 101617. [CrossRef]
6. Salata, S.; Uzelli, T. Are Soil and Geology Characteristics Considered in Urban Planning? An Empirical Study in Izmir (Türkiye). *Urban Sci.* **2023**, *7*, 5. [CrossRef]
7. da Silva, T.; Fleskens, L.; van Delden, H.; van der Ploeg, M. Incorporating soil ecosystem services into urban planning: Status, challenges and opportunities. *Landsc. Ecol.* **2018**, *33*, 1087–1102. [CrossRef]
8. Petrova, S.; Nikolov, B. Soil related ecosystem services in urban areas—A literature review. *Ecol. Balk.* **2023**, *15*, 203–231. Available online: https://eb.bio.uni-plovdiv.bg/wp-content/uploads/2023/10/203-231_eb23304.pdf (accessed on 26 March 2025).
9. Rattan Lal, R.; Negassa, W.; Lorenz, K. Carbon sequestration in soil. *Curr. Opin. Environ. Sustain.* **2015**, *15*, 79–86. [CrossRef]
10. Singh, B.K. (Ed.) *Soil Carbon Storage. Modulators, Mechanisms and Modeling*, 1st ed.; Academic Press: Cambridge, MA, USA, 2018; p. 340.
11. Blanchart, A.; Consalès, J.N.; Séré, G.; Schwartz, C. Consideration of soil in urban planning documents—A French case study. *J. Soils Sediments* **2019**, *19*, 3235–3244. [CrossRef]
12. Pereira, P.; Inácio, M.; Pinto, L.; Kalinauskas, M.; Bogdzevic, K.; Zhao, W. Mapping ecosystem services in urban and peri-urban areas. A systematic review. *Geogr. Sustain.* **2024**, *3*, 491–509. [CrossRef]
13. Yang, J.L.; Zhang, G.L. Formation, characteristics and eco-environmental implications of urban soils—A review. *Soil Sci. Plant Nutr.* **2015**, *61* (Suppl. S1), 30–46. [CrossRef]
14. Ossola, A.; Livesley, S.J. Drivers of soil heterogeneity in the urban landscape. In *Urban Landscape Ecology*, 1st ed.; Francis, R.A., Millington, J.D.A., Chadwick., M.A., Eds.; Taylor & Francis: London, UK, 2016; pp. 19–42.
15. Katikas, L.; Krzywoszynska, A.; Naciph Mora, K.; Roca Vallejo, R. Preliminary assessment of the knowledge gaps related to soil literacy. *Soils Eur.* **2024**, *1*, e118883. [CrossRef]
16. Peake, L. Critical Soil Governance Lessons from Around the World: The Good, The Bad and the Ugly. In Proceedings of the Soil Policy Legacy Report, 22nd World Congress of Soil Science, Glasgow, UK, 24 December 2022. Available online: https://soils.org.uk/wp-content/uploads/2023/02/BSSS_WCSS-Soil-Policy-Legacy-Report_Jan-2023_Final_no-crops-compressed.pdf (accessed on 14 January 2025).
17. Orlova, K.S.; Savin, I.Y. Ecosystem Services Provided by Urban Soils and Their Assessment: A Review. *Eurasian Soil Sci.* **2024**, *57*, 1072–1083. [CrossRef]
18. Ungaro, F.; Maienza, A.; Ugolini, F.; Lanini, G.M.; Baronti, S.; Calzolari, C. Assessment of joint soil ecosystem services supply in urban green spaces: A case study in Northern Italy. *Urban For. Urban Green.* **2022**, *67*, 127455. [CrossRef]
19. Hyun, J.; Kim, Y.J.; Yoo, G. A method for soil quality assessment in the metropolitam greenery: A comprehensive view of ecosystem services and soil functions. *MethodsX* **2023**, *10*, 102102. [CrossRef] [PubMed]
20. Hyun, J.; Kim, Y.J.; Kim, A.; Plante, A.F.; Yoo, G. Ecosystem services-based soil quality index tailored to the metropolitan environment for soil assessment and management. *Sci. Total Environ.* **2022**, *820*, 153301. [CrossRef]
21. Séré, G.; Lothode, M.; Blanchart, A.; Chirol, C.; Tribotte, A.; Schwartz, C. Destisol: A decision-support tool to assess the ecosystem services provided by urban soils for better urban planning. *Eur. J. Soil Sci.* **2024**, *75*, 13557. [CrossRef]
22. Calzolari, C.; Tarocco, P.; Lombardo, N.; Marchi, N.; Ungaro, F. Assessing soil ecosystem services in urban and peri-urban areas: From urban soils survey to providing support tool for urban planning. *Land Use Policy* **2020**, *99*, 105037. [CrossRef]
23. Calzolari, C.; Ungaro, F.; Filippi, N.; Guermandi, M.; Malucelli, F.; Marchi, N.; Staffilani, F.; Tarocco, P. A methodological framework to assess the multiplicity of ecosystem services of soils at regional scale. *Geoderma* **2016**, *261*, 190–203. [CrossRef]
24. Science for Environment Policy. No Net Land Take by 2050? Future Brief 14, 2016. Produced for the European Commission DG Environment by the Science Communication Unit, UWE, Bristol. Available online: https://op.europa.eu/en/publication-detail/-/publication/f1ca6673-23cd-11e6-86d0-01aa75ed71a1/language-en (accessed on 26 March 2025).
25. ARPAE. Climatological Tables. Available online: https://www.arpae.it/it/temi-ambientali/clima/dati-e-indicatori/tabelle-climatiche (accessed on 26 March 2025).

26. Cremonini, L.; Randi, P.; Fazzini, M.; Nardino, M.; Rossi, F.; Georgiadis, T. Causes and Impacts of Flood Events in Emilia-Romagna (Italy) in May 2023. *Land* **2024**, *13*, 1800. [CrossRef]
27. WMO. Manual for estimation of probable maximum precipitation, Operational Hydrology Report 1, World Meteorological Organization Paper n. 332. 1986.
28. SNPA. Consumo di Suolo, Dinamiche Territoriali e Servizi Ecosistemici. Edizione 2024, Report Ambientali SNPA, 43/2024. Available online: https://www.snpambiente.it/temi/suolo/consumo-di-suolo-dinamiche-territoriali-e-servizi-ecosistemici-edizione-2024/ (accessed on 14 January 2025).
29. Regione Emilia-Romagna. Coperture Vettoriali Uso del Suolo di Dettaglio—Edizione 2018. Servizio Statistica, Comunicazione, Sistemi inf. Geografici, Educ. Alla Sostenibilità, Partecipazione, Regione Emilia-Romagna. 2018. Available online: https://geoportale.regione.emilia-romagna.it/download/dati-e-prodotti-cartografici-preconfezionati/pianificazione-e-catasto/uso-del-suolo/2014-coperture-vettoriali-uso-del-suolo-di-dettaglio-edizione-2018 (accessed on 15 January 2025).
30. Tarocco, P. Carta dei Suoli Della Regione Emilia-Romagna, scala 1:50.000. Edizione 2021. Regione Emilia-Romagna, Servizio Geologico Sismico e dei Suoli. Available online: http://mappegis.regione.emilia-romagna.it/gstatico/documenti/dati_pedol/carta_suoli_50k.pdf (accessed on 15 January 2025).
31. USDA. Soil Survey Manual. In *USDA Handbook*; Government Printer: Washington, DC, USA, 1993; Volume 18, pp. 136–140.
32. Decreto Ministeriale del 13/09/1999. Approvazione dei "Metodi Ufficiali di Analisi Chimica del Suolo". Emanato da: Ministro per le Politiche Agricole e Pubblicato su: Gazz. Uff. Suppl. Ordin. n° 248 del 21/10/1999. Available online: https://www.gazzettaufficiale.it/eli/id/2016/04/18/16A02762/sg (accessed on 16 January 2025).
33. MEA. *Millennium Ecosystem Assessment: Ecosystems and Human Well-Being 5*; Island Press: Washington, DC, USA, 2005; Available online: https://www.millenniumassessment.org/documents/document.356.aspx.pdf (accessed on 15 January 2025).
34. Haines-Young, R.; Potschin, M.B. Common International Classification of Ecosystem Services (CICES), V5.1 and Guidance on the Application of the Revised Structure. 2018. Available online: https://cices.eu/content/uploads/sites/8/2018/01/Guidance-V5 1-01012018.pdf (accessed on 15 January 2025).
35. Dominati, E.; Patterson, M.; Mackay, A. A framework for classifying and quantifying the natural capital and ecosystem services of soils. *Ecol. Econ.* **2010**, *69*, 1858–1868. [CrossRef]
36. Ungaro, F.; Calzolari, C. Carta dei Servizi Ecosistemici dei Suoli Della Regione Emilia-Romagna Scala 1:50.000, Note Illustrative (Second Ed.) 2023. Regione Emilia-Romagna, Direzione Generale Cura del Territorio e deLL'AMbiente, Area Geologia, Suoli e Sismica. Available online: https://doi.org/10.13140/RG.2.2.33676.08328 (accessed on 26 March 2025).
37. Menta, C.; Conti, F.D.; Pinto, S.; Bodini, A. Soil Biological Quality index (QBS-ar): 15 years of application at global scale. *Ecol. Indic.* **2018**, *85*, 773–780. [CrossRef]
38. Xue, J.; Su, B. Significant remote sensing vegetation indices: A review of developments and applications. *J. Sens.* **2017**, *2017*, 1353691. [CrossRef]
39. Gorelick, N.; Hancher, M.; Dixon, M.; Ilyushchenko, S.; Thau, D.; Moore, R. Google Earth Engine: Planetary-scale geospatial analysis for everyone. *Remote Sens. Environ.* **2017**, *202*, 18–27. [CrossRef]
40. Klingelbiel, A.A.; Montgomery, P.H. Land capability classification. In *USDA Agricultural Handbook 210*; US Governement Printing Office: Washington, DC, USA, 1961.
41. Regione Emilia-Romagna. Carta Della Capacità d'uso dei Suoli ai Fini Agricoli e Forestali Della Pianura Emiliano-Romagnola in Scala 1:50,000. Servizio Geologico Sismico e dei Suoli, Regione Emilia-Romagna. 2021. Available online: http://geo.regione.emilia-romagna.it/gstatico/documenti/dati_pedol/CAPACITA_USO.pdf (accessed on 16 January 2025).
42. Ungaro, F.; Calzolari, C.; Busoni, E. Development of pedotransfer functions using a group method of data handling for the soil of the Pianura Padano-Veneta region of North Italy: Water retention properties. *Geoderma* **2005**, *124*, 293–317. [CrossRef]
43. Ungaro, F.; Calzolari, C.; Pistocchi, A.; Malucelli, F. Modelling the impact of increasing soil sealing on runoff coefficients at regional scale: A hydropedological approach. *J. Hydrol. Hydromech.* **2014**, *62*, 33–42. [CrossRef]
44. Calzolari, C.; Ungaro, F. Predicting shallow water table depth at regional scale from rainfall and soil data. *J. Hydrol.* **2012**, *414–415*, 374–387. [CrossRef]
45. Wu, J.; Feng, Z.; Gao, Y.; Peng, J. Hotspot and relationship identification in multiple landscape services: A case study on an area with intensive human activities. *Ecol. Indic.* **2013**, *29*, 529–537. [CrossRef]
46. Goovaerts, P. *Geostatistics for Natural Resources Evaluation*; Oxford University Press: New York, NY, USA, 1997; pp. 266–271.
47. Statios. *WinGslib*, Version 1.3; Statios Software and Services: San Francisco, CA, USA, 2000. Available online: http://www.statios.com/Quick/gslib.html (accessed on 26 March 2025).
48. Deutsch, C.V.; Journel, A.G. *GSLIB, Geostatistical Software Library and User's Guide*, 2nd ed.; Oxford University Press: New York, NY, USA, 1998; 369p.
49. QGIS Development Team. QGIS Geographic Information System. Open-Source Geospatial Foundation Project. 2023. Available online: http://qgis.osgeo.org (accessed on 26 March 2025).

50. Regione Emilia-Romagna, Uso del Suolo di Dettaglio: Consultazione On-Line Delle Edizioni di Dettaglio 1976-78, 1994, 2003, 2008, 2014, 2017 e 2020, Integrate Con la Cartografia di Base Della Regione Emilia-Romagna. Available online: https://geoportale.regione.emilia-romagna.it/applicazioni-gis/regione-emilia-romagna/pianificazione-e-catasto/uso-del-suolo/uso-del-suolo-standard (accessed on 24 January 2025).
51. Calzolari, C.; Ungaro, F.; Calogiuri, T.; Marchi, N.; Tarocco, P.; Bazzocchi, S.; Barlotti, M.; Magnani, A.; Marzolo, S.; Lombardo, F. Valutazione dei Servizi Ecosistemici e Stima Degli Impatti Economici e Ambientali Conseguenti al Consumo E all'impermeabilizzazione dei Suoli nei Comuni di Forlì, Carpi e S. Lazzaro di Savena. Deliverable SOS4Life 19 B1.2. 2018. Available online: https://www.sos4life.it/wp-content/uploads/SOS4Life_Stima-degli-impatti-conseguenti-al-consumo-di-suolo-nei-comuni-partner_B.1.2-1.pdf (accessed on 26 January 2025).
52. Kamalova, A.; Gordeev, A.; Galitskaya, P.; Selivanovskaya, S. Assessment of Soil Quality in Urban Green Areas of Two Russian Cities by Means of Chemical and Biological Methods. In *Smart and Sustainable Urban Ecosystems: Challenges and Solutions*; Korneykova, M., Vasenev, V., Dovletyarova, E., Valentini, R., Gorbov, S., Vinnikov, D., Dushkova, D., Eds.; SSC 2022. Springer Geography; Springer: Cham, Germany, 2022. [CrossRef]
53. Tresch, S.; Moretti, M.; Le Bayon, R.C.; Mäder, P.; Zanetta, A.; Frey, D.; Stehle, B.; Kuhn, A.; Munyangabe, A.; Fliessbach, A. Urban Soil Quality Assessment—A Comprehensive Case Study Dataset of Urban Garden Soils. *Front. Environ. Sci.* **2018**, *6*, 00136. [CrossRef]
54. Yingming, M.; Shuxun, S.; Shiqi, L.; Jinlong, J. Spatial distribution of pH and organic matter in urban soils and its implications on site-specific land uses in Xuzhou, China. *Comptes Rendus Biol.* **2014**, *337*, 332–337. [CrossRef]
55. Canedoli, C.; Ferrè, C.; El Khair, D.A.; Padoa-Schioppa, E.; Comolli, R. Soil organic carbon stock in different urban land uses: High stock evidence in urban parks. *Urban Ecosyst.* **2020**, *23*, 159–171. [CrossRef]
56. Díaz-Sanz, J.; Robert, S.; Keller, C. Parameters influencing run-off on vegetated urban soils: A case study in Marseilles, France. *Geoderma* **2020**, *376*, 114455. [CrossRef]
57. Montgomery, J.A.; Klimas, C.A.; Arcus, J.; DeKnock, C.; Rico, K.; Rodriguez, Y.; Vollrath, K.; Webb, E.; Williams, A. Soil Quality Assessment Is a Necessary First Step for Designing Urban Green Infrastructure. *J. Environ. Qual.* **2015**, *45*, 18–25. [CrossRef] [PubMed]
58. Schindelbeck, R.R.; van Es, H.M.; Abawi, G.S.; Wolfe, D.W.; Whitlow, T.L.; Gugino, B.K. Comprehensive assessment of soil quality for landscape and urban management. *Landsc. Urban Plan.* **2008**, *88*, 73–80. [CrossRef]
59. Vrščaj, B.; Poggio, L.; Ajmone Marsan, F. A method for soil environmental quality evaluation for management and planning in urban areas. *Landsc. Urban Plan.* **2008**, *88*, 81–94.
60. Lehmann, A.; Stahr, K. The potential of soil functions and planner-oriented soil evaluation to achieve sustainable land use. *J. Soils Sediments* **2010**, *10*, 1092–1102. [CrossRef]
61. Lehmann, A.; David, S.; Stahr, K. TUSEC—Handbuch zur Bewertung von natürlichen Böden und anthropogenenStadtböden/TUSEC A manual for the evaluation of Natural Soils and Anthropogenic Urban Soils. *Hohenh. Bodenkd. Hefte* **2008**, *86*, 224.
62. Tresch, S.; Moretti, M.; Le Bayon, R.C.; Mäder, P.; Zanetta, A.; Frey, D.; Fliessbach, A. A Gardener's Influence on Urban Soil Quality. *Front. Environ. Sci.* **2018**, *6*, 25. [CrossRef]
63. Mamehpour, N.; Rezapour, S.; Ghaemian, N. Quantitative assessment of soil quality indices for urban croplands in a calcareous semi-arid ecosystem. *Geoderma* **2021**, *382*, 114781. [CrossRef]
64. Séré, G.; Le Guern, C.; Bispo, A.; Layet, C.; Ducommun, C.; Clesse, M.; Schwartz, C.; Vidal-Beaudet, L. Selection of soil health indicators for modelling soil functions to promote smart urban planning. *Sci. Total Environ.* **2024**, *924*, 171347. [CrossRef]
65. Medina-Roldán, E.; Lorenzetti, R.; Calzolari, C.; Ungaro, F. Disentangling soil-based ecosystem services synergies, trade-offs, multifunctionality, and bundles: A case study at regional scale (NE Italy) to support environmental planning. *Geoderma* **2024**, *448*, 116962. [CrossRef]
66. Suleymanov, A.; Abakumov, E.; Polyakov, V.; Kozlov, A.; Saby, N.P.A.; Kuzmenko, P.; Telyagissov, S.; Coblinski, J.A. Estimation and mapping of soil pH in urban landscapes. *Geoderma Reg.* **2025**, *40*, e00919. [CrossRef]

land

MDPI

Article

Evolution of "Production–Living–Ecological" Spaces Conflicts and Their Impacts on Ecosystem Service Values in the Farming–Pastoral Ecotone in Inner Mongolia During Rapid Urbanization

Ziqi Yu [1,2], Xi Meng [1,2,*] and Gongjue Yu [1,2]

1 School of Public Management, Inner Mongolia University, Hohhot 010070, China; ziqi.yu@imu.edu.cn (Z.Y.); yugongjue_icon@163.com (G.Y.)
2 Inner Mongolia Social Governance and Innovation Research Base, Hohhot 010070, China
* Correspondence: mengxi_xi@126.com

Abstract: Rapid urbanization is causing ecological and environmental issues to worsen. The stability of the ecosystem function of the farming–pastoral ecotone (FPE) in Inner Mongolia is essential to ensuring the sustained growth of the nearby cities, acting as a vital ecological safeguard in China's northern regions. This study used the "production–living–ecological" spaces (PLES) spatial dynamics, the rate of change index, and the standard deviation ellipse to examine the spatial and temporal evolution of the PLES in the FPE in Inner Mongolia. This study constructed a spatial conflict index model based on the theory of landscape ecology, and evaluated the ecosystem service value (ESV) of the region and visualized the results of the analysis using the micro-scale of the grid. Finally, the relationship between the ESV and PLES spatial conflicts was determined using a bivariate spatial autocorrelation model. The findings show that: (1) During the 20 years, the maximum ecological spatial change rate reached 0.43%, with the cumulative spatial dynamics of PLES totaling 2.49%. Notably, industrial production space activities experienced the most significant increase, amounting to 277.09%. (2) Regional spatial conflict intensity shows an upward trend from 2000 to 2020, with the average conflict level increasing from 0.53 to 0.56, and high conflict values being concentrated in the east. (3) The ESV pattern in the FPE in Inner Mongolia is characterized by "high ESV in the east and low ESV in the central and western regions", with an overall trend of increasing and then decreasing. A notable negative correlation was observed between ESV and PLES spatial conflicts in the region, with Moran's I indicating values of−0.196, −0.293, and−0.163, respectively. Specifically, low-value–high-conflict zones were predominantly found in other ecological spaces, high-value–low-conflict zones was concentrated in forest ecological spaces, and high-value–high-conflict zones were predominantly concentrated in aquatic ecological spaces. The research findings serve as a crucial scientific foundation for the development of ecological civilization and the sustainable advancement of the FPE in Inner Mongolia.

Keywords: production–living–ecological; spatial conflict; ecosystem service values; urbanization; farming–pastoral ecotone in Inner Mongolia

Academic Editors: Michele Munafò, Luca Congedo and Francesca Assennato

Received: 4 January 2025
Revised: 18 February 2025
Accepted: 18 February 2025
Published: 21 February 2025

Citation: Yu, Z.; Meng, X.; Yu, G. Evolution of "Production–Living–Ecological" Spaces Conflicts and Their Impacts on Ecosystem Service Values in the Farming–Pastoral Ecotone in Inner Mongolia During Rapid Urbanization. *Land* **2025**, *14*, 447. https://doi.org/10.3390/land14030447

1. Introduction

Human production–living–ecological spaces (PLES) are territorial complexes formed by the ongoing interaction between natural ecological processes and socio-economic systems [1]. Spatial conflict is a phenomenon in which multiple requirements contend with scarce spatial resources in the process of using land resources, driven by various interests, resulting in contradictions [2,3]. Ecosystems are necessary and fundamental for human survival, providing supply, regulation, cultural, and ecological functions for human development [4,5]. Land used for agricultural production and ecological maintenance has gradually been replaced by land for industry, mining, transportation, and construction since China's reform and opening up [6,7]. This exacerbated the conflict between the need for urban development and the need for ecological and environmental protection. Conflicts and contradictions in the PLES are intensifying, with significant impacts on ecosystem services [8,9]. Therefore, scientifically revealing the attributes of the geographic development of the PLES, quantitatively assessing the ecosystem services value (ESV), and investigating the influence of the evolution of the PLES geographic conflict on the ESV in a specific region can help to analyze the correlation between the economic benefits and ecological benefits of different types of land in the region (either positive or negative correlations). Depending on the correlation, this can provide an important scientific basis for the optimization of regional PLES and the construction of ecological civilization, as well as a data reference for the future formulation of policies on the direction of regional spatial development. This can provide an important scientific basis for the optimization of the PLES in the region and the construction of ecological civilization.

PLES are intertwined and synergistic, forming a complex system through the interweaving of different human behavioral activities. PLES include production spaces (PS), living spaces (LS), and ecological spaces (ES). Production space mainly that provides industrial and agricultural products, LS caters to the activities of human beings, such as housing and leisure, and ecological space mainly provides ecological services and ecological products [10]. Scholars have performed extensive research on PLES, primarily focusing on its theoretical meaning [11], examining its classification system [12], conducting a thorough assessment of PLES using a range of models [13], optimizing PLES [14], and examining PLES's spatial–temporal evolution and driving mechanisms [15]. Common methods for identifying PLES include land use dynamics and land use transfer matrices [16–18]. The research on PLES has concentrated predominantly on their spatial and temporal evolution, coupling mechanism, and spatial optimization [19]. The ESV method has been widely used to properly evaluate the quantity of ecosystem services, with the ESV method being the most popular since it is straightforward and simple to use, with the added benefit of high comparability [20].

Scholars have conducted a large amount of research on PLES and ESV, but few have investigated the influence of the emergence of the PLES dispute on ESV. The existing research has mainly focused on the influence of land use on the ESV, such as land use changes, using methods such as ecosystem service correlation analyses [21,22], a regression examination of land use changes and ecosystem service drivers [23,24], and the geoprobe model-based quantitative detection of driver interactions [25,26]. The farming–pastoral ecotone (FPE) in Inner Mongolia confronts major difficulties such as land deforestation, grassland degradation, and a limited land-carrying capacity, and its human–land relationship is complicated and rapidly changing. Previous studies on the FPE in Inner Mongolia have focused on land use structure [27,28] and ecological and environmental effects [29], but research on ESV from the perspective of the evolution of PLES spatial conflicts is lacking. The ESV is impacted by the evolution of PLES, and research on these effects can explore the connection between different land types and ESV, offer evidence in favor of China's sustainable land resource use, and strike an equilibrium between ecological value and economic development. Therefore, it is important to investigate the spatial and temporal evolution characteristics of the PLES in the FPE in Inner Mongolia, to reveal the impact of the conflicting evolution of the PLES on the ESV, and to provide citations regarding the ecological protection and sustainable growth of the FPE in Inner Mongolia. This could provide ideas for research focusing on other ecological transition zones, such as the farming–pastoral ecotone. It will also help China to adhere to the sustainable advancement concept of "green mountains are golden mountains" while continuing to promote the development of urbanization and modernization, control the conflicts between development and ecological protection within a reasonable range, and establish a win–win situation in the PLES.

2. Materials and Methods

2.1. Study Area

The FPE in Inner Mongolia is a crucial region for the cooperative growth of both plantations and animal husbandry. It is situated between 109°36′–120°52′ E and 39°35′–45°15′ N, and it has a complex topography and geomorphology, with mountains, plateaus, hills, and plains scattered throughout. These is a temperate continental monsoon climate; the area is situated at the climatic crossroads of a semi-moisture area and a semi-arid area (Figure 1). The FPE in Inner Mongolia consists of 29 counties (banners), including five in Hohhot, two in Baotou, ten in Chifeng, ten in Ulanchab, and two in the Xilingolite League.

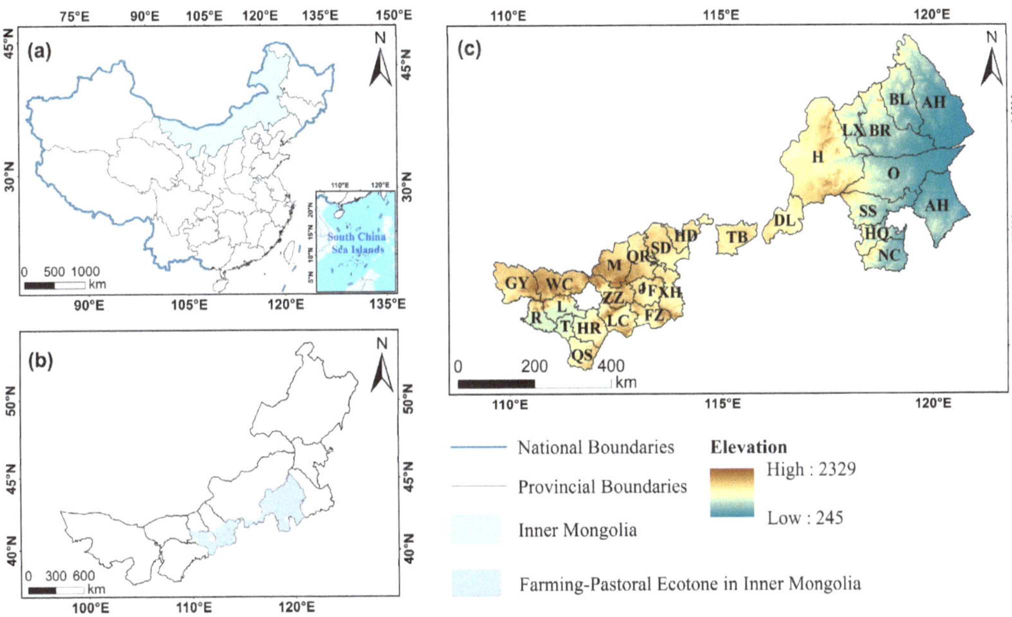

Figure 1. Study area: (**a**) location of the Inner Mongolia in China; (**b**) FPE in the Inner Mongolia; (**c**) The 29 counties (banners) in the FPE in the Inner Mongolia. The abbreviations of county (banners) names are listed in Table S1.

2.2. Data

The Resource and Environmental Science Data Center of the Chinese Academy of Sciences provided the administrative vector boundary data, the land use data at a 30 m spatial resolution, and the DEM data at a 90 m spatial resolution (https://www.resdc.cn/ accessed on 15 January 2023). The socioeconomic data were obtained from the Inner Mongolia Statistical Yearbook and National Compendium of Cost–Benefit Information on Agricultural Products. The average values of yield per hectare, planted area, and unit price of major grain crops such as wheat, corn, and soybean in Inner Mongolia were used as the research benchmarks to obtain the ecological service value equivalent scale of the ecosystems in the FPE in Inner Mongolia.

2.3. Methods

To investigate the evolution of spatial conflicts and their impact on the ESV in the FPE in Inner Mongolia from the spatial perspective of the PLES, this study first collected and processed land use and socioeconomic data, and then analyzed the spatial changes in the PLES, using spatial dynamics, the rate of change index, and the standard deviation ellipse to characterize the spatial changes in the PLES characteristics. Secondly, the spatial conflict composite index (SCCI) measurement model was constructed to reveal the spatial conflict intensity in the spatio-temporal dimension, ESVs were assessed based on the modified model, and finally, the impact of the intensities of PLES conflicts on the ESV was quantitatively measured (Figure 2).

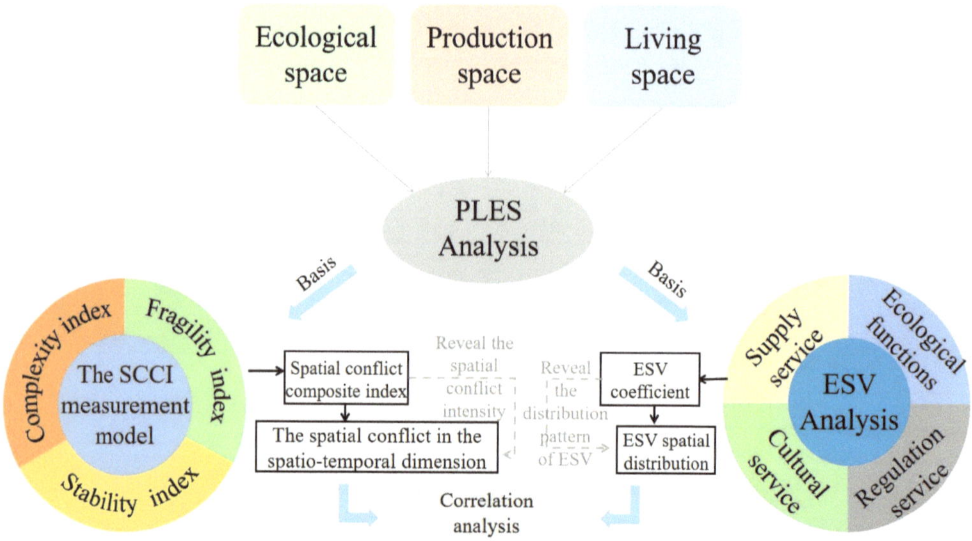

Figure 2. Flowchart revealing the PLES evolution conflicts and their impacts on ESV.

2.3.1. Dynamic Degrees and Rate of Change Index of PLES

(1) PLES dynamic degrees

PLES dynamic degrees refer to the proportional change in the area of each form of PLES over time. This not only reflects the conversion rate of PLES in the research area but also reflects the increase or decrease in a specific category of PLES land use in a specific period of time. It primarily consists of the integrated dynamic degrees of the PLES and the single-dynamic degrees [30,31]. Based on relevant research findings, this study categorized the PLES in the FPE in Inner Mongolia into eight varieties [19] (Table S2). The integrated dynamic degrees of the PLES and the single-dynamic degrees equations are as follows:

$$G = \frac{S_i - S_{ii}}{S_{ii}} \times \frac{1}{T} \times 100\% \quad (1)$$

where G signifies the dynamic degrees index of a certain category of PLES; S_i is the area of the PLES type following the transfer; S_{ii} is the area when no transfer has occurred; T (year) represents the time interval of the study.

$$LC = \left[\frac{\sum_{i=1}^{n} \Delta LU_{i-j}}{2\sum_{i=1}^{n} LU_i} \right] \times \frac{1}{T} \times 100\% \quad (2)$$

where LC (%) is the integrated dynamic degrees of the PLES; LU_i is the region of PLES category i at the study's outset; $\Delta LU_{i\text{-}j}$ (km^2) is the total worth of the region of PLES type i transformed into non-i PLES type in the given time duration T; n is the number of regional PLES types.

(2) PLES change rate index

The PLES change rate index is the intensity of the degree of change in the areas of productive, living, and ecological land utilization relative to the study area over time. It enables an effective comparison of the differences in the changes across PLES types. The equations are as follows:

$$F_i = \frac{\Delta V_i}{A} \times \frac{1}{(t_2 - t_1)} \times 100\% \quad (3)$$

$$\Delta V = \left[LA_{(i,t_2)} - ULA_i \right] + \left[LA_{(i,t_1)} - ULA_i \right] \quad (4)$$

where F_i signifies the rate of change index (%) for PLES type i; ΔV_i signifies the overall alteration in this type of PLES over the study period; ULA_i signifies the area (km^2) of category i PLES types that did not undergo a transformation during the study period; $LA_{(i,t_1)}$ is the area (km^2) of PLES type i at the start of the study period; $LA_{(i,t_1)} - ULA_i$ represents the sum of the areas (km^2) of *ith* PLES types that were transformed to other non-type i PLES categories during the study period; $LA_{(i,t_2)} - ULA_i$ indicates the area added during the study period (km^2); t_1 refers to the start time of the study period; and t_2 corresponds to the end time of the study period.

2.3.2. Standard Deviation Ellipse

Originally established by Lefever in 1926, the standard variation ellipse (SDE) is used to characterize the spatial allocation of geographical components and is currently mostly used to examine the patterns of space development in a variety of fields [32]. The method's main parameters are the core point, long axis, short axis, and direction angle, which, as a spatial analysis tool, can accurately represent the geographic distribution of the study object. The formulas are as follows:

Mean center coordinates:

$$\overline{X_w} = \frac{\sum_{i=1}^{n} w_i x_i}{\sum_{i=1}^{n} w_i}, \overline{Y_w} \frac{\sum_{i=1}^{n} w_i y_i}{\sum_{i=1}^{n} w_i} \quad (5)$$

Azimuth:

$$\theta = \frac{\arctan\left[\left(\sum_{i=1}^{n} x_i'^2 - \sum_{i=1}^{n} y_i'^2 \right) + \sqrt{\left(\sum_{i=1}^{n} x_i'^2 - \sum_{i=1}^{n} y_i'^2 \right)^2 + 4\left(\sum_{i=1}^{n} x_i' y_i' \right)^2} \right]}{2\sum_{i=1}^{n} x_i' y_i'} \quad (6)$$

Axis standard deviation:

$$\delta_x = \sqrt{\frac{\sum_{i=1}^{n} (x_i' \cos\theta - y_i' \sin\theta)^2}{n}}, \ \delta_y = \sqrt{\frac{\sum_{i=1}^{n} (x_i' \sin\theta - y_i' \cos\theta)^2}{n}} \quad (7)$$

where $(\overline{X_w},\overline{Y_w})$ is the weighted centroid; (x_i,y_i) are the spatial center positions; w_i is the weight; and θ is the ellipse direction; x'_i and y'_i are the center-to-mean center coordinate deviations, respectively. The standard deviations along the x and y axes are indicated by the symbols δ_x and δ_y, respectively.

2.3.3. SCCI Measurement Model

This study constructed a geographic conflict composite index (SCCI) of PLES in the FPE in Inner Mongolia focusing on three aspects: spatial complexity, spatial vulnerability, and spatial stability [33] (Table 1). Spatial intricacy is described by the Area-Weighted Mean Patch Fractal Dimension (AWPFD). This illustrates how different human endeavors impact spatial patterns and the evolution of urban geographic phenomena. Higher AWPFD values indicate less human disturbance. The Spatial Vulnerability Index represents the spatial vulnerability. External forces are the primary cause of a land system's susceptibility, and different land types have differing resistance to disturbances. The Spatial Stability Index measures spatial stability. Stability is strongly influenced by the geographic morphology of land use units. The degree of geographic conflict increases and stability decreases with increasing spatial morphological fragmentation.

Table 1. Calculation methods of PLES conflict indices.

Index	Formula		Formula Description
Spatial Complexity Index	$SCI = \sum_{i=1}^{m}\sum_{j=1}^{n}\left[\frac{2\ln(0.25P_{ij})}{\ln(a_{ij})}\right]\left(\frac{a_{ij}}{A}\right)$	(8)	a_{ij} is the patch area; P_{ij} denotes the perimeter of a discrete landscape patch; A represents the aggregate area of specific landscape categories; subscript notation ij corresponds to the jth land cover class within the ith spatial unit cell; m quantifies the total number of assessment grids in the study area; n enumerates the total number of spatial types.
Spatial Fragility Index	$SFI = \sum_{i=1}^{n} F_i \cdot \frac{a_i}{S}$	(9)	F_i is the fragility index of spatial type i; n represents the overall count of spatial categories; a_i refers to the extent of each landscape category within the unit; S refers to the overall size of the spatial unit.
Spatial Stability Index	$SSI = 1 - \frac{PD-PD_{\min}}{PD_{\max}-PD_{\min}}$, $PD = \frac{n_i}{A}$	(10)	n_i is the number of patches of spatial type i within the spatial unit; A is the area of the spatial unit; PD is the plaque density.
Spatial Conflict Composite Index	$SCCI = SCI + SFI - SSI$	(11)	$SCCI$ is the Spatial Conflict Composite Index; SCI, SFI, and SSI are the Spatial Complexity, Fragility, and Stability Indices, respectively.

After comparing grid scales of 1 km × 1 km, 5 km × 5 km, 7 km × 7 km, and 10 km × 10 km, and considering the amount of data and the size of the study area, the final decision was made to analyze the characteristics of the evolution of the PLES conflict based on the 7 km × 7 km grid scale. The results were classified into very low (0.41–0.49], low (0.49–0.53], medium (0.53–0.57], high (0.57–0.63], and very high (0.63–0.72] spatial conflicts.

2.3.4. Ecosystem Services Valuation

The value per unit area conversion factor approach was employed to assess the ESV in FPE in Inner Mongolia [34].

(1) Ecosystem service evaluation indicator construction

Four primary indicators and eleven supplementary indicators of supply, regulation, support, and cultural services were chosen as indicators for the ecosystem services valuation. Referring to the table of ESV coefficients of Xie et al. [35–37], and according to the specific conditions of the study area, the average value of the ESV coefficients of paddy fields and drylands was used for the APS. The FES was calculated as the average of the ESV coefficients of coniferous forests, mixed coniferous forests, broad-leaved forests, and shrubs. GES is calculated as the average of the ESV coefficients of grassland; AES is calculated as the average of the ESV coefficients of aquatic areas; OES is calculated as the average of the ESV coefficients of desert and bare land. Since the IPS, ULS, and RLS have a negative effect on the overall ecosystem of the research region, the ESVs of all three categories of space are recognized as 0 to exclude their effects (Table S3).

(2) Equivalent based on food price correction

The ESV coefficient equivalent factor was updated based on the yield, planted area, per-unit cost, and ecological landscape type of the main agricultural produce in the FPE in Inner Mongolia. The economic value of food crops per unit area of farmland ecosystems in the FPE in Inner Mongolia was determined to be RMB 1256.53 hm^2 based on the planting acreage, production, and the mean cost of agricultural produce for the four main crops, rice, wheat, maize, and soybeans, in Inner Mongolia in the years 2000, 2010, and 2020. This was carried out in order to remove the effect of crop price variations in different years on the total value (Table S4). The equations are as follows:

$$E_a = \frac{1}{7} = \sum_{i=1}^{n} \frac{m_i p_i q_i}{M} \quad (12)$$

where E_a is the financial worth of food production services supplied by the cultivated land ecosystem per unit area (CNY/hm^2); n is the main agricultural produce species in the study region; p_i is the average price of the *ith* crop (CNY/t); q_i is the yield per unit region of the *ith* agricultural produce (t); m_i is the planted area of the *ith* crop (hm^2); M is the planted region of all agricultural produce (hm^2).

(3) ESV Calculation

ESVs were calculated in accordance with the equivalent ecosystem service value per unit of space factors table (Table S4). The equation is as follows:

$$ESV = \sum_{i=1}^{n} \sum_{j=1}^{m} e_{ij} \times E_a \times A_j \quad (13)$$

where ESV is the regional ecosystem service value (10^4 CNY/year), e is the i ecosystem service value equivalent factor for the *jth* ecosystem; and A is the region of the *jth* ecosystem (hm^2).

2.3.5. Bivariate Spatial Autocorrelations

Moran's I index was applied for this study to perform a geographical autocorrelation analysis of ESV and SCCI [38]. A bivariate geographical self-correlation analysis of the effect of PLES evolution on ESV can provide an academic foundation for homeland space refinement by revealing the mechanism of PLES evolution on ESV from the perspective of spatial correlation, identifying the spatial clustering features of the ESV and spatial conflict level and visually expressing them. The equations are as follows:

Bivariate global Moran's index:

$$I = \frac{n\sum_{i}^{n}\sum_{j}^{n} W_{ij} Q_k^i Q_l^j}{(n-1)\sum_{i}^{n}\sum_{j}^{n} W_{ij}} \tag{14}$$

Bivariate local Moran's index:

$$I_S = Q_k^i \sum_{j=1}^{n} W_{ij} Q_l^j \tag{15}$$

$$Q_k^i = \frac{X_k^i - \overline{X}_k}{\partial_K}, Q_l^j = \frac{X_l^j - \overline{X}_l}{\partial_l} \tag{16}$$

where I is the bivariate global Moran index; I_s is the bivariate local Moran index; n is the number of research units; X_k^i is the attribute worth k of research unit i; X_l^j is the worth of attribute l of research unit j; $\overline{X}_k$, $\overline{X}_l$ are the mean worths of properties k and l; ∂_K, ∂_l are the variance in properties k and l; and W is a weight grid that measures the proximity of the study units and can indicate the geographic relationship between research units i and j. The range of values for Moran I is $[-1,1]$. A favorable worth for Moran I signifies a favorable spatial relationship, meaning that the properties of adjacent research units are similar. Conversely, a negative value for Moran I signifies an adverse spatial relationship and a value of 0 signifies that spatial correlation is irrelevant and that spatial randomness is displayed. The intensity of the spatial interdependence increases with the Moran I value's proximity to 1 or -1.

3. Results

3.1. PLES Spatio-Temporal Characterization

The ES is the primary distribution among the PLES of the FPE in Inner Mongolia. The distribution of ES is the most extensive and presented as a sheet. The LS is scattered in the form of points and the PS is distributed in line or band in the PLES of Inner Mongolia (Figure 3). The LS area increased from 4234.14 km^2 in 2000 to 4603.26 km^2 in 2020, accounting for 0.25%. The PS area increased from 43,624.26 km^2 in 2000 to 43,796.36 km^2 in 2020, accounting for an increase of 0.12%. The ES area decreased from 97,937.65 km^2 in 2000 to 97,392.1275 km^2 in 2020, accounting for a decrease of 0.37%.

The proportion of LS increased by 0.25%, the percentage of productive space increased by 0.12%, and the percentage of ES decreased by 0.37%. IPS and ULS had the lowest rate of change index (0.01%), while GES had the highest rate of change index (1.72%) between 2000 and 2010. The category shift is more pronounced between 2010 and 2020, with ES showing the highest rate of change index (3.22%) and LS showing the lowest (0.15%). The transition between categories is very smooth between 2000 and 2020, with ES changing at the highest rate (0.43%) and LS changing at the slowest rate (0.04%). Overall, there were clear regional disparities and more significant spatial changes in the FPE in Inner Mongolia between 2000 and 2020 (Table 2). In 2000–2010 and 2010–2020, the study period was more intense and significant, and there were obvious differences between regions. Throughout the entire research cycle, the degree of ecological space change was more significant. In Inner Mongolia, the changes in land use in the FPE are mainly the mutual transformation of land types, so the intensity of the regional area changes was not high from the initial stage to the end of the study.

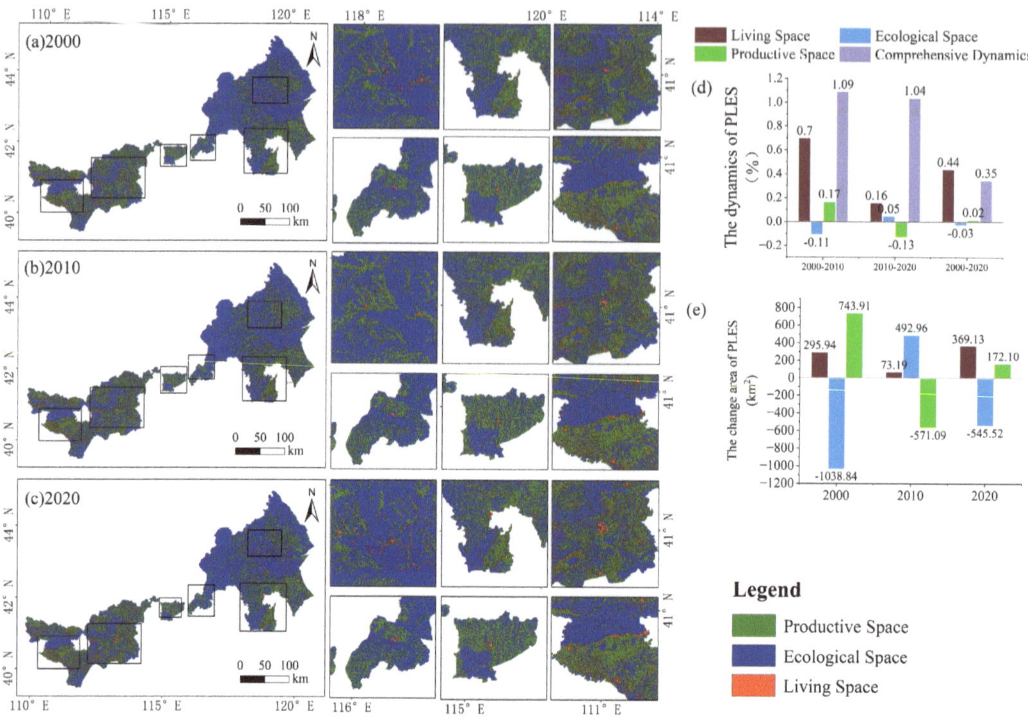

Figure 3. Spatial distribution of PLES in FPE in Inner Mongolia: (a) Spatial distribution of PLES in 2000; (b) Spatial distribution of PLES in 2010; (c) Spatial distribution of PLES in 2020; (d) The dynamic of PLES; (e) The change area of PLES.

Table 2. Index of the change rate of land use type in PLES.

Primary Classification	Secondary Classification	2000–2010 (%)	2010–2020 (%)	2000–2020 (%)
PS	APS	0.89	0.97	0.22
	IPS	0.01	0.03	0.01
ES	FES	0.87	0.87	0.09
	GES	1.72	1.77	0.27
	AES	0.08	0.08	0.01
	OES	0.52	0.50	0.06
LS	ULS	0.01	0.02	0.01
	RLS	0.10	0.13	0.03

The dynamics of PLES within the FPE in Inner Mongolia became increasingly apparent between 2000 and 2020. From the perspective of single-dynamic degrees, the LS had the highest dynamic degrees and the largest increase from 2000 to 2010 (a single-dynamic degree of 0.70% and an increase of 295.94 km^2), followed by the PS (a single-dynamic degree of 0.17% and an increase of 743.19 km^2) and the ES (a single-dynamic degree of −0.11% and a decrease of 1038.48 km^2). LS had the highest single-dynamic degrees (0.16%), with an increase of 73.19 km^2, ES had the second-highest single-dynamic degrees (0.05%), with an increase of 492.96 km^2, and productive space had the lowest single-dynamic degrees (−0.13%), with a reduction of 571.09 km^2 from 2010 to 2020. LS single-dynamic degrees increased by 369.13 km^2 (0.44%) and PS single-dynamic degrees increased by

172.10 km^2 (0.02%) between 2000 and 2020, whereas ES single-dynamic degrees decreased by 545.52 km^2 (−0.03%).

From the perspective of integrated dynamic degrees, the region's integrated dynamic degrees changed more dramatically between 2010 and 2020, with the integrated dynamic degrees during the study period being 0.35%. The integrated dynamic degree was 1.04% in 2000–2010 and the highest integrated dynamic degree was 1.09% in 2010–2020. With the highest single-dynamic degrees among the different land use types during the study period (48.93% for 2000–2010, 38.71% for 2010–2020, and 277.09% for 2000–2020), the IPS has grown significantly over the 20 years. While the single-dynamic degrees for GES were the lowest in the periods 2000 to 2010 and 2000 to 2020, at−1.25% and −0.13%, respectively, the lowest single-dynamic degree between 2010 and 2020 was obtained for FES, at −2.84%, suggesting that the area of GES and FES was reduced more over the 20 years (Figure 3 and Table S5).

The APS and ULS form a southwest-northeast spatial distribution pattern with negligible changes in the spatial standard deviation ellipse change in the FPE at various times (Figure 4). The overall migration trend of APS, FES, GES, and OES is insignificant, and the centers of gravity are roughly overlapping. With reduced directionality and a more dispersed geographical distribution pattern, the standard deviation ellipse's overall oblateness for IPS keeps declining. The IPS experienced a significant increase in its overall size between 2000 and 2010, showing a more visible shift in the spatial pattern of its land use, and then decreased significantly from 2010 to 2020, with more concentrated development and a large horizontal expansion tendency. The total area of GES experienced a decline from 2000 to 2010, but increased dramatically from 2010 to 2020, indicating that the policy of windbreak and sand control, as well as reverting cropland to grassland, plays a significant role. The area of AES rose from 2000 to 2010, and water conservation was successful. However, there was a considerable decline in the area from 2010 to 2020, indicating that the loss of water resources was more severe owing to a variety of circumstances. The overall change in RLS was weakly significant, with an increase in area and a more dispersed distribution from 2000 to 2010, followed by a drop in area from 2010 to 2020, indicating a higher agglomeration of rural living land as urbanization expands.

Overall, APS, FES, GES, and OES migrated southwest to northeast, while IPS, AES, ULS, and RLS migrated northeast to southwest. This is due to the phenomenon of urban development and the uneven distribution of water resources in the FPE of Inner Mongolia. In the study area, Jining District, Tumed Left Banner, and Horinger County, which are located in the southwest, have a higher gross domestic product and higher development level than other areas, so there is a tendency for IPS, AES, ULS, and RLS to migrate to the southwest, whereas the northeast region is less developed, with more unused land being used for agricultural and forestry production, meaning that APS, FES, GES, and OES show a tendency to migrate to the northeast.

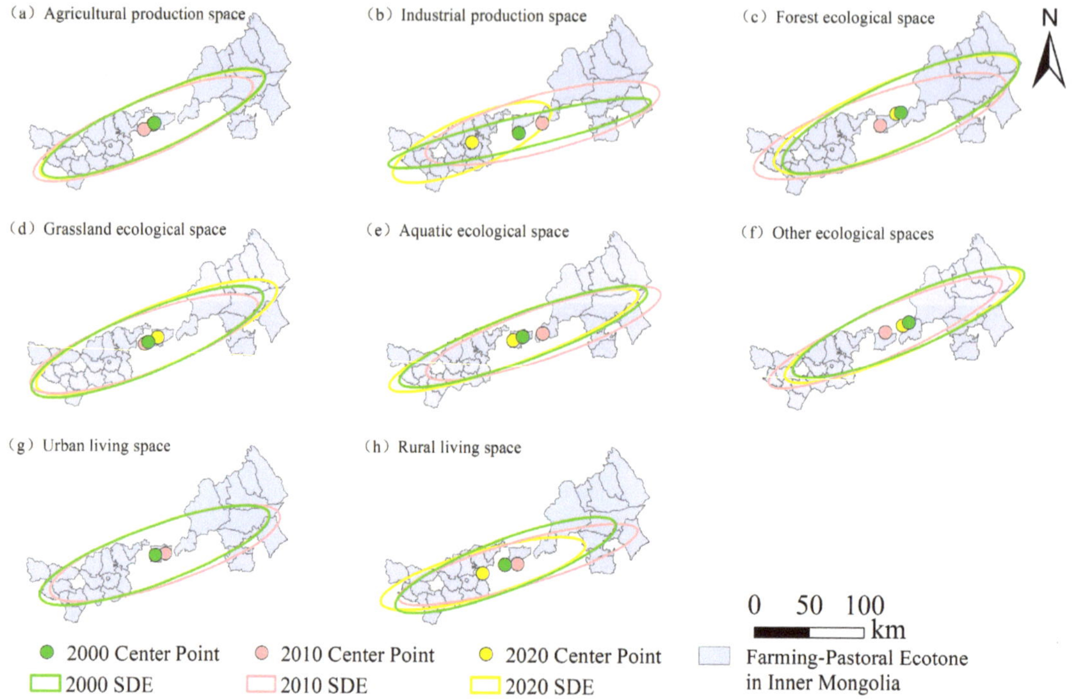

Figure 4. Elliptic variation in the spatial standard deviation of PLES in FPE in Inner Mongolia from 2000 to 2020.

3.2. PLES Spatial Conflict Characterization

There was an upward trend in the average level of spatial conflicts in the FPE in Inner Mongolia. The mean intensity of disputes rose from 0.53 to 0.56, with low and medium spatial conflicts making up over 70% of the total from 2000 to 2020. High- and very high-spatial-conflict areas are concentrated in the eastern region of the FPE located in Inner Mongolia, and the primary land type of high-spatial-conflict areas is other ecological land. Medium-spatial-conflict areas accounted for the greatest proportion of total land, while high-spatial-conflict areas accounted for the smallest proportion, from 2000 to 2020. The percentage of all three types of space conflict increased, except for very low and low spatial conflict, which declined between 2000 and 2010. The proportion of low, high, and high spatial conflicts increased while the proportion of very low and medium spatial conflicts decreased between 2010 and 2020.

In 2000, the east-central region of Ongniud Banner (O) in Chifeng City was the primary location for areas of high spatial conflict, while the central regions of Dolun County (DL) in Xilingol and the southeast region of Ar Horqin Banner (AH) in Chifeng City were the primary locations for stronger spatial conflicts. The southeastern, central, and western regions of the FPE in Inner Mongolia were the primary locations for areas of medium spatial conflicts, with the primary land types of APS and GES. Areas with very low rates of spatial conflicts were primarily found in the FES. In 2010, the east–central area of Ongniud Banner (O) in Chifeng City experienced the most intense spatial conflicts. Many spatial disputes emerged, primarily concentrated in the western half of Hexigten Banner (H) in Chifeng City, with the primary land types being AES and OES. Medium spatial conflicts were primarily found in the western, central, and southeastern regions, with APS and GES being the most common land categories. Very low and low rates of spatial conflicts were primarily concentrated in the eastern region. The high levels of spatial conflicts are expected to increase in 2020, particularly in the east–central region of Ongniud Banner (O) in Chifeng City. The spatial conflict level decreased in the western part of Hexigten Banner (H) in Chifeng City, and high spatial conflicts increased in the western part of the FPE in Inner Mongolia, primarily concentrated in the Tumed Left Banner (L), and Tumed Right Banner (R) in Hohhot (Figure 5).

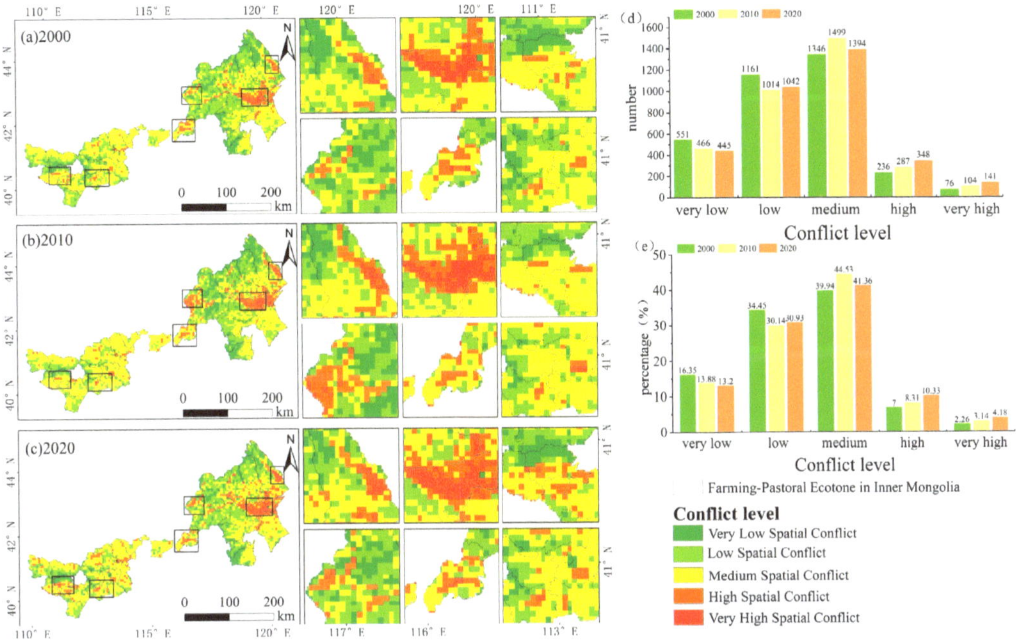

Figure 5. Characteristics of spatial conflicts of PLES in FPE in Inner Mongolia between 2000 and 2020: (**a**) Spatial distribution of conflict levels in 2000; (**b**) Spatial distribution of conflict levels in 2010; (**c**) Spatial distribution of conflict levels in 2020; (**d**) The number of spaces conflicts; (**e**) The proportion of spaces conflicts.

3.3. ESV's Spatio-Temporal Patterns

The ESV (104 CNY/hm^2) in the FPE in Inner Mongolia was divided into five distinct segments based on the natural breaks approach: very low (0–2418.81], low (2418.81–4878.55], medium (4878.55–6921.03], high (6921.03–9530.87], and very high (9530.87–40,658.90] (Figure 6). The ESV exhibited a spatial distribution characterized by lower values in the central and western regions and higher values in the eastern regions. Ar Horqin Banner (AH), Bahrain Left Banner (BL), Hexigten Banner (H), and Ningcheng County (NC) in eastern Chifeng City are the primary locations for the ESV high-value regions situated within the FES and AES regions. Forests and aquifers have good ecological functions, such as water conservation and climate supervisors, which provide high ecosystem service value functions to the FPE in Inner Mongolia.

Very low- and low-value ESV regions were predominantly found in the central, western, and southern regions of APS and OES. The water supply in the APS's ecological supply service is negative, which is detrimental to water conservation efforts. OES consists primarily of land types such as sandy land, saline land, marshy land, and bare rocky gravel land, which have poor ecological functions and hence provide a lower ecosystem service value function for the FPE in Inner Mongolia.

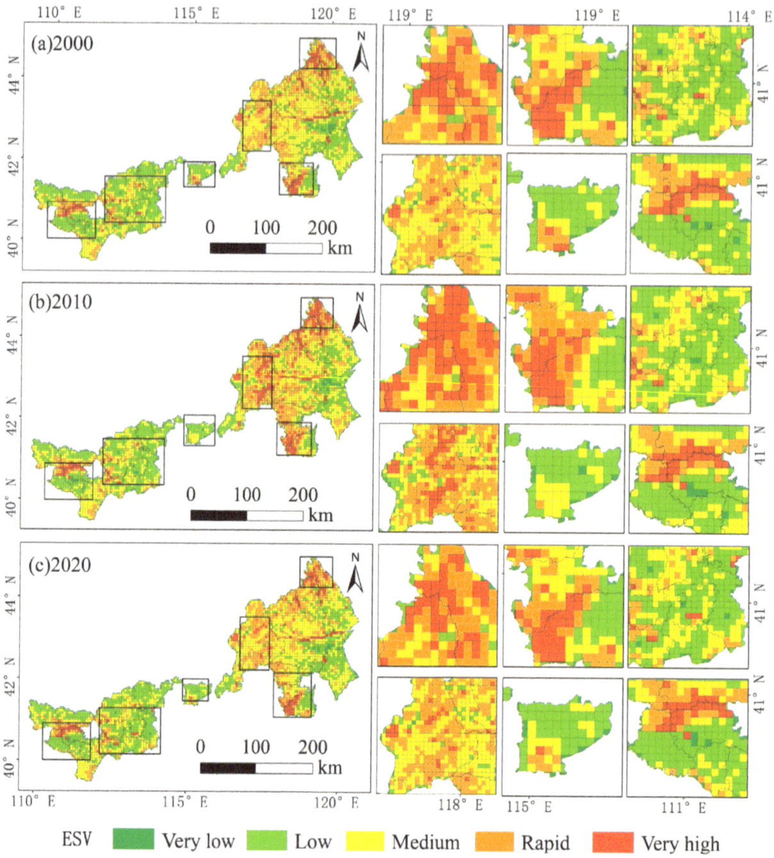

Figure 6. Spatial distribution of ESV levels of the FPE in Inner Mongolia from 2000 to 2020: (**a**) Spatial distribution of ESV levels in 2000; (**b**) Spatial distribution of ESV levels in 2010; (**c**) Spatial distribution of ESV levels in 2020.

Overall, the ESV in the FPE in Inner Mongolia decreased by CNY 0.45 billion over the 20 years. From 2000 to 2010, the implementation of forest land protection policies in the FPE in Inner Mongolia improved, with an increase in the area of forested space in the western, northeastern, and other mountainous areas at higher elevations, increasing the ESV of the FES. Between 2000 and 2020, regulation services accounted for the largest share of the ESV, whereas cultural services represented the smallest proportion. Hydrological regulation had the highest share of the value of secondary individual ecosystem services, while water provisioning had the lowest proportion, with the combined proportion of regulation and ecological functions accounting for 90% of the ESV.

The proportion of climate regulation of the FPE in Inner Mongolia increased from 23% to 24%, with a value increase of CNY 1.775 billion from 2000 to 2010. However, the proportion decreased from 24% to 23%, with a value decrease of CNY 1.832 billion, which represents a decrease of CNY 0.057 billion from 2010 to 2020. Grassland ecological land was the largest contributor to the region's ESV. Its proportion of the ESV dropped from 57.82% in 2000 to 49.05% in 2010 but increased to 57.09% in 2020, reflecting a 0.73% decrease from 2000. The transition from GES to FES resulted in a decline in ESV for GES in 2010. The FPE in Inner Mongolia had the highest proportion of GES in their land use categories, at 48%, 42%, and 47%, respectively, throughout this period (Figure 7).

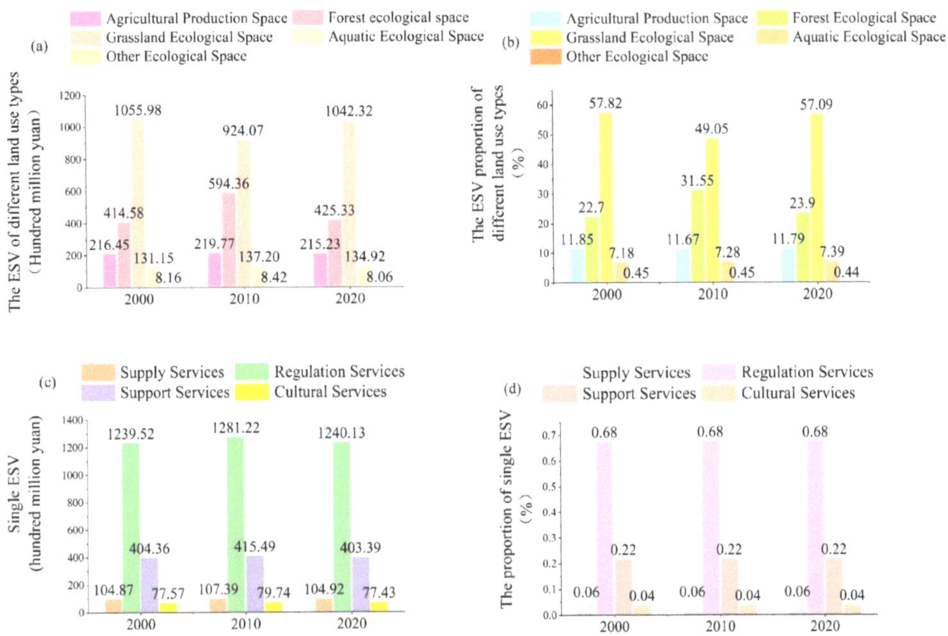

Figure 7. ESV analysis in the FPE in Inner Mongolia from 2000 to 2020: (**a**) The ESV of different land use types; (**b**) The ESV proportion of different land use types; (**c**) Single ESV; (**d**) The proportion of single ESV.

3.4. Effects of PLES Spatial Conflicts on ESV

The bivariate global Moran′I values of the FPE in Inner Mongolia in 2000, 2010, and 2020 were −0.196, −0.293, and −0.163, respectively. An inverse relationship was observed between ESV and PLES spatial conflicts, suggesting that higher levels of PLES spatial conflicts in the study area resulted in a lower ESV, with the strength of this correlation demonstrating an upward trend.

High-value–high-conflict zones exhibited core aggregation and a linear distribution; these were primarily found along the eastern edges of the Ongniud Banner (O), Bahrain Right Banner (BR), and Hexigten Banner (H), as well as the watershed ecological land in Duolun County (DL) in the center and Liangcheng County (LC) and Qahar Right Rear Banner (QR) in the west. This suggests that the aquatic ecological land in the area has a high level of ESV but also a high ecological risk. The distribution of high-value–high-conflict clusters increased in central Duolun County (DL) from 2010 to 2020, while their distribution along the eastern Hexigten Banner (H) and western Qahar Right Front Banner (F) and Shangdu County (SD) showed a decreasing trend. From 2000 to 2010, there was a notable increase in high value–high-conflict clusters, primarily in the eastern Bairin Right Banner (BR) and Hexigten Banner (H).

Low-value–low-conflict zones were primarily found along the boundaries of the FPE in Inner Mongolia, which consists primarily of OES and APS with a low ESV. Because these areas are on the outskirts of urban development, the intensity of land use is low, resulting in a low spatial conflict intensity and creating a low-value–low-conflict zone. During 2000 and 2010, the spread of low-value–low-conflict zones grew dramatically in the eastern Bairin Left Banner (BL) and Linxi County (LX), while between 2010 and 2020, it climbed greatly in western Guyang County (GY).

The low-value–high-conflict zones were primarily concentrated in Ongniud Banner (O) in the east and Duolun County (DL) in the center, with a scattered distribution in the west, where the land use type is primarily OES, which has a low ESV and high ecological risk due to its fragile ecological environment. The distribution of low-value–high-conflict zones in the eastern Hexigten Banner (H) rose dramatically between 2000 and 2010, indicating a decade-long imbalance in ecological and environmental protection in the region.

High-value–low-conflict zones are mainly clustered in a piecemeal manner, distributed in the northeastern Ar Horqin Banner (AH), Bairin Left Banner (BR), Hexigten Banner (H), the southeastern Harqin Banner (HQ), and Ningcheng County (NC), and in the western Wuchuan County (WC), Guyang County (GY), and the borders of Tumed Left Banner (L), and Tumed Right Banner (R). Its ecological environment is in a relatively balanced state, exhibiting the spatial characteristics of a high-value–low-conflict region. The region is primarily an FES, with high vegetation cover, rich ecological resources, better environmental quality, and little influence from human activities (Figure 8).

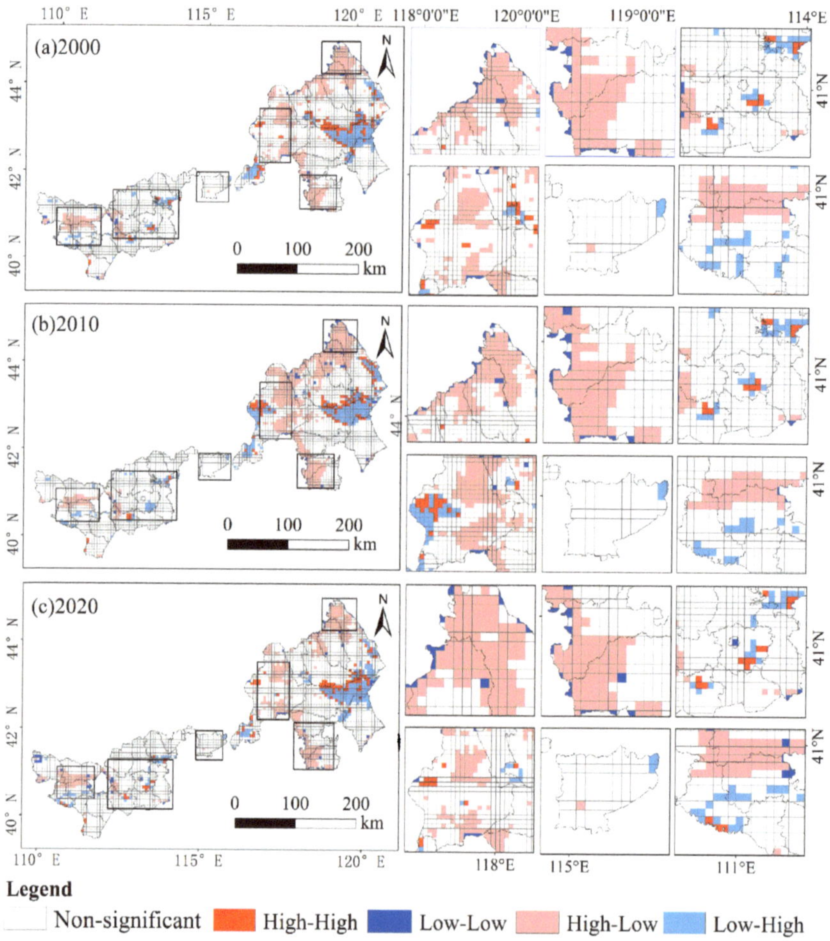

Figure 8. Lisa map of the local autocorrelation of FPE in Inner Mongolia from 2000 to 2020: (**a**) Lisa map of the local autocorrelation in 2000; (**b**) Lisa map of the local autocorrelation in 2010; (**c**) Lisa map of the local autocorrelation in 2020.

4. Discussion

4.1. PLES Conflict Evolution Characterization

OES and AES are the most prevalent space classes, with high and very high spatial conflicts. OES includes sand, gobi, saline, and alkaline land, and the aquatic ecological land includes rivers, lakes, and mudflats.

Numerous factors influenced the changes in the aquatic area of the FPE in Inner Mongolia. The climate of the FPE in Inner Mongolia has been dry and evaporative in recent years; this, along with the standard deviation ellipse, caused the AES area to decrease. Migration to the west increased the degree of conflict in the east. Climate change increased the use of irrigation water and coal mining, among other things, resulting in fewer lakes in Inner Mongolia; some areas of land were desertified and the sand problem has grown worse [39,40]. These regions exhibit significant spatial vulnerability and limited stability, so the composite index of spatial conflict shows high and very high levels of spatial conflict. The majority of medium spatial conflicts were found in APS, including drylands and paddy fields.

In the early period, stability was low and the composite index of spatial conflict showed a medium level because of the growth of agricultural reclamation areas impacted by human activities, extensive planting, and rough cultivation, which cause soil fee declines, severe wind erosion, and sands in arable land [41]. Therefore, the spatial conflict index for these space classes obtained a medium level in the results. FES and GES are the two most common types of area with very low and low spatial conflicts. China's strategy of converting farmland back to forest and grassland [42,43] and the Ministry of Agriculture's recent round of reforms to grassland rights [44] are responsible for the expansion of grassland and forest areas, as well as the improvement in grassland's ecological conditions. Because of the area's comparatively flat topography and uniform vegetation, the soil is shielded from wind and water erosion and serves a crucial water conservation function [45]. Partially located in a nature reserve with little human interference, it has a high degree of stability and low complexity, with a low level of conflict.

4.2. PLES Spatial Conflicts and ESV Correlation Analysis

A distinct negative spatial relationship was observed between ESV and PLES spatial conflicts, suggesting that the region's ecological conditions were suboptimal throughout the study period. The ESV in the FPE in Inner Mongolia showed an overall decreasing trend from 2000 to 2020, while the intensity of spatial conflicts showed an increasing trend. The FPE in Inner Mongolia is a complex ecosystem in which production, living, and ES interact with each other and have an impact on the regional ecosystem. Existing studies indicate that the main factor influencing changes in ESV is land use changes [46–48]. Between 2000 and 2010, the FPE in Inner Mongolia experienced a notable enhancement in regional ecological quality, driven by the continuous implementation of ecological initiatives like the "Three North" shelterbelt program and farmland-to-forest/grassland conversion projects [49]. This period also contributed to a significant increase in ESV and a decrease in the intensity of spatial conflicts, with the high-value–low-conflict areas in the eastern part of the FPE growing significantly. FES and APS exhibited a declining tendency between 2010 and 2020, while GES and IPS showed an increasing trend due to growing urbanization and economic development [29]. This resulted in a decrease in ESV during this period, from CNY 188.383 billion in 2010 to CNY 182.587 billion in 2020, a decrease of CNY 0.45 million from the beginning of the study. Forest land, cropland, and other ecological land make up the majority of the eastern portion of the FPE in Inner Mongolia. The area of mixed cropland and forest land is larger, and the support and regulation services offered are more significant. However, because of the proximity of urban construction land, this area is more impacted by human activity, resulting in a higher pattern of spatial conflicts and greater ecological risks. The ESV of the FPE in Inner Mongolia generally shows a decreasing trend while the increase in human activities also increases the intensity of regional spatial conflicts. The PS and LS of the FPE in Inner Mongolia have been expanding, crowding out the ES due to regional economic development and increasing urbanization.

4.3. Innovations and Limitations

The FPE is a dynamic area that evolved throughout time in response to a particular historical and humanistic context, located between farming and pastoral areas and characterized by a distinct landscape ecology. However, environmental issues are substantial as a result of climate change and human activity.

The ecosystem equilibrium of FPE in Inner Mongolia served as the foundation for the stabilization of China's northern ecological security barrier. Using land use data from 2000 to 2020, we analyzed the land use changes in the FPE in Inner Mongolia from the standpoint of PLES. We studied the spatio-temporal evolution and correlation of ESV and spatial conflicts in the region using a 7 km × 7 km grid. Furthermore, this provided a scientific basis for balancing development and ecological conservation within the Inner Mongolia FPE. Urbanization is causing living and producing spaces to encroach on ES in the Inner Mongolia FPE. The intensity of spatial conflicts in the region increased over the research period, with OES in the east experiencing more intense spatial conflicts. High values were generally found in the FES and AES, while regional ESV was low in the center and west and high in the east. The area of high ESV increased significantly in 2010 as a result of the expansion of forest land, and the high-value–low-conflict areas were primarily found in the FES when looking at the association between ESV and spatial conflict. The ecosystem will unavoidably be impacted by rapid urban expansion; however, the preservation of the natural environment should be fully taken into account while developing land. The future development of the FPE in Inner Mongolia should follow the ecological priority plan and attempt to prevent PS and LS from encroaching on ES. Meanwhile, we should maintain the policy of converting cropland back to forests and grasslands, regulate the expansion speed of construction land, and customize urbanization development patterns to local contexts, alleviating spatial land use tensions. For example, in terms of farming in the FPE, irrigated agriculture and conservation agriculture should be vigorously promoted, and advanced production technology should be introduced while strengthening the construction of windbreak and sand-fixing forests, in order to improve agricultural production efficiency and farmland windbreak and soil-fixing capacity. To improve the ecological barrier function of the cross-border agricultural and animal husbandry zone, when constructing artificial protection forests, the region's native vegetation should be used as a guide and a mix of trees and shrubs should be planted, while ecological industries should be developed in accordance with local conditions. The conflict intensity of the PLES was only examined in this study from the perspectives of complexity, vulnerability, and stability. In the future, social and economic factors can be fully taken into account when building the spatial conflict measurement model, and further research on grid-scale refinement may be performed. The ESV methodology used in the study evaluates the economic value of ecosystem services. Four primary indicators and eleven supplementary indicators of supply, regulation, ecological functions, and cultural services were chosen as the indicators for the ecosystem services valuation. The existing approaches evaluate supply, regulation, ecological functions, and cultural services independently, neglecting their synergistic or conflicting linkages, which resulted in less comprehensive valuation conclusions. Meanwhile, the spatial heterogeneity of socioeconomic factors, as well as other natural elements (e.g., biodiversity and climate change), has an impact on ESV, and such spatial heterogeneity should be fully considered in the future through implementing dynamic economic value coefficients and enriching the quantification of ESV. Future studies might explore the factors that influence spatial variations in PLES.

5. Conclusions

This study established a spatial conflict index model of PLES, quantitatively assessed the ESV in the FPE in Inner Mongolia, and revealed the impact of the evolution of the PLES spatial conflict on the ESV based on the grid scale. The main results and recommendations are as follows:

- The extent of ES declined, whereas LS experienced a significant expansion. The IPS purposes throughout all periods had the highest single-dynamic degrees attitude, while the GES and FES had the lowest single-dynamic degrees attitude. The highest single-dynamic degrees attitude of LS was 0.44%, with the largest increase in the area being 369.13 km^2, while the lowest single-dynamic degrees attitude of ES was −0.03%, with a decrease in the area of 545.52 km^2. Over the past 20 years, there were significant decreases in the grassland and FES.
- APS, FES, GES, and OES migrated southwest to northeast, while IPS, AES, ULS, and RLS migrated northeast to southwest. With the development of urbanization, spatial conflicts have become more serious, and the uncoordinated spatial development among PS, ES, and LS has become more obvious, with other ecological land types predominantly occupying high-value conflict zones and being concentrated in the eastern region.
- The spatial distribution of ESV exhibits a pattern characterized by lower values in the central and western regions, while higher values are observed in the eastern regions. Regulation services had the highest proportion of the ESV, while cultural services had the lowest proportion. Climate regulation in the regulating service is based on the trend of transferring the ecological land in forest and water areas, which showed an initial rise followed by a subsequent decline. ESV is negatively correlated with spatial conflict, and the level of spatial correlation exhibits an initial upward trend followed by a downward shift.

This study provides statistical support and policy recommendations for developing a regional spatial development plan and balancing economic development and environmental conservation in Inner Mongolia in the future. It also offers research concepts and strategies for other similar ecological transition zones.

Supplementary Materials: The following supporting information can be downloaded at: https://www.mdpi.com/article/10.3390/land14030447/s1, Table S1: Full names and abbreviations of 29 counties (banners) in the FPE in Inner Mongolia; Table S2: Classification system of PLES in the FPE in Inner Mongolia; Table S3: Equivalent factors of ecosystem service value per unit area; Table S4: Equivalent factors of ecosystem service value per unit area of FPE in Inner Mongolia; Table S5: The single dynamic degree of different land use type.

Author Contributions: Conceptualization, Z.Y. and X.M.; methodology, Z.Y.; software, G.Y.; validation, Z.Y. and G.Y.; formal analysis, Z.Y.; investigation, X.M.; resources, G.Y.; data curation, X.M.; writing—original draft preparation, Z.Y. and X.M.; writing—review and editing, X.M. and G.Y.; visualization, X.M. and G.Y.; supervision, Z.Y.; funding acquisition, Z.Y. All authors have read and agreed to the published version of the manuscript.

Funding: This research was funded by the Philosophy and Social Science Program of Inner Mongolia Autonomous Region, grant number 2023NDB246.

Data Availability Statement: Data are contained within the article or Supplementary Material.

Acknowledgments: The authors gratefully acknowledge the support of the funding. The authors are also deeply grateful to the editors and reviewers for their critical and constructive comments, which have significantly improved the quality of this manuscript.

Conflicts of Interest: The authors declare no conflicts of interest.

Abbreviations

PLES	Production–living–ecological space
FPE	Farming–pastoral ecotone
ESV	Ecosystem service value
PS	Production space
LS	Living space
ES	Ecological space
APS	Agricultural production space
IPS	Industrial production space
FES	Forest ecological space
GES	Grassland ecological space
AES	Aquatic ecological space
OES	Other ecological space
ULS	Urban living space
RLS	Rural living space

References

1. Hou, Y.; Zhang, Z.; Wang, Y.; Sun, H.; Xu, C. Function Evaluation and Coordination Analysis of Production-Living-Ecological Space Based on the Perspective of Type-Intensity-Connection: A Case Study of Suzhou, China. *Land* **2022**, *11*, 1954. [CrossRef]
2. Yu, B.; Lv, C. The Progress and Prospect of Land Use Conflicts. *Prog. Geogr.* **2006**, *25*, 106–115.
3. Zhou, G.; Peng, J. The Evolution Characteristics and Influence Effect of Spatial Conflict: A Case Study of Changsha-Zhuzhou-Xiangtan Urban Agglomeration. *Prog. Geogr.* **2012**, *31*, 717–723.
4. Costanza, R.; d'Arge, R.; de Groot, R.; Farber, S.; Grasso, M.; Hannon, B.; Limburg, K.; Naeem, S.; O'Neill, R.V.; Paruelo, J.; et al. The value of the world's ecosystem services and natural capital. *Nature* **1997**, *387*, 253. [CrossRef]
5. Song, J.; Zhang, Z.; Chen, L.; Wang, D.; Liu, H.; Wang, Q.; Wang, M.; Yu, D. Changes in ecosystem services values in the south and north Yellow Sea between 2000 and 2010. *Ocean Coast. Manag.* **2021**, *202*, 105497. [CrossRef]
6. Wan, L.; Ye, X.; Lee, J.; Lu, X.; Zheng, L.; Wu, K. Effects of urbanization on ecosystem service values in a mineral resource-based city. *Habitat Int.* **2015**, *46*, 54–63. [CrossRef]
7. Wang, Z.; Xu, M.; Lin, H.; Qureshi, S.; Cao, A.; Ma, Y. Understanding the dynamics and factors affecting cultural ecosystem services during urbanization through spatial pattern analysis and a mixed-methods approach. *J. Clean. Prod.* **2021**, *279*, 123422. [CrossRef]
8. Li, J.; Sun, W.; Yu, J. Change and regional differences of production-living-ecological space in the Yellow River Basin: Based oncomparative analysis of resource-based and non-resource-based cities. *Resour. Sci.* **2020**, *42*, 2285–2299. [CrossRef]
9. Wang, A.; Liao, X.; Tong, Z.; Du, W.; Zhang, J.; Liu, X.; Liu, M. Spatial-temporal dynamic evaluation of the ecosystem service value from the perspective of "production-living-ecological" spaces: A case study in Dongliao River Basin, China. *J. Clean. Prod.* **2022**, *333*, 130218. [CrossRef]
10. Huang, J.; Lin, H.; Qi, X. A literature review on optimization of spatial development pattern based on ecological-production-living space. *Prog. Geogr.* **2017**, *36*, 378–391.
11. Liu, Y. On the logical structures, checks and balances mechanisms and development principles of "production-living-ecological" spaces. *Hubei Soc. Sci.* **2016**, *3*, 5–9.
12. Zhang, H.; Xu, E.; Zhu, H. An ecological-living-industrial land classification system and its spatial distribution in China. *Resour. Sci.* **2015**, *37*, 1332–1338.
13. Wang, D.; Jiang, D.; Fu, J.; Lin, G.; Zhang, J. Comprehensive Assessment of Production-Living-Ecological Space Based on the Coupling Coordination Degree Model. *Sustainability* **2020**, *12*, 2009. [CrossRef]
14. Wang, D.; Fu, J.; Jiang, D. Optimization of Production-Living-Ecological Space in National Key Poverty-Stricken City of Southwest China. *Land* **2022**, *11*, 411. [CrossRef]
15. Wang, J.; Sun, Q.; Zou, L. Spatial-temporal evolution and driving mechanism of rural production-living-ecological space in Pingtan islands, China. *Habitat Int.* **2023**, *137*, 102833. [CrossRef]

16. Zhang, B.; Feng, Q.; Li, Z.; Lu, Z.; Zhang, B.; Cheng, W. Land Use/Cover-Related Ecosystem Service Value in Fragile Ecological Environments: A Case Study in Hexi Region, China. *Remote Sens.* **2024**, *16*, 563. [CrossRef]
17. Huang, J.; Zhang, Y.; Zhang, J.; Qi, J.; Liu, P. Study on the ecological environment quality and its driving factors of the spatial transformation of production-living-ecological space in Baishan City. *Sci. Rep.* **2024**, *14*, 18709. [CrossRef]
18. Xu, H.; Zhang, F.; Li, W.; Shi, J.; Johnson, B.A.; Tan, M.L. Spatial-temporal pattern of change in production-living-ecological space of Nanchong City from 2000 to 2020 and underlying factors. *Environ. Monit. Assess.* **2024**, *196*, 94. [CrossRef]
19. Fu, J.; Bu, Z.; Jiang, D.; Lin, G.; Li, X. Sustainable land use diagnosis based on the perspective of production-living-ecological spaces in China. *Land Use Policy* **2022**, *122*, 106386. [CrossRef]
20. Costanza, R.; de Groot, R.; Sutton, P.; van der Ploeg, S.; Anderson, S.J.; Kubiszewski, I.; Farber, S.; Turner, R.K. Changes in the global value of ecosystem services. *Glob. Environ. Change-Hum. Policy Dimens.* **2014**, *26*, 152–158. [CrossRef]
21. Solomon, N.; Segnon, A.C.; Birhane, E. Ecosystem Service Values Changes in Response to Land-Use/Land-Cover Dynamics in Dry Afromontane Forest in Northern Ethiopia. *Int. J. Environ. Res. Public Health* **2019**, *16*, 4653. [CrossRef] [PubMed]
22. Xiao, J.; Dai, J.; Fang, X.; Li, L.; Chen, L. Multiple Scenario Simulation of Ecosystem Service Value in Xuzhou City Based on PLUS Model. *China Land Sci.* **2024**, *38*, 125–134.
23. Li, J.; Bai, Y.; Alatalo, J.M. Impacts of rural tourism-driven land use change on ecosystems services provision in Erhai Lake Basin, China. *Ecosyst. Serv.* **2020**, *42*, 101081. [CrossRef]
24. Pan, N.; Guan, Q.; Wang, Q.; Sun, Y.; Li, H.; Ma, Y. Spatial Differentiation and Driving Mechanisms in Ecosystem Service Value of Arid Region:A case study in the middle and lower reaches of Shule River Basin, NW China. *J. Clean. Prod.* **2021**, *319*, 128718. [CrossRef]
25. Zhou, Y.; Chen, T.; Wang, J.; Xu, X. Analyzing the Factors Driving the Changes of Ecosystem Service Value in the Liangzi Lake Basin-A GeoDetector-Based Application. *Sustainability* **2023**, *15*, 15763. [CrossRef]
26. Huang, M.; Yue, W.; Fang, B.; Feng, S. Scale response characteristics and geographic exploration mechanism of spatial differentiation of ecosystem service values in Dabie Mountain area, central China from1970to2015. *Acta Geogr. Sin.* **2019**, *74*, 1904–1920.
27. Dang, H.; Rong, L.; Li, Y.; Zhao, M. Spatiotemporal evolution characteristics and influencing factors of productionliving-ecological spaces in the farming-pastoral ecotone:Taking Hohhot of Inner Mongolia as an example. *Arid Zone Res.* **2023**, *40*, 1698–1706.
28. Wang, Y.; Gao, J.; Jin, Y.; Cao, B.; Wang, Y.; Zhang, X.; Zhou, J. Habitat Quality of Farming-pastoral Ecotone in Bairin Right Banner, Inner Mongolia Based on Land Use Change and InVEST Model From 2005 to 2015. *J. Ecol. Rural Environ.* **2020**, *36*, 654–662.
29. Yang, S.; Yang, H.; Zhao, G. Eco-environmental effects and differentiation mechanism of land use transition in the agro-pastoral belt in Inner Mongolia: A perspective of production-living-ecological sapces. *J. Arid Land Resour. Environ.* **2024**, *38*, 22–30.
30. Jian, Z.; Luo, H.; Shan, N. A study on the spatial and temporal evolution and carbon effects of production-living-ecological in Xinjiang under carbon peak and carbon neutrality goals. *Arid Zone Res.* **2024**, *41*, 1238–1248.
31. Man, W.; Liu, M.; Wang, Z.; Hao, Y.; Xiang, H.; Wei, S.; Mao, D.; Jia, M.; Ren, C. Remote sensing investigation of grassland change in Northeast China during 1990~2015. *China Environ. Sci.* **2020**, *40*, 2246–2253.
32. Zhang, Y.; Jiang, P.; Cui, L.; Yang, Y.; Ma, Z.; Wang, Y.; Miao, D. Study on the spatial variation of China's territorial ecological space based on the standard deviation ellipse. *Front. Environ. Sci.* **2022**, *10*, 982734. [CrossRef]
33. Jensen, D.; Baird, T.; Blank, G. New landscapes of conflict: Land-use competition at the urban-rural fringe. *Landsc. Res.* **2019**, *44*, 418–429. [CrossRef]
34. Zhao, F.; Liu, X.; Zhao, X.; Wang, H. Effects of production-living-ecological space changes on the ecosystem service value of the Yangtze River Delta urban agglomeration in China. *Environ. Monit. Assess.* **2023**, *195*, 1133. [CrossRef]
35. Xie, G.; Lu, C.; Leng, Y.; Zheng, D.; Li, S. Ecological assets valuation of the Tibetan Plateau. *J. Nat. Resour.* **2003**, *18*, 189–196.
36. Xie, G.; Zhang, C.; Zhang, C.; Xiao, Y.; Lu, C. The value of ecosystem services in China. *Resour. Sci.* **2015**, *37*, 1740–1746.
37. Xie, G.; Zhang, C.; Zhang, L.; Chen, W.; Li, S. Improvement of the Evaluation Method for Ecosystem Service Value Based on Per Unit Area. *J. Nat. Resour.* **2015**, *30*, 1243–1254.
38. Saputra, W.; Giyarsih, S.R.; Muhidin, S. Spatial analysis of slum areas on the riverbanks of Palembang City using the Anselin Local Moran's I analysis. *Geojournal* **2023**, *88*, 6523–6538. [CrossRef]
39. Feng, S.; Liu, X.; Zhao, W.; Yao, Y.; Zhou, A.; Liu, X.; Pereira, P. Key Areas of Ecological Restoration in Inner Mongolia Based on Ecosystem Vulnerability and Ecosystem Service. *Remote Sens.* **2022**, *14*, 2729. [CrossRef]
40. Guo, X.; Chen, R.; Thomas, D.S.G.; Li, Q.; Xia, Z.; Pan, Z. Divergent processes and trends of desertification in Inner Mongolia and Mongolia. *Land Degrad. Dev.* **2021**, *32*, 3684–3697. [CrossRef]
41. Liu, Q.; Tong, Y. The effects of land use change on the eco-environmental evolution of farming-pastoral region in Northern China: With an emphasis on Duolun County in Inner Mongolia. *Acta Ecol. Sin.* **2003**, *23*, 1025–1030.
42. Wang, S.; Xing, X.; Wu, Y.; Guo, X.; Li, M.; Ma, X. Restoration of vegetation in the Yellow River Basin of Inner Mongolia is limited by geographic factors. *Sci. Rep.* **2024**, *14*, 14922. [CrossRef] [PubMed]

43. Zhao, L.; Jia, K.; Liu, X.; Li, J.; Xia, M. Assessment of land degradation in Inner Mongolia between 2000 and 2020 based on remote sensing data. *Geogr. Sustain.* **2023**, *4*, 100–111. [CrossRef]
44. Tan, S.; Ye, Z.; Du, H. Impacts and mechanism of the new round of grassland tenure confirmation on grassland ecology: Take the pastural area of Inner Mongolia as an example. *Resour. Sci.* **2024**, *46*, 610–620.
45. Harmon, D.D.; Rayburn, E.B.; Griggs, T.C. Grassland Ecology and Ecosystem Management for Sustainable Livestock Performance. *Agronomy* **2023**, *13*, 1380. [CrossRef]
46. Wang, Y.; Dai, E.; Yin, L.; Ma, L. Land use/land cover change and the effects on ecosystem services in the Hengduan Mountain region, China. *Ecosyst. Serv.* **2018**, *34*, 55–67. [CrossRef]
47. Zhang, L.; Zhang, H.; Xu, E. Information entropy and elasticity analysis of the land use structure change influencing eco-environmental quality in Qinghai-Tibet Plateau from 1990 to 2015. *Environ. Sci. Pollut. Res.* **2022**, *29*, 18348–18364. [CrossRef]
48. Wang, Y.; Zhang, X.; Peng, P. Spatio-Temporal Changes of Land-Use/Land Cover Change and the Effects on Ecosystem Service Values in Derong County, China, from 1992-2018. *Sustainability* **2021**, *13*, 827. [CrossRef]
49. Xu, W.; Rao, L. Impacts of land use and climate change on ecosystem services in agro-pastoral ecotone. *Environ. Sci.* **2023**, *44*, 5114–5124.

land

MDPI

Article

Integration of Climate Change and Ecosystem Services into Spatial Plans: A New Approach in the Province of Rimini

Denis Maragno [1,*], Federica Gerla [1,2,*] and Francesco Musco [1]

1 Departments of Architecture and Arts, Università Iuav di Venezia, 30135 Venezia, Italy; francesco.musco@iuav.it
2 National PhD in Earth Observation, Department of Civil and Environmental Engineering, Sapienza Università di Roma, 00185 Rome, Italy
* Correspondence: dmaragno@iuav.it (D.M.); fgerla@iuav.it or federica.gerla@uniroma1.it (F.G.)

Abstract: This study presents a spatial methodology for integrating climate change (CC) risks and ecosystem service (ES) assessments into strategic spatial planning, applied to the Metropolitan Plan of the Province of Rimini (Emilia-Romagna, Italy). The proposed approach combines IPCC-aligned climate vulnerability analysis with ecosystem service mapping based on the methodology developed by CREN. Climate risks, including urban heat islands, droughts, and urban floods, were assessed using satellite-derived indices such as Land Surface Temperature (LST), Vegetation Health Index (VHI), and hydraulic modeling. For ESs, nine key services were evaluated and mapped by integrating land use, forest cover, and habitat data with biophysical modulation factors (e.g., slope, carbon stock, infiltration capacity). The results highlight priority areas where climate adaptation and ecological functions converge, enabling targeted interventions. This integrated workflow offers a replicable and scalable planning tool to support evidence-based decision-making at the metropolitan level. Its adoption is recommended by other local and regional authorities to strengthen the climate and ecological responsiveness of spatial planning instruments.

Keywords: climate change; ecosystem services; urban planning; climate adaptation; urban adaptation; remote sensing; earth observation

Academic Editors: Luca Congedo, Francesca Assennato and Michele Munafò

Received: 19 March 2025
Revised: 17 April 2025
Accepted: 23 April 2025
Published: 25 April 2025

Citation: Maragno, D.; Gerla, F.; Musco, F. Integration of Climate Change and Ecosystem Services into Spatial Plans: A New Approach in the Province of Rimini. *Land* **2025**, *14*, 934. https://doi.org/10.3390/land14050934

1. Introduction

Growing attention has been devoted to climate change (CC) due to its significant impacts on urban systems, ecosystems, and public health—collectively referred to as climate impacts (CIs) [1,2]. Climate projections for urban areas highlight the need for integrated planning activities to reduce greenhouse gas emissions and enhance urban resilience to future climate impacts [3,4]. Therefore, it is important to prioritize CC mitigation and adaptation in urban agendas and revise conventional urban planning approaches. Local governments play a vital role in this transformation, as they can define strategies and measures to reduce climate-altering gas emissions and adapt their territories through local planning tools, as outlined in the Paris Agreement (2015). Besides climate issues, it is essential to consider ecosystem properties to ensure quality of life and environmental well-being. Evaluating ecosystem services (ESs) is gaining consensus, especially in urban and regional planning, as a tool for assessing ecosystem supply and biodiversity [5,6]. By mapping ESs at various scales, it is possible to increase awareness of the role ecosystems play in supporting human well-being and reinforce the interconnection between urban and natural systems. However, this evaluation approach presents several limitations, primarily due to the scarcity of high-quality spatial data [7,8]. The connection between

CC adaptation and ESs has become increasingly significant. Nature-based solutions and green infrastructure are now widely recognized as essential components of planning strategies. This marks a significant shift, particularly in Italy, where green spaces have traditionally been evaluated based solely on their size rather than on their contribution to ecological integrity and socio-economic resilience [9–11]. To effectively address climate impacts through ecosystem services, planning must adopt an integrated, cross-sectoral, and multiscale perspective. Incorporating ecosystem services into planning instruments and initiatives is essential at all levels, from regional to local. Regional and metropolitan strategies should be defined to enable local implementation, following guidelines that endorse the principle of subsidiarity [12].

This paper outlines the methodological approach to developing the Plan for the Metropolitan Territory of the Province of Rimini (PTAV)[1], in accordance with Regional Urban Planning Law No. 24/2017[2] [13–15], with scientific support provided by Iuav University of Venice during the development phase. The PTAV provides the knowledge and strategies that municipalities are required to adopt and implement in the drafting of local-scale plans, with the aim of promoting community well-being and addressing environmental and climate emergencies[3]. It aims to define a long-term, resilient vision for territorial development, aligning public interest with urban and environmental planning decisions. This shared approach enhances governance capacity and supports more informed and effective decision-making. It emphasizes the importance of assessing the territory and its evolutionary processes (Regional Law 24/17, Art. 22). By incorporating the Climate Adaptation Planning Methodology [16], the PTAV aims to reduce climate vulnerabilities and develop local resources, focusing on ecosystem services (ESs). The plan provides a strategic opportunity for the province and its municipalities to collaborate around a shared vision, pooling their respective strengths and resources to pursue common objectives. Integrating climate change adaptation and ecosystem services into planning strategies makes the plan both innovative and dynamic, transforming the urban knowledge framework into a diagnostic tool for evaluating planning effectiveness over time. Moreover, the spatial knowledge derived from assessments of climate vulnerability[4] and ecosystem services offers essential support to metropolitan and local authorities in advancing socio-economic and environmental goals, particularly through ecological transition and enhanced climate resilience. The paper outlines the response to the legal requirement to assess and consider climate impacts at the provincial scale for the territory of Rimini, integrating them into strategic planning processes. It also reflects on the added value of incorporating both climate risks and ecosystem services into spatial planning as a means to support sustainable development and climate adaptation strategies.

This contribution, therefore, aims to answer the following research questions:

- RQ1: How do we consider climate impacts within planning processes?
- RQ2: How should ESs be assessed in relation to the competencies of planning instruments? What are the implications for database structure and knowledge frameworks?
- RQ3: How can ESs be considered within strategic planning processes? What are the benefits? What does it imply in monitoring?

2. Materials and Methods

The methodological approach adopted is strongly related to the methodology suggested by the IPCC [2]. From this methodology, the operational steps involved in developing climate vulnerability assessments are, in fact, defined.

This study produced sensitivity maps for each climate impact and adaptive capacity maps for each ecosystem service (ES) to guide the PTAV planning process, incorporating adaptation and ES issues. Sensitivities and adaptive capacities were assessed separately to better define strategic plans. This approach enriched the planning process with new spatial information, enhancing and improving existing adaptive capacities where needed.

Hazard characterizes an event potentially impacting a specific territory. The authors integrated climate change (CC) and ecosystem services (ESs) into the spatial planning process. For CC, they first defined future climate hazards specific to the Province of Rimini, then identified potential impacts, focusing on heat waves and intense rainfall events. As for ESs, the Emilia-Romagna Region, in line with the European Directive on environmental sustainability and Italian Law No. 221/2015, enacted Regional Law 24/2017 to promote green economy measures and reduce the overexploitation of natural resources. This legal framework established the basis for surveying and assessing ESs provided by environmental systems. Nevertheless, an integrated workflow was needed to combine climate impact and ecosystem service assessments in a coherent framework to inform PTAV decision-making. This section outlines the methodological approach adopted for assessing climate impacts and ecosystem values within the Province of Rimini's planning process. The workflow structure is illustrated in Figure 1.

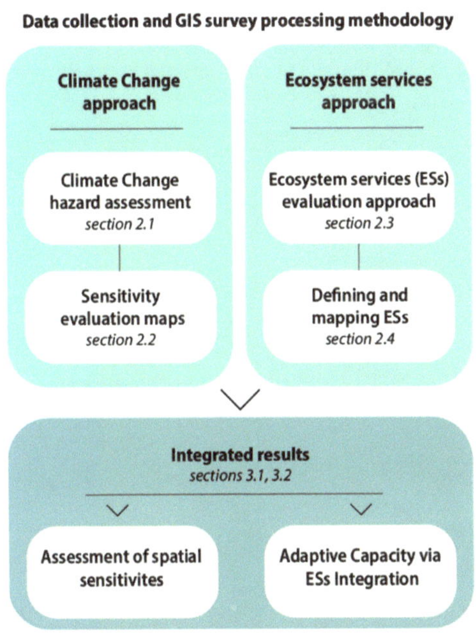

Figure 1. Workflow of the methodology.

Starting from the spatial knowledge framework, the methodological approach integrates climate and ecosystem analyses into a unified strategy that addresses two emerging themes in planning processes in an innovative and dynamic manner.

2.1. *Geographical and Climatic Context: Current and Expected Hazard Scenarios*

This section describes the climate change hazard assessment, considering past climate scenarios and trends regarding temperature and precipitation (Figure 2).

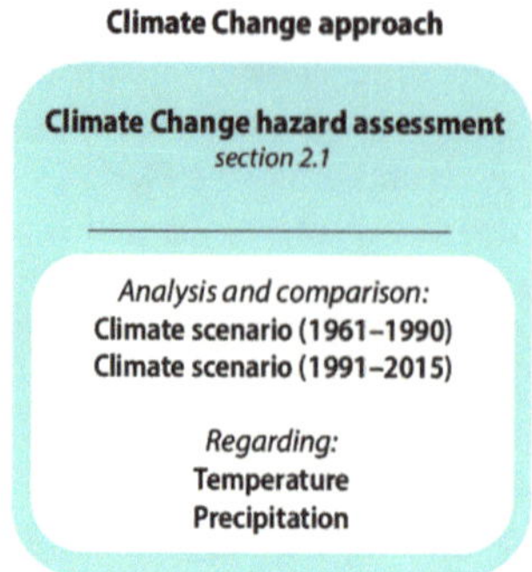

Figure 2. Workflow of the geographical and climatic context.

The Province of Rimini is located along the southernmost stretch of Emilia-Romagna's coastline, with a hinterland consisting of a flat area in the north that extends to the sea to the east in the areas of Rimini and Riccione. To the west, a hilly and mountainous Apennine belt makes up the area (Figure 3). The Province of Rimini has a warm temperate climate that is stably humid, with hot summers and a reduced diurnal temperature range, thanks to the influence of the Adriatic Sea.

Figure 3. Localization of Rimini Province: study area.

The studies carried out by ARPAE[5], comparing the recent climate (1991–2015) with that of the 30-year reference period 1961–1990, show evident changes attributable to CC [17]. According to the "IdroMeteoClima Emilia-Romagna Report" [18], carried out by the Arpa Emilia-Romagna Climate Observatory, in 2020, there was a temperature deviation of about +0.5 °C on the recent climate (1991–2015) and +1.5 °C on the 1961–1990 climate (Figure 4a–c).

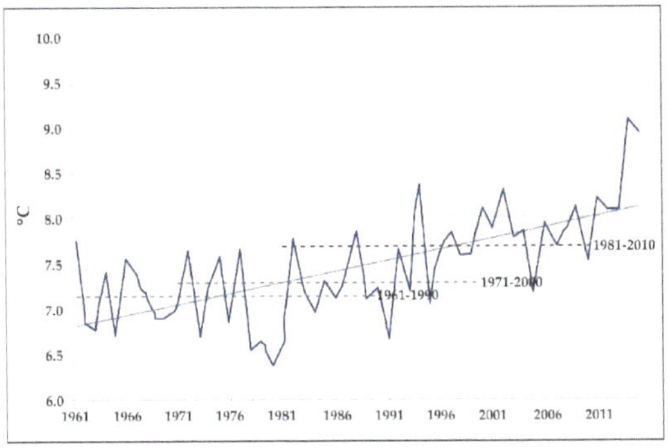

(a)

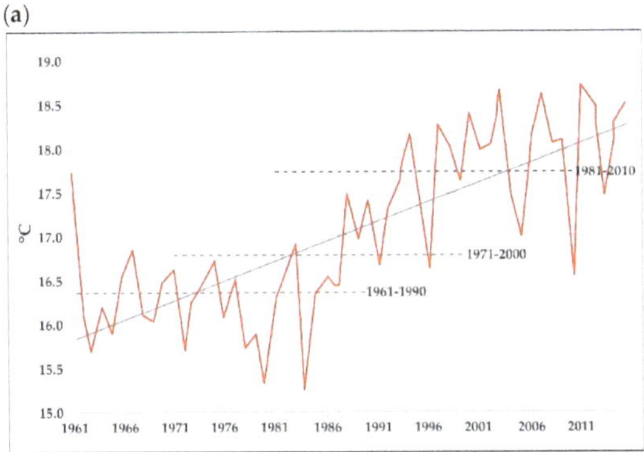

(b)

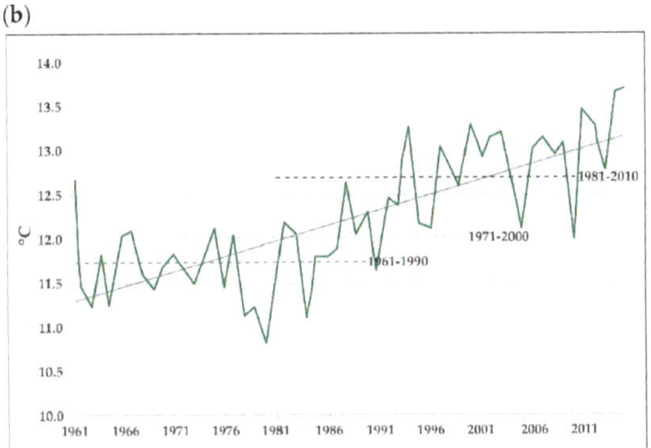

(c)

Figure 4. (**a**–**c**) Graph of historical trends and temperature (°C) minimum (**a**), maximum (**b**), and average (**c**) between 1961 and 2015 in Emilia-Romagna (Atlante climatico dell'Emilia-Romagna 1961–2015).

Overall, 2020 was, on average, the fifth warmest year after 2014, 2015, 2018, and 2019. It was also the mildest year ever since 1961 for average winter temperatures, with few days of frost and anomalous late spring frosts, which caused severe damage to fruit crops in advanced phenological development. As far as rainfall is concerned, no significant changes in annual precipitation patterns have been observed. However, shifts in the regime are evident, with prolonged droughts and more frequent extreme events (Figure 5).

In recent years, significant rainfall deficits have characterized the phenomenon over almost the entire region compared to 1961–1990. The negative anomalies have also been very intense, with values as low as −300 mm in Emilia-Romagna in 2020. November

2020, usually the wettest month, presented the lowest level of rainfall in the last 60 years, with deviations of about 70% less than the climatic reference values (1961–1990). The year 2021 was a drought year: −235 mm of precipitation compared to the reference climate (1991–2020), with negative anomalies over much of the regional territory [19].

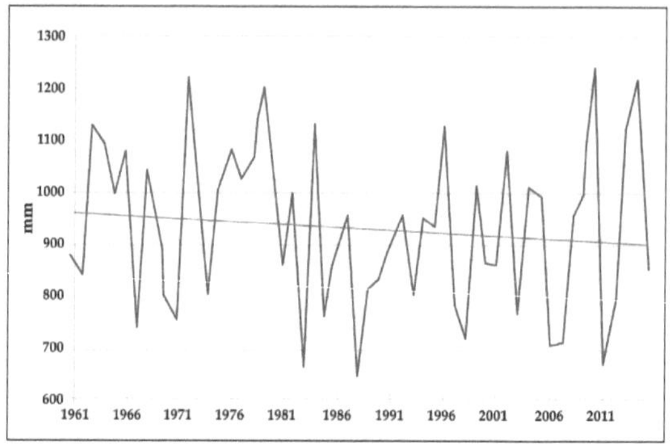

Figure 5. Graph of historical trends and trends in annual precipitation (mm) between 1961 and 2015 in Emilia-Romagna (Atlante climatico dell'Emilia-Romagna 1961–2015).

Rainfall events become more polarized, tending toward the extremes precisely with more intense and concentrated events occurring than in the past [20–22].

The analysis of climate trends made it possible to identify hazards for the territory, from which it was possible to identify possible climate impacts. Data were collected and processed for each climate impact in a (Geographic Information System) environment to assess the territory's sensitivities and vulnerabilities (considering E.S.s as adaptive capacity) [2]. From the analyses performed, the authors decided to consider urban heat islands, droughts, and urban flooding as climate impacts in the PTAV.

2.2. Sensitivity Assessment Maps in GIS: Heat, Drought, and Floods

The following steps outline the main components of the climate impact stress assessment. The climate impacts considered are urban heat island, water stress conditions, and floods (Figure 6).

The phenomenon of the urban heat island describes elevated temperatures in urban areas relative to their rural surroundings [23–25]. During the day, artificial surfaces store more thermal energy (heat) than the surrounding agricultural and forest areas [26]; thus, during the night, the temperature difference amplifies. While the rural system cools rapidly, the urban fabric takes much longer, resulting in a slow and gradual release of heat due to the thermal inertia that characterizes the materials of the built space.

Ongoing CC requires planning processes to be supported by spatial survey tools that can identify and localize urban areas particularly vulnerable to such phenomena [27]. The present contribution develops its assessment and mapping of phenomena related to excessive heat on the Land Surface Temperature (LST) calculation. This term is used to describe a method that measures the temperature of surfaces. The LST thus emerges as a significant parameter as it provides insight into the thermal impact caused by climate variations [28].

The operation behind the LST is developed in a GIS environment, starting with the emissivity of the Earth's surface and the brightness temperature of the Earth's surface. The authors calculate the brightness temperature and land surface emissivity by processing data from the Landsat 8 mission[6] [29].

The survey produced raster images where each pixel expresses the surface temperature for every 30 m of land area. The processed images were identified based on the climate survey carried out in the previous phase, processing data from warmer days subject to heat waves. At the provincial level, the analysis of the LST allowed the distribution of ground temperatures greater than 30 degrees to be identified (Figure 7).

Determining water stress conditions at the provincial level requires adequate spatial information. To this end, the authors employed remote sensing techniques and calculated the VHI (Vegetation Health Index)[7]. Assuming that a constant and intense increase in soil temperature, combined with a prolonged period of no rainfall, harms vegetation vigor and causes significant vegetation stress, the VHI bases itself on the inverse correlation between LST (reflected soil temperature) and NDVI (presence of chlorophyll in the leaf).

The VHI is derived from the relationship between the Temperature Condition Index (TCI) and the Vegetation Condition Index (VCI). The calculation of the TCI uses temperature data obtained from the LST, while that of the VCI is based on NDVI vegetation data reflecting soil moisture conditions. The VCI indicates standardized values (in %) reflecting vegetative stresses related to low moisture content, while the TCI reflects values (in %) of vegetative stresses related to high temperatures. The VHI is, therefore, a representative indicator of the general health of the vegetation present in an area at the time of satellite data acquisition. It is a balanced estimate between the observation period's thermal state and plant tissue's moisture content [30]. In detail, the VHI manifests increasing drought conditions[8], below threshold value 40, where some territories present a high vegetative stress status [31,32]. The spatialization of the VHI enables the authors to identify water stress gradients that, when adequately correlated with specific context information, can indicate the potential propensity of a specific forest or tree type to drought stress and fire exposure. The VHI allowed us to investigate the behavior of areas of high ecosystem value concerning the possible specific impact of drought. The authors considered the relationship whereby the impact of drought decreases the provision of ecosystem services (ESs) in the long term; thus, to maintain the provision of these services, the assessment enabled the identification of areas most susceptible to climate impacts.

Two specific time frames were analyzed: one following a rainfall event and one during a period of prolonged drought, both identified through satellite imagery. This comparison made it possible to observe and identify two main behaviors, distinguishing poorly performing areas in which there is strong ecological fragmentation (in red in Figure 8) and areas that still maintain a positive characterization but which should be alerted to as they have threshold values at a minimum (in yellow in Figure 8).

Given that climatic variations follow a worsening trend, it is essential to prioritize action in these areas where ecological continuity must be ensured to guarantee the provision of ESs.

The last impact assessment concerned the floods (Figure 9). In the context of settlements, one of the most widespread impacts resulting from intense and concentrated rainfall events is urban flooding or flash floods. When heavy rain falls in built-up areas and flows into underground drainage networks, it causes localized water overloads, leading to temporary flooding [33]. CC exacerbates the inefficiency of drainage systems, emphasizing the need to analyze, represent, and interpret surface runoff paths and the diverse relationships between urban and rural areas. The study investigated the spatial correlation between surface runoff and land use patterns to address this territorial vulnerability. This helped identify the most vulnerable urban structures and their ability to retain rainwater, facilitating a rethinking of land use planning by designing new transformations into a climate resilience system. This work calculates the potential impact of flooding by simulating rainwater runoff, according to the Maragno et al. (2021) methodology [34].

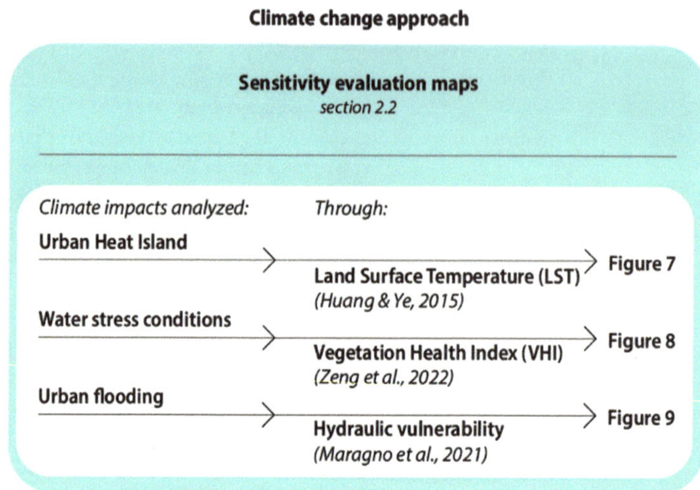

Figure 6. Climate impacts considered in the sensitivity assessment (Section 2.2) and related indicators: LST for urban heat island (Huang & Ye, 2015) [29], VHI for water stress (Zeng et al., 2022) [32], and hydraulic vulnerability for urban flooding (Maragno et al., 2021) [34].

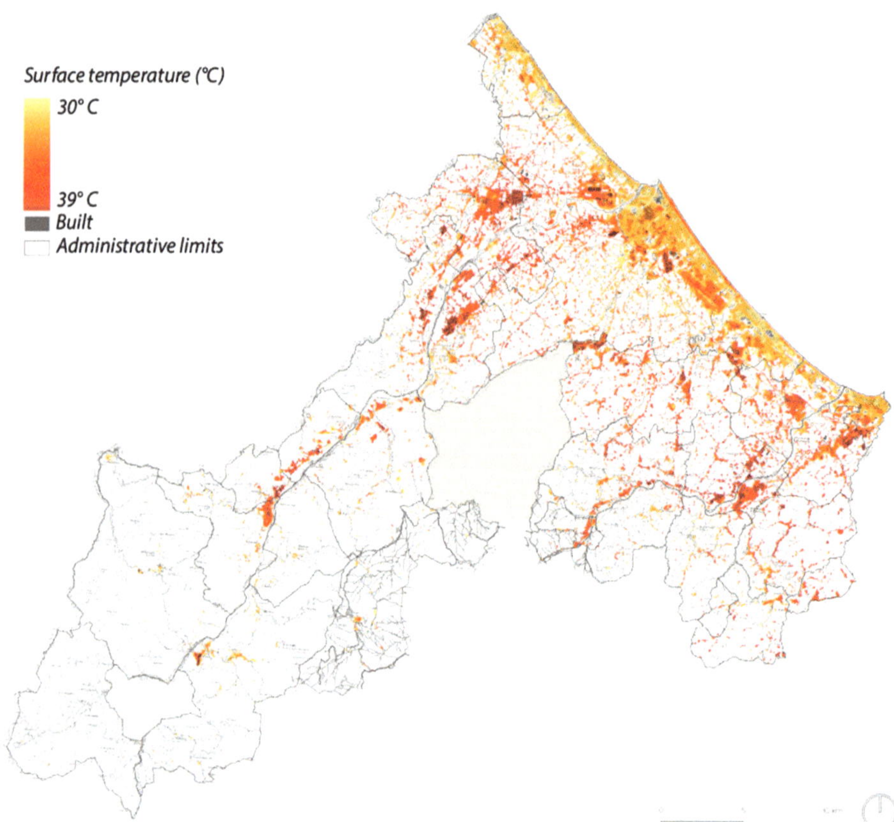

Figure 7. Heat waves: areas vulnerable to high temperatures (based on Copernicus data and Huang and Ye (2015) [29] Land Surface Temperature methodology).

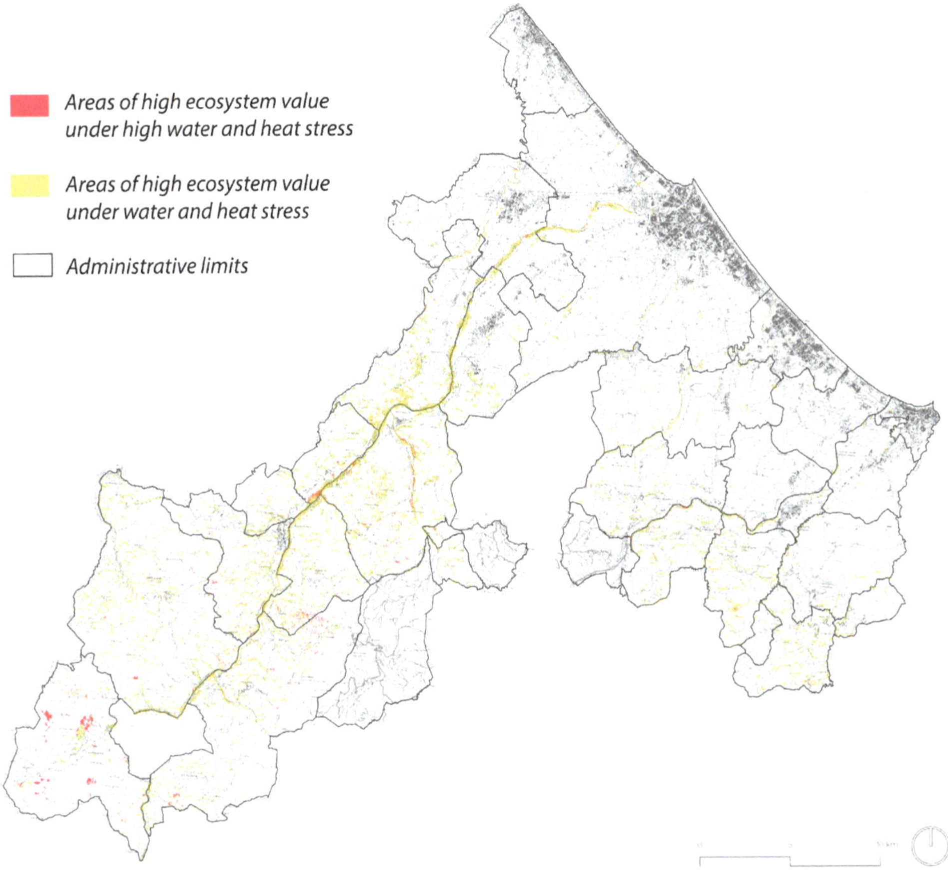

Figure 8. Drought map: vegetation component subjected to water and heat stress (based on Copernicus data and Kogan (1995) [30] Vegetation Health Index methodology).

Consequently, the delineation of inflow and outflow areas is associated with the morphology of the land and its hydraulic response, which allows for the spatial estimation of the surface runoff related to land uses. The study employs a spatial index of hydraulic vulnerability derived from a combination of soil sealing levels (determined using the Curve Number procedure, a parameter of the surface runoff training model equation) and land morphologies (such as slopes, depressions, elevations, and low-lying areas). The methodology allows for an assessment of surface runoff dynamics and estimated hydraulic impacts based on changes in land cover, monitored through the European Corine Land Cover project (CLC 2018). This research facilitates the creation of different scenarios of surface runoff, measured considering specific rainfall reactivity indicators (such as H 30 mm), which are spatially associated with the corresponding soil saturation volume. The methodology returns a trend that describes the different hydraulic performances of the study area.

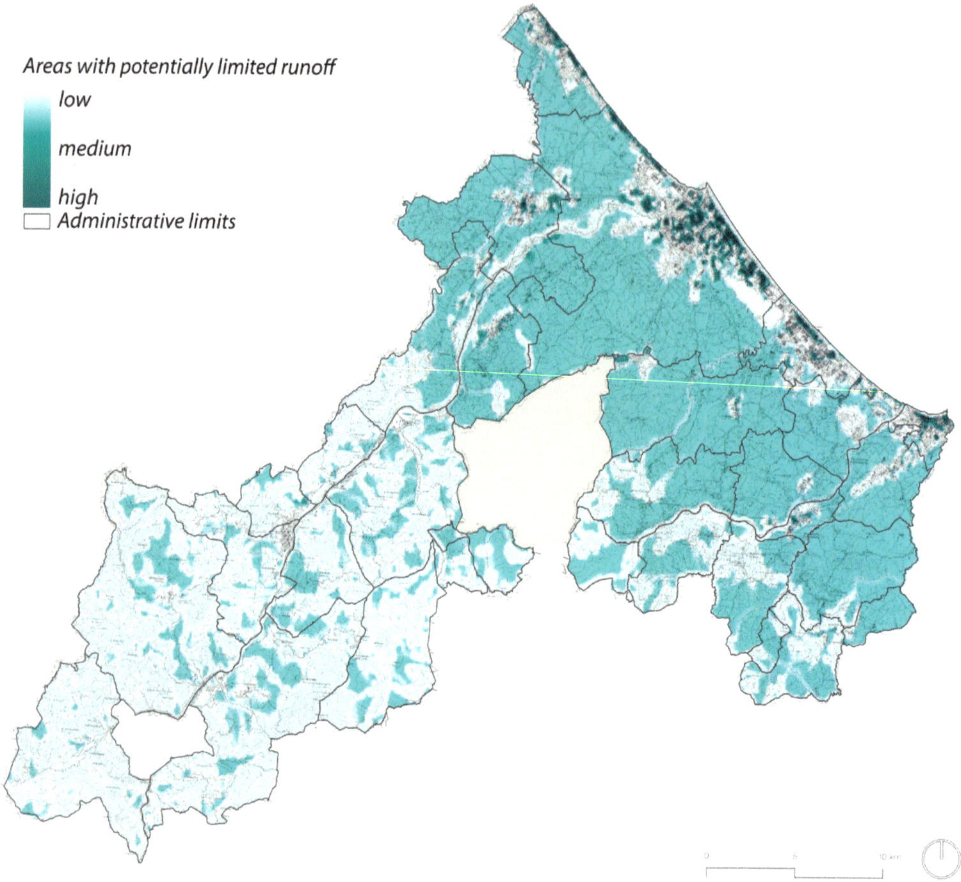

Figure 9. Vulnerabilities from surface runoff (based on Copernicus data and Maragno, Pozzer (2021) [34] Urban Flooding assessment methodology).

2.3. Ecosystem Services Evaluation Approach

The Centre for Natural Ecological Research (CREN) developed a methodology for assessing ESs. This methodological approach represents a valuable tool for decision-makers in territorial planning, providing criteria for evaluating ecosystem services. It aims to integrate scientific knowledge of ecological relationships within a complex framework encompassing social, political, and value aspects. The long-term goal is to protect and preserve ecosystems [35]. The authors obtained the data and information necessary for the analyses from the Emilia-Romagna Region and ISPRA[9] Database (Figure 10).

The first step in this methodology is creating a digital map of the environmental system to initiate the evaluation of ESs. This map is a fundamental cognitive system for further analyses within a GIS environment.

The map incorporates three spatial information layers covering the entire province:

1. Land Use Layer (updated to 2020): this layer provides information about how land uses within the province;

2. Forest Layer (updated to 2014): this layer focuses on the province's distribution and characteristics of forests. It provides valuable insights into the forest ecosystem and its potential services;
3. Habitat Layer (updated to 2020): this layer highlights the presence and distribution of different habitats within the province, offering crucial information about the diversity and richness of local ecosystems.

The environmental system map is the basic level required for the calculation of the following ESs:

- ES 1: Protection from extreme events;
- ES 2: Regulation of the microclimate;
- ES 3: CO_2 regulation;
- ES 4: Erosion control;
- ES 5: Agricultural production;
- ES 6: Forest production;
- ES 7: Water purification;
- ES 8: Regulation of the hydrological regime;
- ES 9: Recreational service.

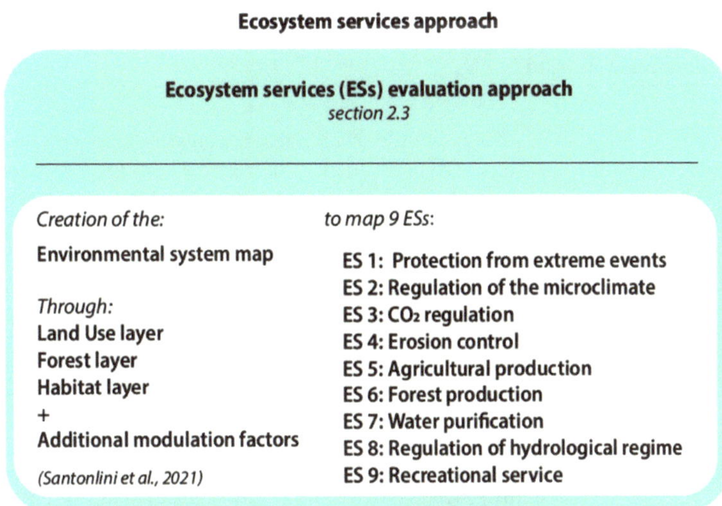

Figure 10. ESs evaluation framework (Section 2.3), based on the integration of land use, forest, and habitat layers with additional factors to map nine ESs (Santolini et al., 2021) [35].

The authors chose these ESs because they connect to the ecological support and regulatory functions fundamental to delivering all other services. Secondly, the choice was due to the study area's physical, geomorphological, and settlement conformation, where different ecological and structural elements coexist but are pivotal to much of the regional territory. In the adopted methodology, each ecosystem service is influenced by specific environmental variables, referred to as "additional modulation factors", which are inherently defined within the methodological framework. Therefore, the second step involves integrating the additional modulation factors into the environmental system map, which changes depending on the ES evaluated (Table 1).

Table 1. Additional modulation factors for each ES (elaboration based on CREN methodology).

Modulation Factors	ES 1	ES 2	ES 3	ES 4	ES 5	ES 6	ES 7	ES 8	ES 9
Forest cover (%)	X		X				X	X	
Slope (class)	X				X	X	X	X	
Current increase in forest biomass (m^3/ha)						X			
Influence of road infrastructure (m)		X			X		X		
Organic carbon stock in soil 0–100 cm (Mg/ha)			X						
Land Capability Classification (LCC) (class)					X				
Evapotranspiration coefficient (KC) (index)								X	
Deep Water Infiltration (WAR) (index)								X	
Aquifers in rock clusters								X	
Purification capacity (BUF) (index)							X		
Current erosion (RUSLE) (Mg/ha/year)				X					
Distance to urban centers (m)									X
Distance to the road network (m)									X
Distance to trails and cycle paths (m)									X
Distance to protected areas (m)									X

Integrating the environmental system map with the additional modulation factors results in the "Evaluation Matrix" for each ES evaluated. Each assessment matrix has within its values established by the methodology itself, called "weights", which depend on the land use of the environmental system map and the modulation factors.

For example, in the case of ES1: protection from extreme events, the weights used, in addition to those derived from the environmental system map, are associated with forest cover and slope modulation factors. The assignment of the weights (from 0 irrelevant to 5 very relevant) determines how much the presence of the modulation factors affects the delivery of a specific ES.

2.4. *Defining and Mapping Ecosystem Services in GIS*

Following the described methodological approach, the performance level of nine ESs in the Province of Rimini was assessed and mapped (Figure 11).

The analysis employed ecological phenomena that support and regulate the ecosystem's functioning to estimate crucial thresholds for utilization. Such an approach aids decision-making regarding when, where, and to what extent action should be taken to ensure the stable provision of ESs and the protection of biodiversity within a territory.

Assessing ESs' direct and indirect contributions to human well-being is a crucial step in determining the "minimum critical dimension of impact needed to safeguard the collective function of the asset over time, including social utility and resulting well-being" [35].

This assessment enables the development and promotion of strategies and interventions for ES preservation or enhancement during the planning phase [36]. For this reason, the nine ESs, once processed, were synthesized into a single information layer by raster summation. The sum of the individual ecosystem service maps allows the development of an ecosystem value map that describes ESs' overall trend and distribution in the study area. In this way, the summary map allows for understanding the areas with a higher concentration of ESs.

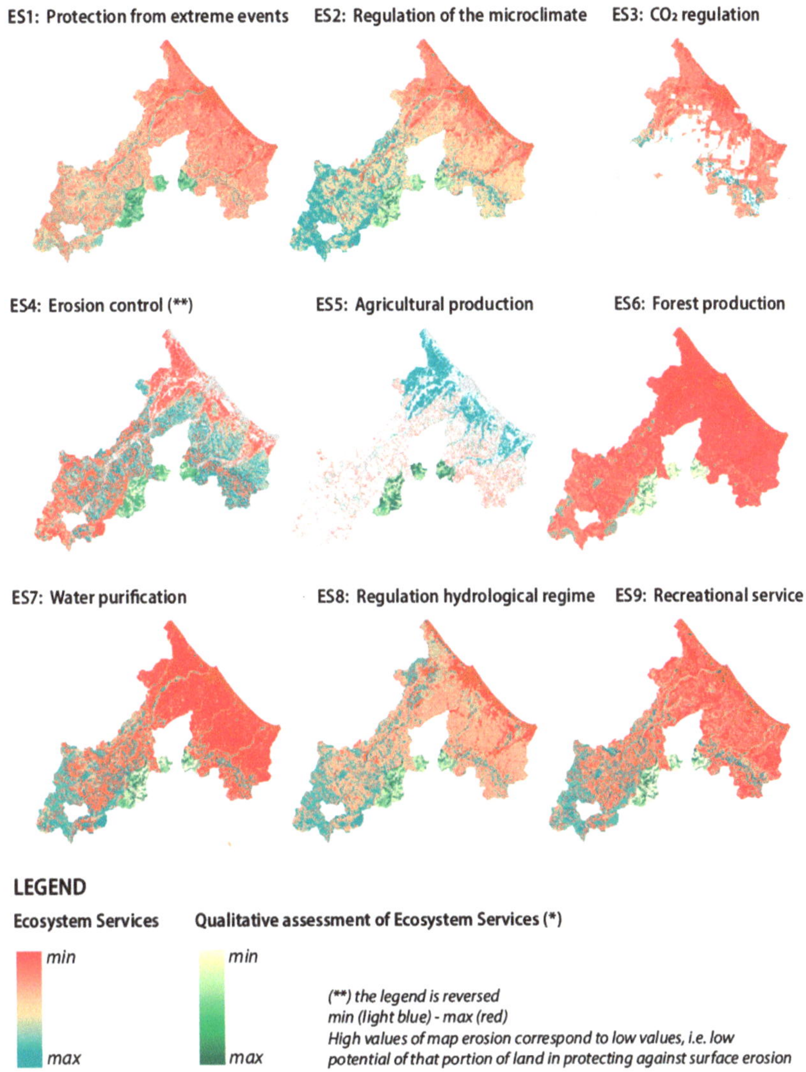

Figure 11. Nine ESs (*) The gradient in green on the map depicts two municipalities recently annexed to the province of Rimini and for which the information layers required for ecosystem assessments were not available. For this reason, while waiting for these information layers, an assessment was carried out on a qualitative basis. (**) In the ES "Erosion control" the legend is reversed.

3. Results

The resulting maps locate significant impacts and disruptions. The maps identify priority areas where action is necessary to address climate impacts and to protect and enhance ecosystem values. Thus, in this section, climate-related impacts and ecosystem values within the study area will be identified and spatially integrated to highlight areas where both aspects coexist and where climate change-related pressures and ecosystem values are simultaneously present (Figure 12).

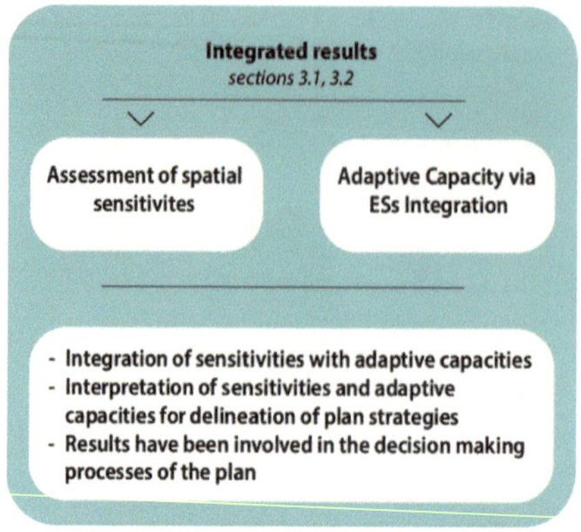

Figure 12. Workflow of the integrated results.

3.1. Integrated Assessment of Spatial Sensitivities

The results of the climate impact assessment maps highlight the areas in the entire area of the Province of Rimini that are susceptible to high temperatures and flash floods. It is observable that these areas are concentrated in the municipalities alongside the coastline (Figure 13).

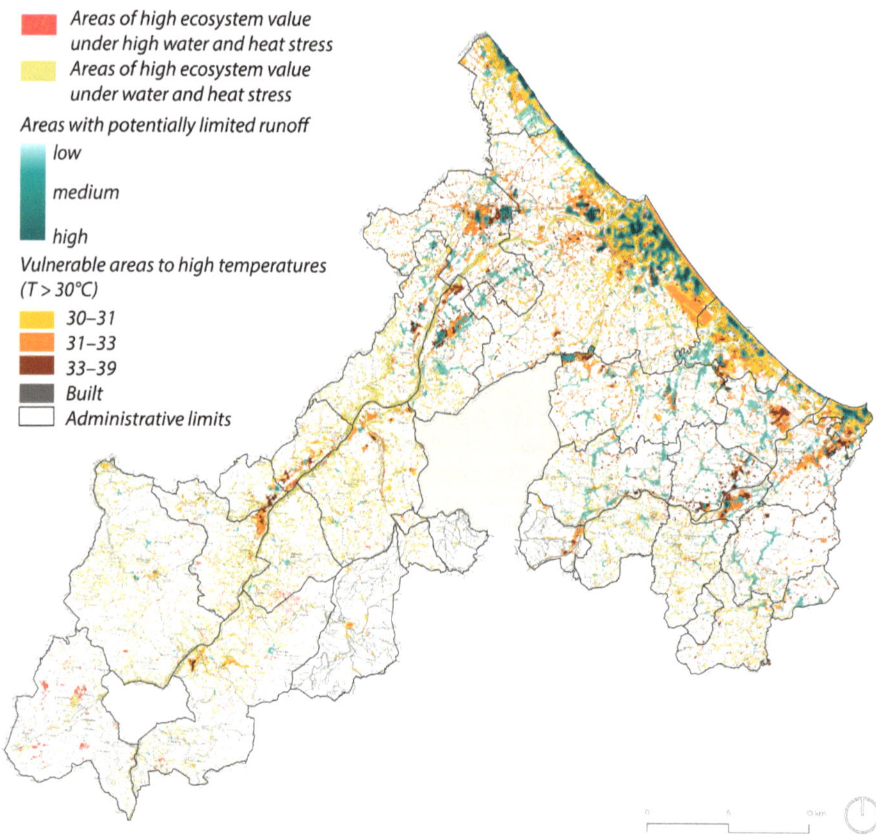

Figure 13. Climate impact sensitivity map.

This result draws attention to the urban areas affected by the phenomenon by allowing the adoption of some adaptation strategies to mitigate the temperature range [37]. The map (Figure 14) reveals that the municipalities with the biggest spots vulnerable to high temperatures are Cattolica, Riccione, Bellaria-Igea Marina, Rimini, and Misano Adriatico municipalities, characterized by urbanization focusing on tourism.

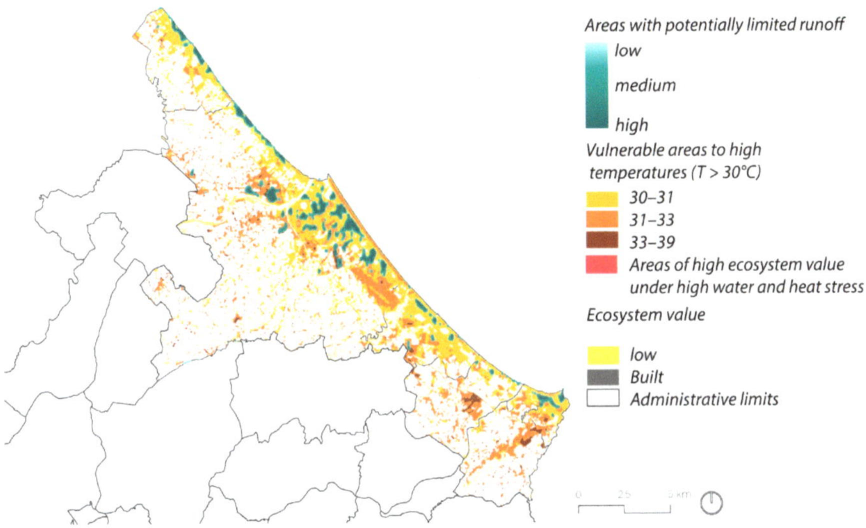

Figure 14. Climate impact sensitivity map: focus on coastal areas.

Considering only the portion of the territory affected by these climatic vulnerabilities, this extends over 6000 hectares in the Province of Rimini and reports Land Surface Temperatures above the 30 °C threshold during the hottest days of the year. These areas have been classified into three levels, reflecting different priorities for intervention: first, second, and third levels (Table 2).

Table 2. Classification of vulnerable areas to high temperatures (T > 30 °C).

Coastal Municipalities	Total Surface of the Municipalities (km^2)	Urban Surface Vulnerable to High Temperature (km^2)	Third-Level Priority (Class 30–31 °C on Urban Areas Surface)	Second-Level Priority (Class 31–33 °C on Urban Areas Surface)	First-Level Priority (Class 33–39 °C on Urban Areas Surface)
Bellaria-Igea Marina	18.08	6.57	5.5 km^2 (83.7%)	1.06 km^2 (16.1%)	0.01 km^2 (0.2%)
Cattolica	6.06	4.3	2.8 km^2 (65.6%)	1.3 km^2 (30.9%)	0.2 km^2 (3.5%)
Misano	22.36	6.01	2 km^2 (33.4%)	3.4 km^2 (56.1%)	0.6 km^2 (10.5%)
Riccione	17.45	10.62	7.5 km^2 (70.4%)	2.9 km^2 (28.1%)	0.2 km^2 (1.5%)
Rimini	135.27	39.7	21.1 km^2 (53%)	17.3 km^2 (43.5%)	1.3 km^2 (3.5%)

In contrast to the coastal area and Conca Valley regions, the Marecchia Valley shows a less severe presence of heat islands. The presence of vegetation on the slopes localizes the phenomenon primarily along the valley floor, where urban areas and spaces dedicated to human activities, such as industrial and commercial zones, are concentrated and interconnected by central infrastructure nodes. Like urban heat islands, the risk of surface runoff reveals a spatial reality that significantly impacts the densely built areas of coastal cities. The spatial representation of climatic stress, linked to land use patterns, highlights significant hydraulic vulnerability in territorial contexts characterized by high population density and extensive pavement. It can be observed that the coastal communities of Bellaria-Igea Marina, Rimini, Cattolica, and, to a lesser extent, Riccione and Misano are particularly prone to flooding events caused by heavy rainfall. Three levels of priority of intervention have also been distinguished here (Table 3). The tendency for flooding in the coastal area is thus more pronounced in complex urban areas lacking permeable zones or green spaces. Conversely, rural areas with low urbanization rates in the Marecchia and Conca Valleys exhibit less evident data concerning these phenomena. This trend is therefore connected to the presence of vegetated areas having a rather high health status, as seen in the map. The ecological component, unconnected to the urban and industrial areas, maintains a high conservation status and simultaneously helps mitigate the effects and consequences of heat islands and flooding. Also, in the map, it is possible to observe how there is, in the areas described, a temperature-calming effect due to the evapotranspiration mechanisms carried out by the vegetation and how the properties of the land, combined with the hypogeal systems of vegetation, contribute to a decrease in surface runoff. The study of extreme weather events highlights the need to establish a spatial correlation between surface runoff, heat, and land use. Furthermore, the results demonstrate how these impacts have adverse consequences for the economic well-being of the territories. The increased vulnerability of coastal areas to CC may also affect the attractiveness and tourism value of the region shortly.

Table 3. Classification of areas with potentially limited runoff.

Coastal Municipalities	Total Surface of the Municipalities (km^2)	Urban Flooding Total Surface (km^2)	Third-Level Priority	Second-Level Priority	First-Level Priority
Bellaria-Igea Marina	18.08	10.56	10.1 km^2 (95.9%)	0.2 km^2 (1.8%)	0.26 km^2 (2.3%)
Cattolica	6.06	1.71	1.3 km^2 (74.9%)	0.3 km^2 (17.5%)	0.1 km^2 (7.6%)
Misano	22.36	13.45	13.4 km^2 (99.6%)	0.05 km^2 (0.4%)	0 km^2 (0%)
Riccione	17.45	4.5	4.3 km^2 (94.3%)	0.2 km^2 (4.8%)	0.04 km^2 (0.9%)
Rimini	135.27	94.87	90.9 km^2 (95.8%)	2.5 km^2 (2.7%)	1.4 km^2 (1.5%)

3.2. *Identification of Adaptive Capacities Through the Integration of ESs*

In the provincial territory of Rimini, the condition of ecosystem components is highly diverse, both in terms of individual ecosystem services and across different areas of the territory (Table 4; Figure 15).

On average, there is a low presence of ESs in the coastal zone (Figure 16), and, therefore, in this area, particular attention should be given to existing and newly planned settlements and infrastructure to protect and safeguard existing ESs and enhance their extent, including through dedicated projects and investments.

Table 4. Classification of ESs according to (5) classes.

Coastal Municipalities	Total Surface of the Municipalities (km^2)	Low	Medium–Low	Medium	Medium–High	High
Bellaria-Igea Marina	18.08	10.74 km^2 (59.4%)	6.15 km^2 (33%)	1.17 km^2 (6.5%)	0.2 km^2 (1.1%)	0 km^2 (0%)
Cattolica	6.06	5.1 km^2 (84.2%)	0.4 km^2 (6.9%)	0.3 km^2 (4.7%)	0.06 km^2 (0.9%)	0.2 km^2 (3.3%)
Misano	22.36	10.62 km^2 (47.5%)	8.75 km^2 (39.1%)	2.72 km^2 (12.2%)	0.27 km^2 (1.2%)	0 km^2 (0%)
Riccione	17.45	12.65 km^2 (72.5%)	3.09 km^2 (17.7%)	1.33 km^2 (7.6%)	0.38 km^2 (2.2%)	0 km^2 (0%)
Rimini	135.27	68.15 km^2 (50.4%)	49.89 km^2 (36.88%)	15.15 km^2 (11.2%)	1.64 km^2 (1.2%)	0.44 km^2 (0.32%)

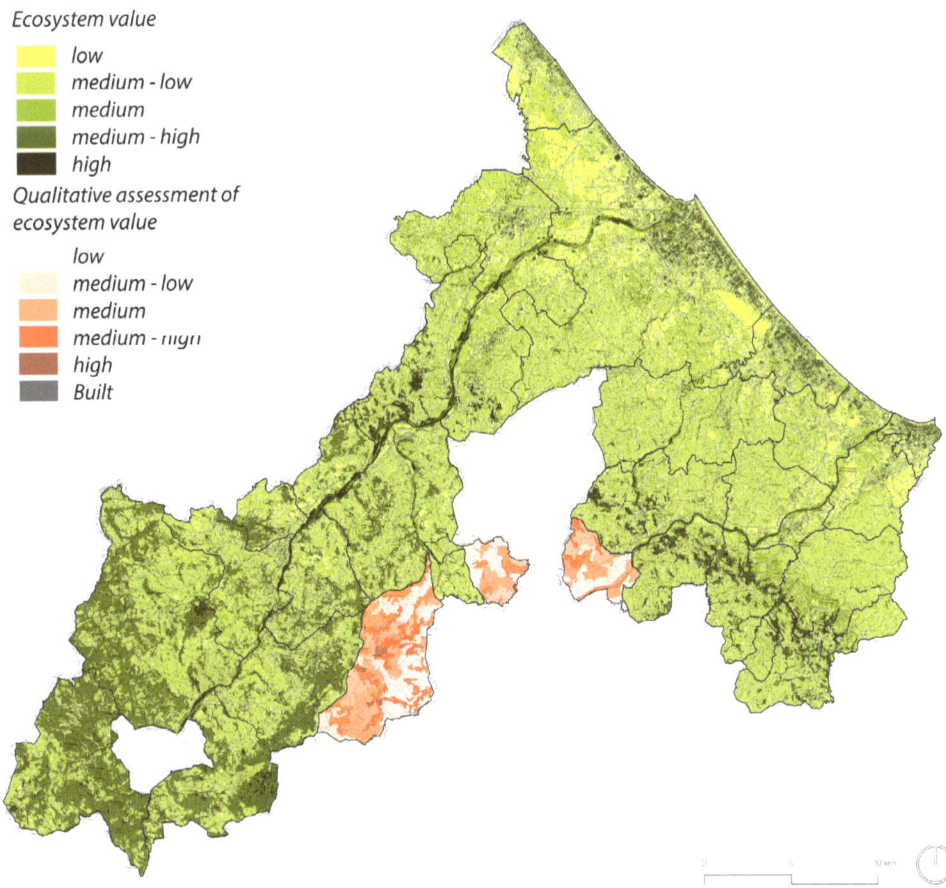

Figure 15. Ecosystem value map. The gradient in orange on the map depicts two municipalities recently annexed to the Province of Rimini and for which the information layers required for ecosystem assessments were not available. For this reason, while waiting for these information layers, an assessment was carried out on a qualitative basis.

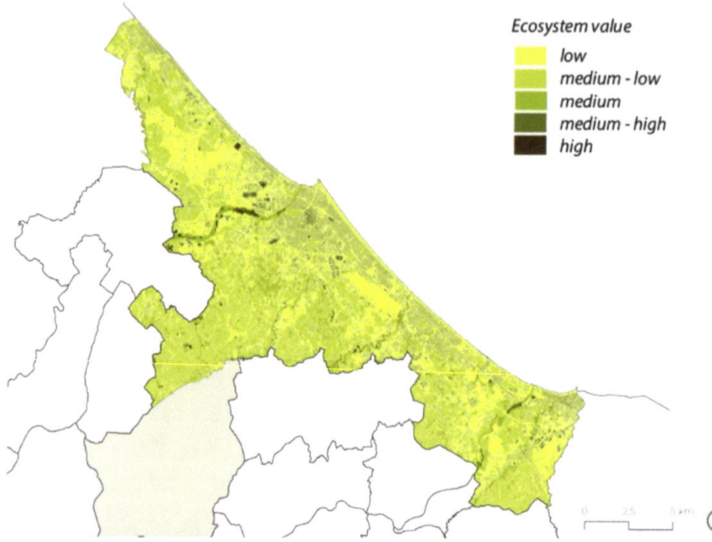

Figure 16. Ecosystem value map—focus on coastal areas.

There are diversified situations for different ESs in the lowland and early Apennine areas. Therefore, the need emerges to pay contextual and particular attention to ecosystem services (ESs) and the type of project/investment under consideration. Finally, in the Apennine area, ESs have a general state of well-being, with exceptions related to specific areas. However, this area also demonstrates the need for special attention to maintain the current ESs.

The maps produced make it possible to enrich the urban digital heritage of the Province of Rimini. In addition to mapping the vulnerabilities to CC and ESs of the study area, the cognitive layers produced aim to support and guide the plan in an integrated manner, digitally aggregating spatial information related to social and economic issues.

From the integration of the two maps obtained, areas where there is a coexistence of both CC impacts and ES values have been identified and localized at a territorial level. Integrating the two themes proved fundamental for identifying priority areas for intervention and directing the strategies to be undertaken at the planning level.

4. Discussion

The location of the main impacts makes it possible to determine the areas to intervene as a matter of priority to cope with climatic impacts and protect and enhance ecosystem values, which are also understood, but not limited to, as adaptive capacity.

The integration of the processed information layers has supported the identification of all those natural or semi-natural areas where ecosystem values are still high. In these areas, it becomes a priority to maintain and protect the presence of the ecological components, which help to moderate the criticality caused by climate impacts while increasing the quality of life. The strengthening of the provincial and regional ecological network system, the valorization of the primary and transversal ecological linking elements, and the connection with urban green areas, therefore, emerge as indispensable aspects, also identifying the peri-urban coastal sphere (Figure 17) as a priority sphere of action both for the strengthening of ecosystem performance and for the reconnection and creation of continuous and interconnected green nets. In the case of the coastal peri-urban sphere, it becomes necessary to protect the currently existing ecosystem values while also contributing to their enhancement to calm the effects of high temperatures and limited runoff.

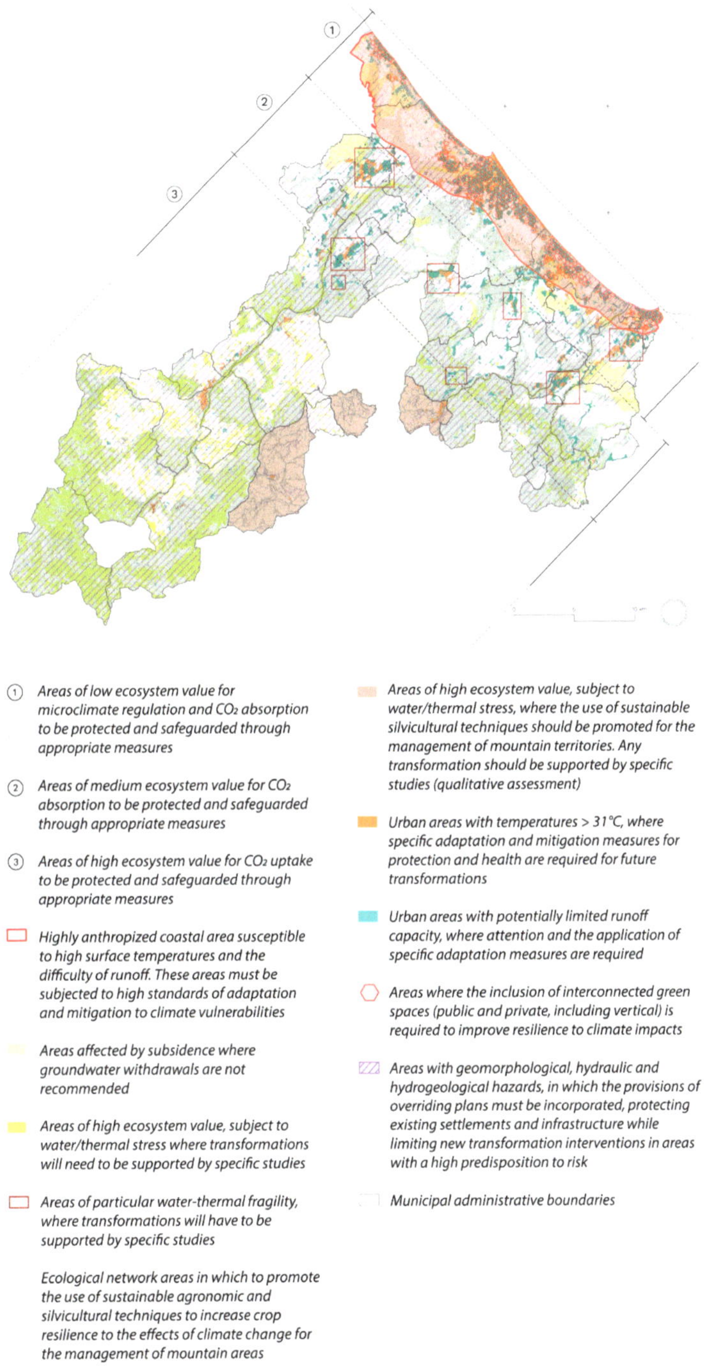

Figure 17. Map of strategies where the synthesis map of ecosystem services and the climate change map were integrated.

The results from the methodological approach described fit into various phases of drawing up the wide-area plan, helping to innovate the contents and nature of the planning instruments. Firstly, the elaborations make it possible to increase the knowledge related to the territory by using innovative interpretative keys: climatic and ecosystemic.

This contributes to expanding the knowledge framework by enabling a more cross-cutting analysis of the territory. In addition to the environmental and socio-economic components, indispensable elements to be analyzed within an urban context, information is added concerning the interactions that these elements may have, in positive or negative terms, with the ecosystem values present in the territory and consequently with the possible climatic impacts that may occur.

In addition to the purely cognitive phase, the results also contribute to outlining objectives and strategies with a strong interest in the issues of climate change and ecosystem services. In this way, the plan orients toward proposals and policies to cope with the impacts of climate change while protecting and enhancing the ecosystem values in the territory.

The elaborations on CC contribute to outlining actions aimed at security and increasing territorial resilience. Securing infrastructures and urban areas while respecting the environmental characteristics, matrices, and elements of the territory enhances its development potential and helps mitigate the vulnerabilities exposed by the climate crisis.

These results translate into indications for municipalities aimed at the following:

- Incorporating and integrating knowledge about the territory's risks, considering both the traditional and known framework of risks (hydrogeological, geomorphological, seismic risk) and risks linked to critical climatic conditions. The objective is manifold: on the one hand, to increase the level of awareness of territorial and climatic risks in such a way as to defend and protect existing settlements and infrastructures while limiting new transformation interventions in areas with a high predisposition to risk;
- Identifying critical issues caused by climate impacts and integrating this information into the planning, design, and assessment processes. When planning a transformation of the territory, every municipality must use the information layers related to critical climatic conditions provided by the province to assess the feasibility of the planned interventions. When intending to transform an area located in portions of the territory with a high predisposition to one or more criticalities related to temperature, runoff, water, and thermal stress of vegetation, the municipality must carry out a climate impact assessment. In this way, the municipalities, studying the specifics of their territory, identify and assess the areas most exposed to climatic criticalities and where to act to mitigate impacts while avoiding land transformations with negative consequences;
- Restoring ecosystem and ecological values by integrating them into adaptation strategies and climate compensation devices when a land transformation occurs.

Elaborations on ESs contribute to environmental protection and enhancement policies. The value provided by ESs is recognized as a fundamental resource for life and health protection [38]. Ecosystem valuation, therefore, becomes a key element in estimating land quality and differentiating environmental improvement actions. The ecosystem value becomes no longer a negligible element for defining urban planning and the admissibility conditions of territorial and urban transformations.

Guidance has also been developed for municipalities to conduct ecosystem assessments aimed at the following:

- Considering the ES assessment within ordinary environmental and territorial sustainability assessment procedures. ESs are considered indispensable for assessing the quality of the territory, which is why their protection becomes increasingly important;

- Strengthening the ecological network system by integrating ecosystem services and green and blue infrastructure in such a way as to reinforce ecosystem performance while promoting local green plans integrated with general planning.

It is important to underline that the methodology remains strongly dependent on data availability; in the absence of adequate data, analyses concerning both climate and ecosystem values cannot be effectively carried out. While one of the main constraints in climate-related assessments lies in the need for clean, error-free satellite imagery, the limitations affecting ecosystem service evaluations are of a different nature. The regional methodology adopted for mapping ecosystem services (ESs) and applied in this study relies extensively on pre-existing information layers that cannot be independently produced (e.g., organic carbon maps, forest cover data, and other informational layers produced by the Emilia-Romagna Region). Consequently, it depends on the timely updating of datasets by external actors—such as regional authorities, research institutes, and scientific organizations—posing considerable challenges for replicating the approach in other territorial contexts.

Moreover, the lengthy timeframes required for generating and updating these data layers limit the adaptability and responsiveness of the methodology. For this reason, the present contribution should not be interpreted as a final product but rather as a preliminary step in the ongoing process of updating and innovating spatial planning practices. In this respect, the qualitative approach adopted for the newly annexed municipalities (Figure 13) represents our starting point for the development of a standardized and replicable methodology applicable to other Italian contexts.

5. Conclusions

This paper presents the analysis developed during the formulation of the Territorial Plan for the Province (PTAV) of Rimini, with a specific focus on the integration of climate change impacts and ecosystem services into spatial planning. These themes are innovative within urban planning as they are typically not addressed in standard planning instruments. The climate change assessment focused on the impacts of heat and extreme precipitation, while the analysis of ecosystem services addressed multiple functions such as microclimate regulation, protection against extreme events, CO_2 regulation, erosion control, agricultural and forest productivity, water purification, hydrological regulation, and recreation.

The proposed methodology integrates climate change and ecosystem services into planning processes from two complementary perspectives: cognitive and strategic (RQ1–RQ2). The cognitive perspective enhances territorial knowledge regarding climate change impacts, increasing the information within the knowledge framework, while the strategic one incorporates climate change impacts into decision-making tools for adaptation planning. Similarly, for ecosystem services, the approach answers "how", "where", and "to what extent" to act to ensure a stable provision of services and biodiversity protection.

Innovation is achieved by enriching and dynamically updating the knowledge framework, making it responsive and operational through the integration of new datasets and planning tools. As a result, climate change and ecosystem services offer substantial contributions to the formulation of planning guidelines at the provincial scale, reinforcing the role of existing protected areas while also identifying additional zones for future protection (RQ3). These dimensions are often excluded from ordinary planning instruments, and the present work proposes concrete steps to align spatial planning with emerging environmental priorities. By producing integrated maps that highlight the spatial overlap between climate vulnerabilities and ecosystem values, the PTAV enables the formulation of climate-resilient planning strategies. The methodology transforms the knowledge framework into a diagnostic and adaptive tool capable of supporting the continuous evaluation of planning effectiveness. The strong interconnection between ecological and climatic dimensions marks a significant advancement in spatial planning, providing practical guidance for public administrations engaged in the ecological transition, territorial resilience, and sustainable development.

Author Contributions: Conceptualization, D.M., F.G. and F.M.; methodology, D.M. and F.G.; software, D.M. and F.G.; validation, D.M. and F.G.; formal analysis, D.M. and F.G.; investigation D.M.; resources, D.M. and F.G.; data curation, D.M. and F.G.; writing—original draft preparation, D.M. and F.G.; writing—review and editing, D.M. and F.G.; visualization, D.M., F.G. and F.M.; supervision, D.M.; project administration, F.M.; funding acquisition, F.M. All authors have read and agreed to the published version of the manuscript.

Funding: The authors acknowledge funding provided by the Province of Rimini.

Data Availability Statement: Data supporting the results of this study are available upon request from the corresponding author [Denis Maragno].

Acknowledgments: We are grateful to Katia Federico for meaningful discussions on the topic and to the reviewers for helpful comments. DM would like to thank the IUAV University working group "MERGE: integrated remote sensing for land monitoring and management" for exchanging ideas.

Conflicts of Interest: The authors declare no conflicts of interest.

Abbreviations

The following abbreviations are used in this manuscript:

CC	Climate change
CI	Climate impact
ES/ESs	Ecosystem service/ecosystem services
LST	Land Surface Temperature
VHI	Vegetation Health Index
NDVI	Normalized Difference Vegetation Index
TCI	Thermal Condition Index
VCI	Vegetation Condition Index
CREN	Centro Ricerche Ecologiche Naturalistiche
CLC	Corine Land Cover
ARPAE	Agenzia Regionale per la Protezione Ambiente dell'Emilia-Romagna
ISPRA	Istituto Superiore per la Protezione e la Ricerca Ambientale
PTAV	Piano Territoriale di Area Vasta
GIS	Geographic Information System

Notes

1 The Plan for the Metropolitan Territory of the Province of Rimini (PTAV) is the new general spatial planning tool developed by the province in accordance with Regional Law No. 24/2017. The process began with the drafting of preliminary documents and a public consultation phase, which led to the formulation and publication of the Draft Plan. After its adoption by the Provincial Council, the plan entered the approval phase and is currently under review by the Regional Urban Planning Committee. The PTAV is the instrument that enables provinces to identify the strategic assets of the territorial system in collaboration with their municipalities. The plan aims to promote communities' well-being and address current and future environmental and climate emergencies.

2 Regional Law n. 24, 21 December 2017 "Regional regulation on the protection and use of land" and Article 21: Ecological and environmental endowments 1. The ecological and environmental endowments of the territory consist of the set of spaces, works, and interventions that contribute, together with the infrastructure for the urbanization of settlements, to combat climate change and its effects on human society and the environment, to reduce natural and industrial risks and to improve the quality of the urban environment; [...] 2. The urban and ecological–environmental quality strategy shall provide for the determination of the need for ecological and environmental endowments and the performance requirements to be met by them, coordinating with the

climate change mitigation and adaptation policies established at the European, national, and regional levels and incorporating the indications of sectoral planning [. . .].

3 The Italian spatial planning system is structured across multiple institutional levels, reflecting the distribution of competences among the state, regions, provinces (or metropolitan cities), and municipalities. At the national level, the state defines general guidelines through framework laws and coordination tools. While it plays a limited role in direct planning, it is essential in ensuring consistency and the protection of national interests. Regions are responsible for drafting Regional Territorial Plans (PTR), which are hierarchically superior to provincial and municipal plans, as they establish strategic objectives, guidelines, and binding rules that subordinate instruments must comply with. Provinces and Metropolitan Cities adopt Provincial Territorial Coordination Plans (PTCP) or Wide-Area Territorial Plans (PTAV), which link regional strategies with local needs. At the municipal level, planning is carried out through general urban plans that must align with higher-level plans, ensuring vertical consistency within the system. In recent years, the Italian planning framework has progressively evolved toward the integration of spatial planning with environmental sustainability and climate adaptation, also in response to European guidelines and national regulations aimed at limiting land take.

4 It is crucial to distinguish between the terminology used in regulatory contexts at regional/provincial levels and those in climate assessments (IPCC) due to existing regulatory gaps. Hazard refers to the origin of risk, like heat waves or heavy rain. In this paper, vulnerability is defined as a system's propensity to sustain damage from event-induced stresses, calculated as the difference between sensitivity (physical, morphological, functional, and organizational factors weakening a system) and adaptive capacity.

5 The "Regional Agency for Prevention, Environment and Energy of Emilia-Romagna" (ARPAE) is structured on a territorial basis and is committed to environmental protection, prevention, the management of energy resources, and the development of forecasting models aimed at supporting sustainable development.

6 To make the most effective use of the potential of Landsat-8 imagery, satellite data must be chosen by assessing temperatures during the most intense heat periods in parallel. A set of satellite imagery is then selected based on four criteria: (i) year of acquisition, (ii) month of acquisition, (iii) daily mean temperature, and (iv) absence of significant cloud cover. The selection evaluates orbital moments of acquisition with fewer clouds in the atmosphere.

7 This index is used for various purposes in environmental analysis. For example, it is used to conduct fire risk assessments, to investigate the presence of ground vegetation fraction and its health status, to calculate leaf area in estimating a plant's CO_2 sequestration rate, and to monitor crop and pasture productivity.

8 Drought measurement can be performed according to different indices. Among the most widely used and internationally recognized ones is the SPI (Standardized Precipitation Index). This is a standardized indicator for detecting and assessing precipitation deficit (drought) at different time scales. The SPI makes it possible to quantify the water surplus or deficit concerning the climatology of the area under consideration.

9 ISPRA (Istituto Superiore per la Protezione e la Ricerca Ambientale) is an Italian public research institute focused on environmental protection, including marine, environmental emergencies, and research.

References

1. Cissé, G.; McLeman, R.; Adams, H.; Aldunce, P.; Bowen, K.; Campbell-Lendrum, D.; Clayton, S.; Ebi, K.L.; Hess, J.; Huang, C.; et al. Health, Wellbeing, and the Changing Structure of Communities. In *Climate Change 2022: Impacts, Adaptation and Vulnerability; Contribution of Working Group II to the Sixth Assessment Report of the Intergovernmental Panel on Climate Change*; Cambridge University Press: Cambridge, UK, 2022; pp. 1041–1170. [CrossRef]
2. IPCC. *Climate Change 2022: Impacts, Adaptation and Vulnerability*; Pörtner, H.O., Roberts, D.C., Tignor, M., Poloczanska, E.S.S., Mintenbeck, K., Alegría, A., Craig, M., Langsdorf, S., Löschke, S., Möller, V., et al., Eds.; Cambridge University Press: Cambridge, UK, 2022. [CrossRef]
3. Musco, F.; Maragno, D.; Bertin, M. Pianificare l'adattamento al cambiamento climatico come gestione di una macro-emergenza locale. *Territorio* **2019**, *138*, 138–144. [CrossRef]
4. Sands, P. The United Nations Framework Convention on Climate Change. *Rev. Eur. Community Int. Environ. Law* **1992**, *1*, 270. [CrossRef]
5. Salzman, J.; Bennett, G.; Carroll, N.; Goldstein, A.; Jenkins, M. The global status and trends of Payments for Ecosystem Services. *Nat. Sustain.* **2018**, *1*, 136–144. [CrossRef]
6. Bouwma, I.; Schleyer, C.; Primmer, E.; Winkler, K.J.; Berry, P.; Young, J.; Vadineanu, A. Adoption of the ecosystem services concept in EU policies. *Ecosyst. Serv.* **2018**, *29*, 213–222. [CrossRef]
7. Hasan, S.S.; Zhen, L.; Miah, M.G.; Ahamed, T.; Samie, A. Impact of land use change on ecosystem services: A review. *Environ. Dev.* **2020**, *34*, 100527. [CrossRef]
8. Liu, M.; Wei, H.; Dong, X.; Wang, X.C.; Zhao, B.; Zhang, Y. Integrating Land Use, Ecosystem Service, and Human Well-Being: A Systematic Review. *Sustainability* **2022**, *14*, 6926. [CrossRef]

9. Frantzeskaki, N.; Mahmoud, I.H.; Morello, E. Nature-based solutions for resilient and thriving cities: Opportunities and challenges for planning future cities. In *Nature-Based Solutions for Sustainable Urban Planning: Greening Cities, Shaping Cities*; Springer: Berlin/Heidelberg, Germany, 2022; pp. 3–17.
10. Kim, G.; Kim, J.; Ko, Y.; Eyman, O.T.G.; Chowdhury, S.; Adiwal, J.; Son, Y. How Do Nature-Based Solutions Improve Environmental and Socio-Economic Resilience to Achieve the Sustainable Development Goals? Reforestation and Afforestation Cases from the Republic of Korea. *Sustainability* **2021**, *13*, 12171. [CrossRef]
11. Pietta, A.; Bagliani, M.; Crescini, E. L'Italia si adatta? La definizione delle politiche di adattamento al cambiamento climatico alla scala regionale. *Riv. Geogr. Ital.* **2022**, *2*, 13801. [CrossRef]
12. Yan, S.; Growe, A. Regional Planning, Land-Use Management, and Governance in German Metropolitan Regions. The Case of Rhine–Neckar Metropolitan Region. *Land* **2022**, *11*, 2088. [CrossRef]
13. Regione Emilia-Romagna. Strategia Unitaria di Mitigazione e Adattamento per i Cambiamenti Climatici. 2017. Available online: https://ambiente.regione.emilia-romagna.it/.../strategia-regionale-per-i-cambiamenti-climatici (accessed on 2 February 2022).
14. Regione Emilia-Romagna. Legge Regionale 21 Dicembre 2017, n. 24. Disciplina Regionale Sulla Tutela E L'uso Del Territorio. Titolo II: Disposizioni Generali sulla Tutela e l'uso del Territorio. Capo III: Sostenibilità Ambientale E Territoriale Dei Piani. Art. 21: Dotazioni Ecologiche E Ambientali. 2017. Available online: https://demetra.regione.emilia-romagna.it/al/articolo?urn=er:assemblealegislativa:legge:2017;24&dl_t=text/xml&dl_a=y&dl_id=10&pr=idx,0;artic,0;articparziale,1&anc=tit2 (accessed on 10 October 2022).
15. Regione Emilia-Romagna. Legge Regionale 21 Dicembre 2017, n. 24. Disciplina Regionale Sulla Tutela E L'uso Del Territorio. Titolo II: Disposizioni Generali sulla Tutela e l'uso del Territorio. Capo III: Sostenibilità Ambientale E Territoriale Dei Piani. Art. 22: Quadro Conoscitivo. 2017. Available online: https://demetra.regione.emilia-romagna.it/.../articolo (accessed on 10 October 2022).
16. Lucertini, G.; Di Giustino, G.; dall'Omo, C.F.; Musco, F. An innovative climate adaptation planning process: iDEAL project. *J. Environ. Manag.* **2022**, *317*, 115408. [CrossRef]
17. ARPAE Emilia-Romagna. *Atlante climatico 1961–2015*; Regione Emilia-Romagna: Bologna, Italy, 2017; ISBN 978-88-87854-44-2.
18. ARPAE Emilia-Romagna. Dati Ambientali Emilia-Romagna. Clima. 2021. Available online: https://webbook.arpae.it/clima/ (accessed on 19 February 2022).
19. ARPAE Emilia-Romagna. Rapporti Idro-Meteo-Clima Annuali. 2022. Available online: https://www.arpae.it/it/temi-ambientali/meteo/report-meteo/rapporti-annuali (accessed on 20 February 2022).
20. Persiano, S.; Ferri, E.; Antolini, G.; Domeneghetti, A.; Pavan, V.; Castellarin, A. Changes in seasonality and magnitude of sub-daily rainfall extremes in Emilia-Romagna (Italy) and potential influence on regional rainfall frequency estimation. *J. Hydrol. Reg. Stud.* **2020**, *32*, 100751. [CrossRef]
21. Pavan, V.; Antolini, G.; Barbiero, R.; Berni, N.; Brunier, F.; Cacciamani, C.; Torrigiani Malaspina, T. High resolution climate precipitation analysis for north-central Italy, 1961–2015. *Clim. Dyn.* **2019**, *52*, 3435–3453. [CrossRef]
22. Antolini, G.; Auteri, L.; Pavan, V.; Tomei, F.; Tomozeiu, R.; Marletto, V. A daily high-resolution gridded climatic data set for Emilia-Romagna Italy during 1961–2010. *Int. J. Climatol.* **2016**, *36*, 1970–1986. [CrossRef]
23. Zhu, K.; Blum, P.; Ferguson, G.; Balke, K.D.; Bayer, P. The geothermal potential of urban heat islands. *Environ. Res. Lett.* **2010**, *5*, 044002. [CrossRef]
24. Oleson, K.W.; Bonan, G.B.; Feddema, J.; Jackson, T. An examination of urban heat island characteristics in a global climate model. *Int. J. Climatol.* **2011**, *31*, 1848–1865. [CrossRef]
25. Oke, T.R. City size and the urban heat island. *Atmos. Environ.* **1973**, *7*, 769–779. [CrossRef]
26. Marando, F.; Heris, M.P.; Zulian, G.; Udías, A.; Mentaschi, L.; Chrysoulakis, N.; Maes, J. Urban heat island mitigation by green infrastructure in European Functional Urban Areas. *Sustain. Cities Soc.* **2022**, *77*, 103564. [CrossRef]
27. Cueva, J.; Yakouchenkova, I.A.; Fröhlich, K.; Dermann, A.F.; Dermann, F.; Köhler, M.; Saha, S. Synergies and trade-offs in ecosystem services from urban and peri-urban forests and their implication to sustainable city design and planning. *Sustain. Cities Soc.* **2022**, *82*, 103903. [CrossRef]
28. Hulley, G.; Shivers, S.; Wetherley, E.; Cudd, R. New ECOSTRESS and MODIS land surface temperature data reveal fine-scale heat vulnerability in cities: A case study for Los Angeles County, California. *Remote Sens.* **2019**, *11*, 2136. [CrossRef]
29. Huang, C.; Ye, X. Spatial modeling of urban vegetation and land surface temperature: A case study of Beijing. *Sustainability* **2015**, *7*, 9478–9504. [CrossRef]
30. Kogan, F. Application of vegetation index and brightness temperature for drought detection. *Adv. Space Res.* **1995**, *15*, 91–100. [CrossRef]
31. Masroor, M.; Sajjad, H.; Rehman, S.; Singh, R.; Rahaman, M.H.; Sahana, M.; Avtar, R. Analysing the relationship between drought and soil erosion using vegetation health index and RUSLE models in Godavari middle sub-basin, India. *Geosci. Front.* **2022**, *13*, 101312. [CrossRef]

32. Zeng, J.; Zhang, R.; Qu, Y.; Bento, V.A.; Zhou, T.; Lin, Y.; Wang, Q. Improving the drought monitoring capability of VHI at the global scale via ensemble indices for various vegetation types from 2001 to 2018. *Weather. Clim. Extrem.* **2022**, *35*, 100412. [CrossRef]
33. Özerol, G.; Dolman, N.; Bormann, H.; Bressers, H.; Lulofs, K.; Böge, M. Urban water management and climate change adaptation: A self-assessment study by seven midsize cities in the North Sea Region. *Sustain. Cities Soc.* **2020**, *55*, 102066. [CrossRef]
34. Maragno, D.; dall'Omo, C.F.; Pozzer, G.; Musco, F. Multi-risk climate mapping for the adaptation of the Venice metropolitan area. *Sustainability* **2021**, *13*, 1334. [CrossRef]
35. Santolini, R.; Morri, E.; Pasini, G. *Mappatura e Valutazione dei Servizi Ecosistemici*; CREN: Geneva, Switzerland, 2021.
36. Gómez-Baggethun, E.; Barton, D.N. Classifying and valuing ecosystem services for urban planning. *Ecol. Econ.* **2013**, *86*, 235–245. [CrossRef]
37. Wang, S.; Hu, M.; Wang, Y.; Xia, B. Dynamics of ecosystem services in response to urbanization across temporal and spatial scales in a mega metropolitan area. *Sustain. Cities Soc.* **2022**, *77*, 103561. [CrossRef]
38. Pinto, L.V.; Inácio, M.; Ferreira, C.S.S.; Ferreira, A.D.; Pereira, P.E. Ecosystem services and well-being dimensions related to urban green spaces—A systematic review. *Sustain. Cities Soc.* **2022**, *85*, 104072. [CrossRef]

land

Article

Construction and Optimization of Ecological Security Patterns Based on Ecosystem Services in the Wuhan Metropolitan Area

Beiling Chen [1], Jianhua Zhu [1,2,3,4,*], Huayan Liu [1], Lixiong Zeng [1,2,3,4], Fuhua Li [1], Zhiyan Xiao [5] and Wenfa Xiao [1,2,3,4]

1 Ecology and Nature Conservation Institute, Chinese Academy of Forestry, Key Laboratory of Forest Ecology and Environment of National Forestry and Grassland Administration, Beijing 100091, China
2 Co-Innovation Center for Sustainable Forestry in Southern China, Nanjing Forestry University, Nanjing 210037, China
3 Scientific and Technological Collaborative Innovation Centre for the Yangtze River Economic Belt Ecological Conservation, Beijing 100091, China
4 Forestry and Grassland Carbon Sink Research Institute, Beijing 100091, China
5 Wuhan Forestry Station, Wuhan Garden and Forestry Bureau, Wuhan 430023, China
* Correspondence: zhucool@caf.ac.cn

Abstract: Rapid urbanization has affected ecosystem stability, and the construction of ecological security patterns (ESPs) can rationally allocate resources and achieve ecological protection. Priority evaluation of critical areas can maximize the benefits of ecological protection, which is crucial for sustainable urban development. However, most prior studies have focused on assessing individual elements of the ESP, rarely considering both the protection priority of ecological sources and corridors. We constructed ESPs for the Wuhan Metropolitan Area (WMA) from 2000 to 2020 and evaluated the priority of ecological sources and corridors for protection. The findings indicated that high-level ecological sources exhibited higher overall landscape connectivity and ecosystem service values with lower patch fragmentation. The average area proportions of primary, secondary, and tertiary ecological sources in 2000, 2010, and 2020 were 41.11%, 23.03%, and 29.86%, respectively. High-level ecological corridors had shorter lengths and offered higher comprehensive ecosystem service values. The total length of secondary corridors exceeded that of primary corridors by 1951.19 km, 650.39 km, and 2238.18 km in 2000, 2010, and 2020, respectively. Primary corridors, which connected fragmented and isolated sources, should have their ecological land percentage increased to enhance connectivity. Secondary corridors connected two independent and distant sources, providing the basis for ecological protection in the intervening area, whose surrounding habitats should be protected. This study identifies the ecological protection priority and offers a theoretical basis and practical reference for balancing urban development with ecological protection.

Keywords: ecological security pattern; priority assessment; ecosystem services; remote sensing; Wuhan Metropolitan Area

Citation: Chen, B.; Zhu, J.; Liu, H.; Zeng, L.; Li, F.; Xiao, Z.; Xiao, W. Construction and Optimization of Ecological Security Patterns Based on Ecosystem Services in the Wuhan Metropolitan Area. *Land* **2024**, *13*, 1755. https://doi.org/10.3390/land13111755

Academic Editors: Michele Munafò, Luca Congedo and Francesca Assennato

Received: 20 September 2024
Revised: 23 October 2024
Accepted: 24 October 2024
Published: 25 October 2024

1. Introduction

Rapid urbanization and irrational human activity have affected ecosystem stability and seriously threatened regional ecological security [1]. Ecological problems such as environmental degradation, soil erosion, and increased landscape fragmentation are becoming increasingly prominent [2,3]. Constructing an ecological security pattern (ESP) and enhancing ecological management and protection have remained pivotal approaches to balancing regional development and ecological security [4,5]. However, inevitable conflict exists between urban development and ecological protection, and prioritizing ecological security at the expense of economic growth may not represent an optimal solution. Therefore, effectively allocating resources to maximize ecological security within limited constraints has become a focus in the discourse on urban sustainable development.

ESP refers to a spatial configuration consisting of patches and key locations that are essential for critical ecological processes and ecosystem services [6,7]. Its primary goal is to identify critical areas for ecological security and to optimize the spatial arrangement of ecosystem components through targeted human interventions to enhance regional ecosystem resilience [8]. Constructing and optimizing ESPs can effectively promote ecological connectivity and maintain ecosystem stability, playing a crucial role in regional environmental protection [9,10]. Currently, the widely accepted framework for regional ESP identification involves "ecological sources determination—resistance surface creation—corridors identification" [11,12].

Ecological sources represent the key sites essential to regional ecological functions and processes [13,14]. Two main techniques exist for identifying ecological sources, the first step of ESP construction. Large-scale habitat patches like forests and grasslands, or nature reserves and scenic spots, are chosen as ecological sources by the first technique, direct identification. Despite being relatively convenient, this approach ignores the internal differences in land-use types and changes in ecosystem services, greatly influenced by human subjectivity [15,16]. Another is the indirect identification method, which creates an extensive index system of evaluation considering factors such as ecosystem service function and ecological sensitivity [17,18]. This approach is more comprehensive and objective but more complex and may be affected by trade-offs in ecosystem services [19,20]. Each of these methods has its advantages and disadvantages and should be chosen based on the specific conditions of the research area.

The foundation for determining ecological corridors is the resistance surface construction, which reflects the resistance to species migration between different ecological patches. Most early studies assigned ecological resistance coefficients based on land cover and expert experience. This approach often disregarded the internal distinctions among land-use types and failed to account for human factors [21,22]. Therefore, many studies have considered economic, demographic, and transportation factors, introducing indicators of human activity intensity to modify the basic ecological resistance surface [23,24]. Other studies have used various indicators to construct ecological resistance surfaces. Gao et al. [25] used the habitat quality method to construct resistance surfaces and identified the ESP of Changzhou City. Zhang et al. [26] integrated the ecological environment and economic characteristics, including vegetation cover, distance to water bodies, and GDP, into resistance surface construction. In general, the methods for constructing ecological resistance surfaces lack standardization. The widely used method is modifying the resistance surface based on specific factors.

Ecological corridors are vital channels for materials and energy to move, influencing the integrity of ecosystem service functions and the connectivity of regional landscapes. Circuit theory and the minimum cumulative resistance (MCR) model are the main methods for identifying ecological corridors [27,28]. The MCR model can identify important nodes of corridors but fails to recognize corridor width [29,30]. By analogizing ecological flow to electric currents with stochastic random walk characteristics, circuit theory overcomes these shortcomings [24,31]. Because of its ability to forecast intricate landscape movement patterns and identify critical ecological corridors and key nodes, circuit theory has been extensively utilized in ESP construction [24,32].

The Wuhan Metropolitan Area (WMA) is an inland urban agglomeration located in the middle reaches of the Yangtze River, characterized by its numerous rivers, lakes, and abundant wetland resources [33,34]. However, with the acceleration of urbanization in WMA, ecological issues such as cropland erosion, lake reclamation, and soil degradation are being exacerbated [35,36]. These problems exacerbate the imbalance between urban development and ecological security in WMA, and how to maximize the benefits of ecological protection by using limited resources is an urgent problem. Prioritized management of key areas can effectively enhance the overall function of regional ecosystems at lower costs, thereby maximizing the integrated ecological, economic, and social benefits [37,38]. Lu et al. [39] prioritized the construction and protection of ecological corridors in WMA based on the quality, function, and structure of ecological networks. Zeng et al. [35] identified prioritized protected areas and different development plans for WMA based on landscape connectivity and protected area effectiveness. However, most previous studies have typically focused on grading individual elements within the ESP, rarely comprehensively considering the protection priority of ecological sources and corridors. Such approaches often fail to fully capture the complexity and multidimensionality of ecological protection needs. Therefore, this study improved the existing ESP identification framework and optimized and constructed the ESP of WMA in 2000, 2010, and 2020. We systematically evaluated the ecological protection priority of ecological sources and corridors and deeply explored the multidimensional perspectives of ecological protection.

2. Materials and Methods

2.1. Study Area

The Wuhan Metropolitan Area (WMA) consists of 9 cities, including Wuhan, with a total area of 57,947 km^2 and a population of more than 31.8 million, situated in the eastern part of Hubei province, China (Figure 1) [40]. As a strategic region for the rise of central China, WMA is a comprehensive reform pilot for building an environmentally friendly society and resource-saving society in China [41]. Its landforms are mainly plains and hills, ranging from the Dabie Mountain Range in the northeast to the Mufu Mountain Range in the south (Figure 1c). WMA has a subtropical monsoon climate, with high temperatures, frequent rainfall, and distinct seasons, with an average annual temperature of 16.37 °C and an average annual rainfall of 1102 mm in 2020. Land-use types in the area are dominated by forest and cropland, accounting for 49.13% and 25.20% of the total area. As urbanization progresses, problems such as cropland degradation, reduction of ecological space, and the heat island effect are posing many challenges to regional ecological security [33]. Therefore, it is necessary to construct and optimize the ESP and explore the ecological protection strategy that maximizes benefits at minimal cost, seeking the optimal solution to balance urban economic development and ecological security.

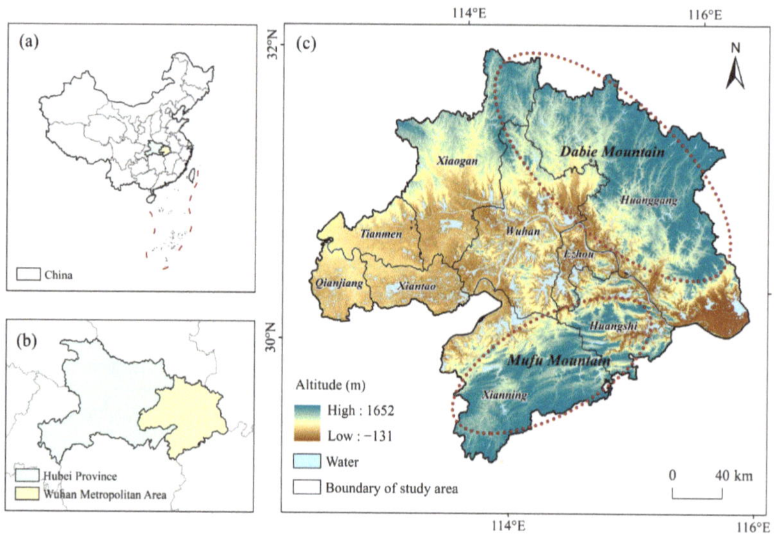

Figure 1. Location of the Wuhan metropolitan area in China (**a**), Hubei Province (**b**), and its altitude (**c**).

2.2. Data Source

Relevant data for the research area in 2000, 2010, and 2020 were gathered (Table 1). Land use was categorized into cropland, forest, shrubland, grassland, water, wetland, construction land, and bare land. All spatial datasets were projected to the WGS_1984_Albers and resampled to a 30 m resolution. Meteorological data were obtained by spatial interpolation using the Anusplin package in RStudio 1.1.463.

Table 1. Pertinent data for the research area.

Data	Resolution or Scale	Source
Land use	30 m	GlobeLand30 (https://www.tianditu.gov.cn, accessed on 12 November 2023)
Meteorological data	-	China Meteorological Data Service Center (http://data.cma.cn/, accessed on 15 November 2023)
DEM	30 m	ASTER Global DEM (http://lpdaac.usgs.gov/, accessed on 18 November 2023)
Soil data	1:1,000,000	Soil Database of China for Land Surface Modeling (http://vdb3.soil.csdb.cn/, accessed on 19 November 2023)
Nighttime light data	1000 m	An improved time-series DMSP-OLS-like data (1992–2022) in China by integrating DMSP-OLS and SNPP-VIIRS [42]
NDVI	250 m	Land Processes Distributed Active Archive Center (http://lpdaac.usgs.gov/, accessed on 20 November 2023)
Statistics data	-	Hubei Province Statistical Yearbook (https://tjj.hubei.gov.cn/, accessed on 20 November 2023)

2.3. *Methods*

Ecosystem services are crucial indicators for assessing the ecological environment and protection, contributing to the sustainable development of urban areas [43,44]. We applied the entropy weight method to integrate ecosystem services and used the landscape connectivity index to identify ecological sources. By integrating circuit theory, the gravity model, and Centrality Mapper, we propose an ESP optimization method to comprehensively evaluate the ecological protection priority of ecological sources and corridors. The research framework for ESP construction and optimization consists of the following five steps (Figure 2): (1) assessing ecosystem services; (2) identifying the ecological sources through the entropy weight method and landscape connectivity analysis; (3) constructing resistance surface and modifying it; (4) using circuit theory to construct the ESP; and (5) optimizing the ESP using the gravity model and Centrality Mapper.

2.3.1. Ecosystem Services Assessment

WMA has been experiencing rapid urbanization in recent years, resulting in significant urban expansion encroaching on cropland, forest, shrubland, grassland, water, and wetland [45]. Six ecological services were chosen for this research in response to these urbanization trends, including water conservation, soil conservation, carbon sequestration, habitat quality, food supply, and ecological recreation. Table 2 shows the methods and references for evaluating the six ecosystem services. Methods for assessing the different ecosystem services are presented in the Supplementary Materials.

Table 2. Methods of six ecological services assessment.

Ecological Service	Method	Reference
Water conservation	InVEST's Seasonal Water Yield Module	Sahle et al. [46]
Soil conservation	Revised Universal Soil Loss Equation (RUSLE)	Liu et al. [47]
Carbon sequestration	Carnegie-Ames-Stanford Approach (CASA)	Peng et al. [7]
Habitat quality	InVEST's Habitat Quality Module	Li et al. [48]
Food supply	Statistical data obtained by spatial allocation using NDVI	Zhang et al. [49]
Ecological recreation	Recreation potential index based on landscape indicators	Schirpke et al. [50]

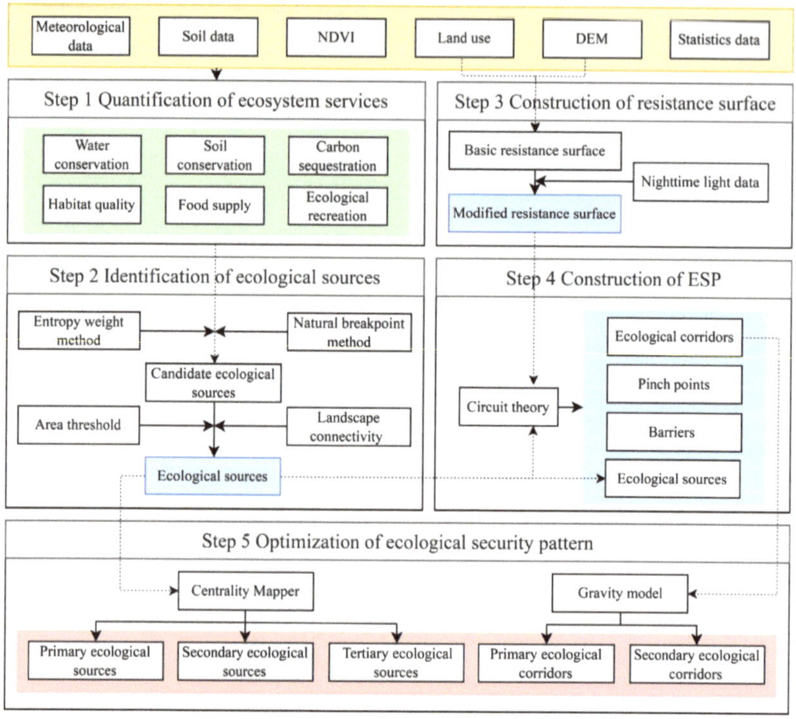

Figure 2. Methodology and framework for constructing and optimizing ESP.

2.3.2. Ecological Source Identification

The key ecological patches that supply ecosystem services and sustain ecological processes are known as ecological sources [51]. Ecological sources were determined by assessing ecosystem services and landscape connectivity. The entropy weight method is a technique independent of evaluator preferences, relying only on the entropy of the original index [52]. Therefore, the entropy weight method was used to weigh and combine six ecosystem service functions and quantify their relative importance. This importance was ranked across five levels, with the highest-ranked patches selected as candidate ecological sources [15].

The ecological sources were identified by considering landscape connectivity to maintain their integrity. Indicating how much the landscape promotes or impedes ecological flow, landscape connectivity is a crucial factor affecting biodiversity, ecosystem stability, and integrity [53]. Possible connectivity (*PC*) is an ideal index for evaluating landscape connectivity, with values ranging from 0 to 1, increasing as connectivity improves [16]. The *dPC* (%), derived from the *PC* value, represents the contribution of patches to overall landscape connectivity. The specific calculation formulas are as follows:

$$PC = \frac{\sum_{i=1}^{n} \sum_{j=1}^{n} a_i \times a_j \times p_{ij}}{A_L^2} \tag{1}$$

$$dPC_i = 100 \times \frac{PC - PC_{i-remove}}{PC} \tag{2}$$

where n indicates the number of nodes overall, A_L denotes the entire study area, a_i and a_j indicate patch i and patch j areas, respectively; p_{ij} denotes all path final connectivity maximum between patch i and patch j, and $PC_{i-remove}$ indicates the potential connectivity of the remaining patches following the patch i removal.

ArcGIS 10.4.1 and Conefor Sensinode 2.6.0 were used to calculate PC and dPC, and ecological patches whose area was less than the minimum patch area threshold were eliminated. Ultimately, we chose patches as ecological sources that were bigger than 2 km^2 and had $dPC > 0.5$.

2.3.3. Resistance Surface Construction

Ecological resistance surface represents the difficulty faced by species migrating between different patches [7]. Using expert experience in assigning resistance values to the corresponding land-use type is a common approach [54]. However, this approach does not adequately describe the resistance since it ignores variations in the same land-use type [55]. Therefore, human and natural factors should be included for comprehensive consideration.

In this study, the indicators used to construct the basic resistance surface—land-use type, altitude, and slope—were selected based on the conditions of WMA (Table 3). The weightings for each resistance factor were determined based on relevant literature [56]. In addition, Nighttime Light Intensity (NLI) was incorporated to optimize the ecological resistance surface, as it better reflects the impact of human activities on species movement [57]. The calculation method is as follows:

$$R_i = \frac{NL_i}{NL_a} \times R \quad (3)$$

where R_i indicates the pixel i's value of modified resistance, NL_a represents land-use type a's average NLI, NL_i denotes the pixel i's NLI, while R denotes the basic resistance coefficient.

Table 3. Resistance factors of ecological security pattern in WMA.

Resistance Factors	Classification	Resistance Value	Weights
Land use	Forest	1	0.6
	Shrubland	10	
	Grassland	10	
	Cropland	30	
	Water	50	
	Wetland	50	
	Bare land	300	
	Construction land	500	
DEM (m)	<100	1	0.2
	100–200	40	
	200–400	60	
	400–700	80	
	>700	100	
Slope (°)	<5	1	0.2
	5–10	40	
	10–20	60	
	20–30	80	
	>30	100	

2.3.4. ESP Construction

An ESP is composed of ecological sources, corridors, and nodes, serving as essential for biodiversity conservation and integrity of the ecosystem, which is advantageous for maintaining ecosystem service functions and safeguarding ecological security [10]. High-frequency areas of ecological flow are referred to as pinch points, having a high possibility for species migration. Pinch points often form when surrounding ecological resistance compresses the corridors in a relatively narrow area [58]. Preservation and restoration in this region should be prioritized, as the loss or deterioration of pinch points will decrease landscape connectivity. Areas where species are prevented from migrating between ecological sources are known as barriers. The removal of barriers is essential to enhance connectivity and protect and restore ecosystems [59].

This study identified corridors, pinch points, and barriers based on circuit theory using the Linkage Mapper 2.0.0 toolbox in ArcGIS. Circuit theory is based on the principle that electrons in a circuit exhibit random walks [60]. Circuit theory allows researchers to integrate all potential routes between sources and predict species migration probabilities by analyzing ecological processes [61]. In circuit theory, ecological sources and landscapes are represented as circuit nodes and conductive surfaces, respectively, while material or energy is analogous to electrons. In this framework, the potential, probability, and degree of obstruction in the flow of ecological information between sources are represented by voltage, current, and resistance, respectively.

2.3.5. Optimization of the ESP

Ecological sources and corridors are the crucial components of an ESP, essential to sustaining ecological processes. We optimized the ESP by determining their relative importance. The Centrality Mapper in Linkage Mapper can be utilized to calculate the centrality of ecological sources, thereby determining their significance and contribution to maintaining the connectivity of the regional ecological network [62]. Thus, we assessed ecological sources by Centrality Mapper and referred to relevant studies [26] to classify them utilizing the natural break point method into three levels: primary, secondary, and tertiary. The interaction strength across ecological corridors, directly proportional to their importance, was assessed using the gravity model [63]. Based on the gravity model, we subdivided ecological corridors into two levels: primary (interaction force $\geq$ 100) and secondary (interaction force < 100) [64]. The following is the computation method:

$$G_{ij} = \frac{\left(\frac{1}{P_i} \times lnS_i\right)\left(\frac{1}{P_j} \times lnS_j\right)}{\left(\frac{L_{ij}}{L_{max}}\right)^2} = \frac{(L_{max})^2 lnS_i \times lnS_j}{(L_{ij})^2 P_i \times P_j} \quad (4)$$

where G_{ij} and L_{ij} represent the interaction and the corridor cumulative resistance value between source i and j; P_i, P_j, S_i, and S_j denote the value of resistance and area of the sources, respectively; while L_{max} represents the total ecological corridors' maximum cumulative resistance.

3. Results

3.1. Land-Use Change

The land-use changes in WMA are depicted in Figure 3. From 2000 to 2020, cropland, forest, and water were the dominant land-use types, comprising around 85% of the region. Cropland was primarily located in the central and western plains. Water was distributed along the Yangtze River through the plains, primarily in the central region. Forest was predominantly found in the northeast and south, notably in the Dabie and Mufu Mountains. Construction land was concentrated in Wuhan City, with the remainder scattered across other urban areas.

Over the past 20 years, cropland decreased the most significantly by 2.5%, and construction land increased the most by 2.3% (Figure 4). Most new construction land was converted from cropland (Table A1). Forest, shrubland, grassland, wetland, and bare land have generally declined, mainly due to the conversion of forest to cropland, shrubland and grassland to forest, and wetland and bare land to water. Water increased from 2000 to 2010, primarily converted from cropland and wetland (Table A2), and decreased from 2010 to 2020, mostly converted to cropland (Table A3).

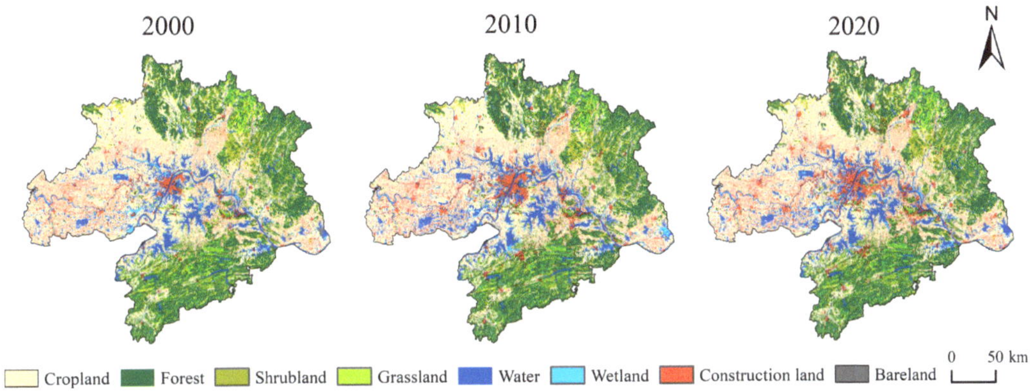

Figure 3. Spatial distribution of land use in WMA from 2000 to 2020.

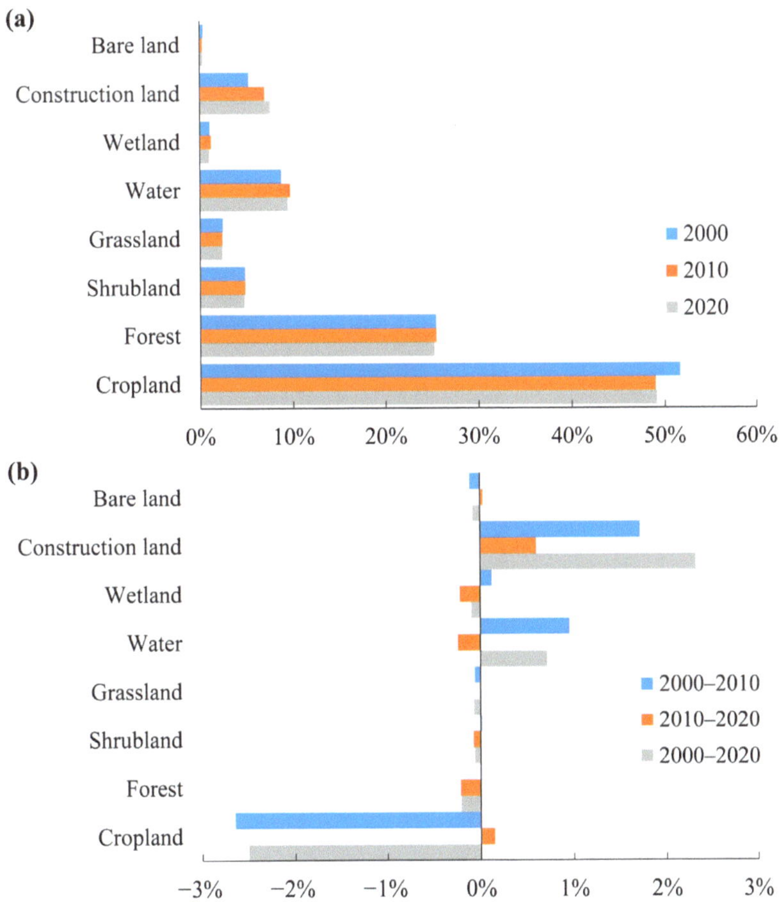

Figure 4. (**a**) Ratio of different land use. (**b**) Changes in different land use.

3.2. Ecosystem Service Change

The spatial patterns of the six ecosystem services differed significantly over time (Figure 5). From 2000 to 2020, the high values of soil conservation, carbon sequestration, ecological recreation, and habitat quality services were predominantly concentrated in the southern and northeastern areas, with high vegetation cover and low human disturbance. The soil conservation value in 2020 reached 813.88 t/hm^2, representing a 17.4% decrease compared to 2000. Carbon sequestration in 2020 was 568.61 gC/m^2, a 22.2% increase over the past 20 years. Ecological recreation and habitat quality services had average values of 0.41 and 0.62 in 2020, increasing by 2.5% and decreasing by 1.6%, respectively. The total water conservation in 2020 was 3.45×10^{10} m^3, an increase of 15.8% from 2000. The high-value areas were mainly located in the eastern and southern parts of the study area, where vegetation is rich and water storage capacity is high. Differences in climatic conditions have contributed to slight shifts in the spatial distribution of the high-value regions across different years. Food supply increased by approximately 436% over the past 20 years and reached 3.31×10^{11} RMB in 2020, with the high-value areas mainly in the central and western croplands.

3.3. Identification and Change of Ecological Sources

The ecological patches were determined through a comprehensive analysis of ecosystem services and landscape connectivity. Figure 6 shows that both the area and number of patches below the threshold initially increase rapidly and then stabilize as the minimum area threshold for ecological patches increases. From 2000 to 2020, the turning points of total patch area and number in different periods were close to the threshold of 2 km^2. When the 2 km^2 threshold was applied, numerous unselected patches emerged. However, these patches were small and had a negligible impact on the spatial distribution of ecological sources (Table 4). Therefore, we chose 2 km^2 as the threshold in this research.

Table 4. Patches with the area below the threshold between 2000 and 2020.

Year	Evaluation Index	Ecological Patches (Area < 2 km^2)	All Ecological Patches	The Proportion of Ecological Patches (Area < 2 km^2)
2000	Total area (km^2)	987.04	6917.86	14.27%
	Number	29,005	29,170	99.43%
2010	Total area (km^2)	847.09	7836.10	10.81%
	Number	13,752	13,945	98.62%
2020	Total area (km^2)	1697.34	11,942.42	14.21%
	Number	29,422	29,843	98.59%

Figure 7 displays the spatial patterns of ecological sources during the three periods. In 2000, 58 ecological sources covered an area of 5179.98 km^2. In 2010, 49 ecological sources spanned 6037.47 km^2. By 2020, the total area and number had increased to 7711.83 km^2 and 78 in 2020. These sources were primarily located in the forested areas of the northeast and south. These regions had high vegetation cover, a well-preserved ecological environment, and high biodiversity. They also exhibited superior soil conservation, carbon sequestration, and water conservation capabilities.

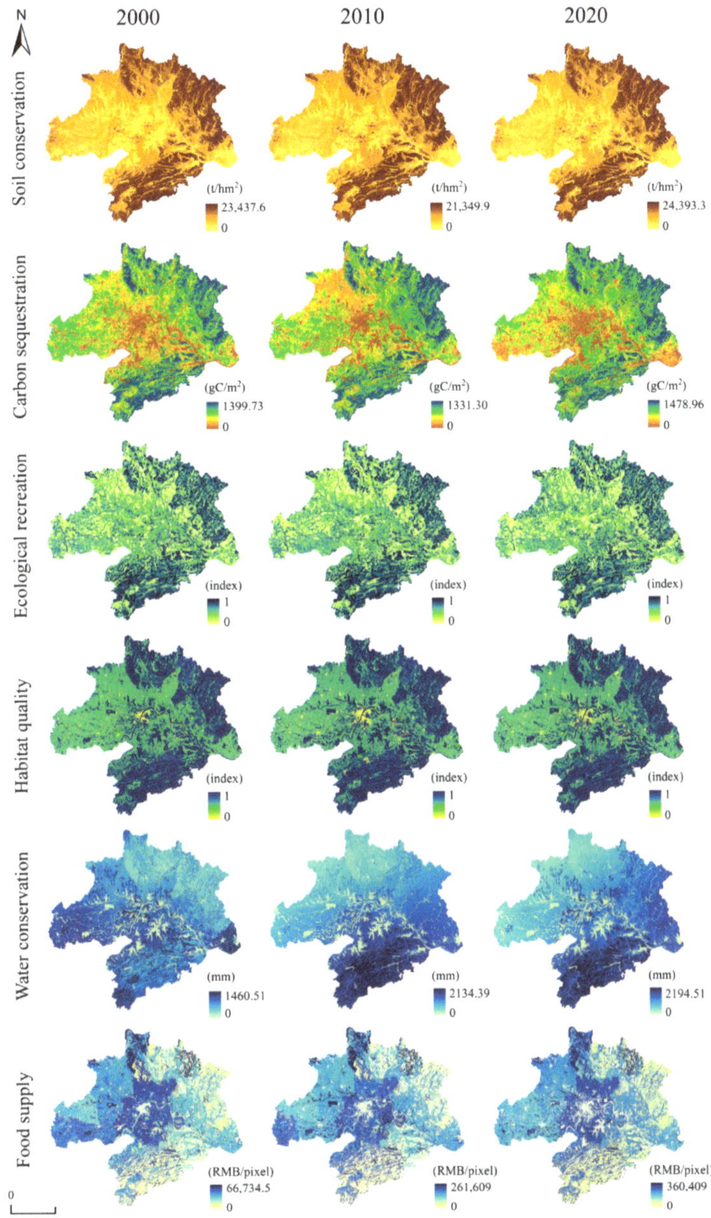

Figure 5. The changes in ecosystem services between 2000 and 2020.

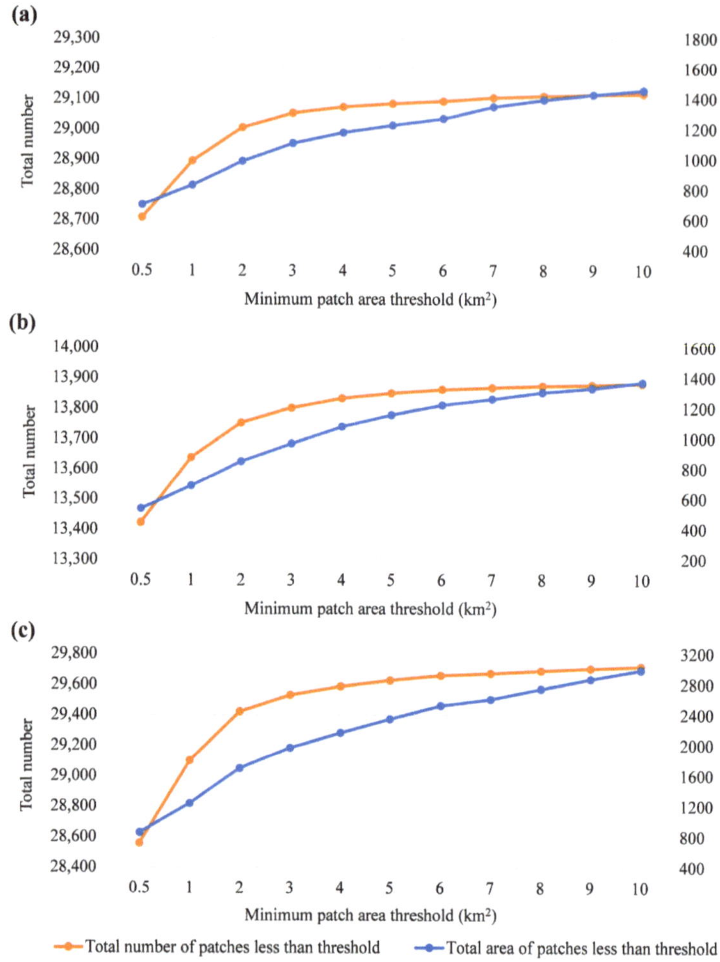

Figure 6. Patches less than the threshold for minimum patch area in (**a**) 2000, (**b**) 2010, and (**c**) 2020.

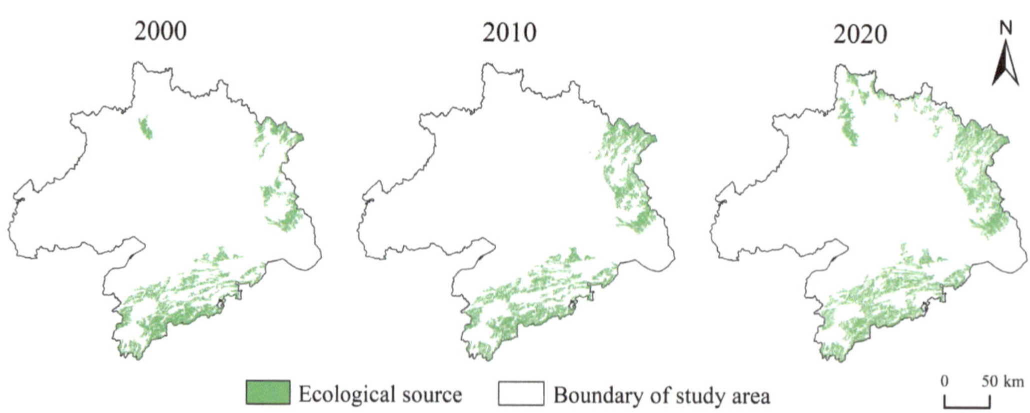

Figure 7. Spatial distribution of ecological sources in WMA from 2000 to 2020.

3.4. Resistance Surface Change

Figure 8 shows that the basic resistance surface values ranged from 0 to 340, with average values of 37.34, 42.16, and 43.86 for the three respective periods. Considering the influence of human disturbances on the ecological resistance surface, we optimized the resistance surface using NLI. The average values of the modified resistance surface for 2000, 2010, and 2020 were 39.72, 47.41, and 50.44, respectively.

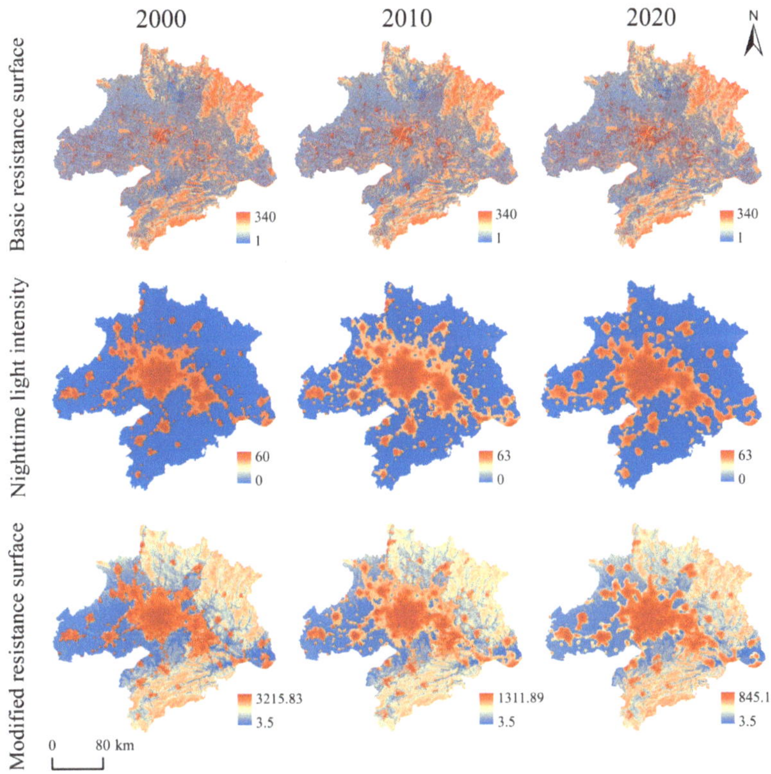

Figure 8. Basic resistance surface, NLI, and modified resistance surface in WMA from 2000 to 2020.

The modified resistance surface was generally similar to the basic resistance surface but showed local variations, particularly in the expansion of the middle and high resistance value areas in the central. High-resistance regions were predominantly concentrated in the central urban area of Wuhan and surrounding urban centers, where construction land was the dominant land-use type. Although the northeastern and southern regions generally had good ecological conditions, higher resistance values were observed due to elevation and slope.

3.5. Spatio-Temporal Change of the ESP

The ESP of WMA is depicted in Figure 9. In 2000 and 2010, there were 154 and 122 ecological corridors, with a total length of 2914.71 km and 1281.96 km, respectively. By 2020, the corridor numbers had increased to 178, while the total length had increased to 3475.39 km. They were primarily found in the east and south and later extended to the north. Predominant land-use types in those regions were forest, shrubland, and grassland.

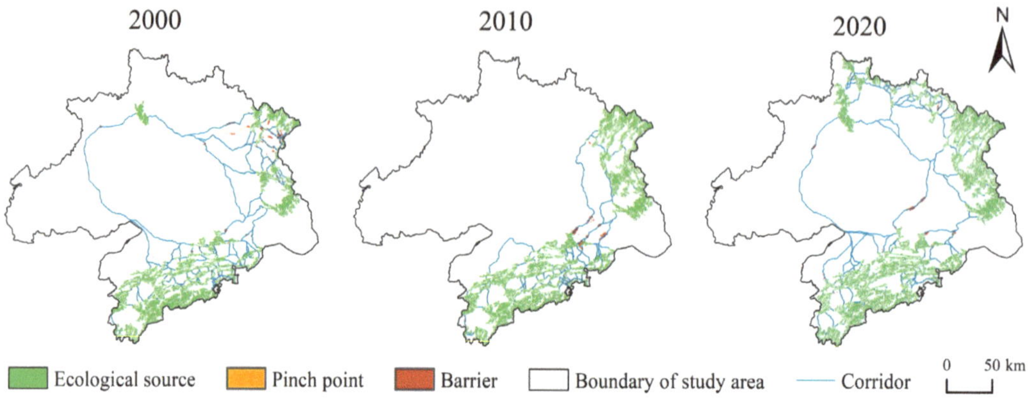

Figure 9. Spatial distribution of the ESP in WMA from 2000 to 2020.

From 2000 to 2020, we identified 244.9 km^2, 157.69 km^2, and 170.46 km^2 of ecological barriers, respectively. These barriers were primarily located on the fringes of ecological sources and corridors. The area of the pinch points was 236.58 km^2, 97.33 km^2, and 197.42 km^2 in 2000, 2010, and 2020, respectively. They were mainly found in the middle of long ecological corridors and close to the junction of ecological sources and corridors. Generally, the distributions of barriers and pinch points were spatially aligned with the corridors, primarily in the northeast and southeast of WMA.

3.6. Spatio-Temporal Change of Optimized ESP

Figure 10 shows the optimized ESP for WMA. In 2000, there were 8 primary, 20 secondary, and 30 tertiary ecological sources. By 2010, these numbers decreased to 5, 16, and 28, respectively. They increased to 9 primary, 21 secondary, and 48 tertiary sources in 2020. The three-year average area proportions of primary, secondary, and tertiary ecological resources were 41.11%, 23.03%, and 29.86%, respectively. From 2000 to 2010, primary and secondary corridors decreased from 107 and 47 to 91 and 31. However, from 2010 to 2020, they increased again to 141 and 37. The total length of secondary corridors exceeded that of primary corridors by 1951.19 km, 650.39 km, and 2238.18 km in 2000, 2010, and 2020, respectively.

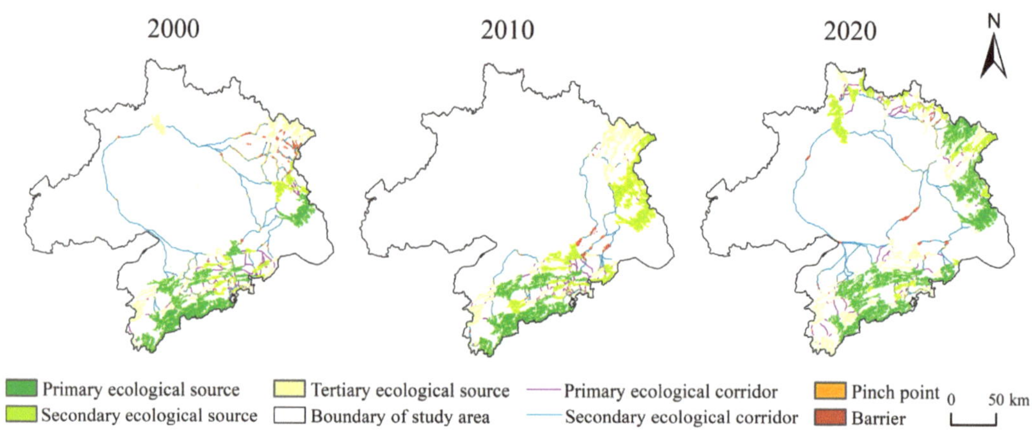

Figure 10. Spatial distribution of the optimized ESP in WMA from 2000 to 2020.

Spatially, primary ecological sources were widespread and primarily located in the southern part of WMA. Secondary ecological sources consisted mainly of small fragmented sources between primary ecological sources or large sources far from the primary ones. Tertiary ecological sources were scattered primarily in peripheral areas, such as the northern and eastern boundaries, exhibiting significant fragmentation. Primary corridors were concentrated in the south and extended north and were characterized by a short and dense configuration. Secondary corridors were predominantly situated in the central and eastern regions, appearing long and dispersed.

4. Discussion

4.1. Priority Assessment of Ecological Protection

Constructing and optimizing the ESP is crucial for maintaining ecosystem functions, ensuring ecological security, and coordinating ecosystem protection with regional high-quality development [53,65]. Implementing ecological protection requires the rational allocation of resources, and it is more efficient to prioritize protection in key areas [22]. However, earlier research has primarily focused on prioritizing single elements within ESPs and has rarely considered both ecological sources and corridors simultaneously [35,39,66]. In contrast, this study integrates the priority of ecological sources and corridors in constructing the ESP, thus addressing ecological protection from multiple dimensions.

The high-level ecological sources in WMA were mainly distributed in the northeast and south. In contrast, low-level ecological sources were primarily in the north and surrounding these sources. This distribution characteristic is consistent with earlier research [35]. Compared to previous studies, this research also emphasizes the prioritization of ecological corridors for comprehensive ecological protection. The high-level corridors were mainly distributed in the northeast and south, while the low-level corridors were concentrated in the central region. This distribution is consistent with previous studies [39]. As a result of incorporating ecosystem services in ecological sources identification, the ecological corridors are not as widely distributed across WMA as in that study.

By counting the ecological sources in 2000, 2010, and 2020, we found that the higher-level ecological sources were characterized by a larger average area and *dPC*, except that secondary sources had a smaller average area than tertiary sources in 2000. Higher-level ecological sources exhibited greater landscape connectivity, lower levels of fragmentation, and higher ecosystem service value [26,67]. These findings are consistent with the conclusions obtained from the previous study [35]. Between 2000 and 2020, the primary land-use types within ecological corridors were cropland, forest, and water. The proportion of cropland in primary corridors was smaller than in secondary corridors, while the proportion of forests was higher. Moreover, the average integrated ecosystem service value in primary corridors was higher than in secondary corridors. Therefore, higher-level ecological corridors demonstrated better environmental conditions, lower ecological resistance, and more efficient ecological flows [68,69].

4.2. Analysis and Recommendations Based on Optimized ESP

We determined the ecological protection priority according to various priorities and proposed corresponding protection measures. Ecological sources are patches that provide essential ecological functions and maintain ecological security, which are crucial for promoting regional sustainable development [26]. Primary ecological sources were mainly intact forest patches, providing abundant resources and critical ecological services. It is necessary to prioritize their protection and preservation, establish nature reserves, and prohibit over-exploitation to ensure their connectivity and integrity [19,64]. Secondary and tertiary ecological sources were more fragmented and had the potential for expansion. Ecological restoration efforts should be strengthened to enhance the overall quality of these patches, improving their structure and connectivity [9]. In the central region with scarce ecological resources, parks and green spaces of appropriate scale could be constructed to enhance the connectivity between sources and cope with the ecological pressure brought by urbanization [70].

Ecological corridors are vital channels for connecting ecological sources, and strengthening their construction can help increase landscape connectivity and ecological protection [71,72]. Primary corridors were mainly short-distance corridors that connected fragmented and isolated sources, reducing fragmentation and enhancing ecological network integrity. Therefore, the protection and construction of these corridors should be prioritized to minimize anthropogenic interference and safeguard the stability of ecosystems [73,74]. Secondary corridors were long-distance corridors that connected two independent sources far from each other, providing a basis for ecological safety protection in the intermediate areas. The land-use composition should be optimized to ensure habitat quality and to reduce the negative impacts of nearby cropland and construction land [75]. Additionally, it is necessary to connect corridors to urban green spaces in urban centers and install stepping stones to reduce the resistance of long-distance corridors and improve landscape connectivity [26,76].

Barriers refer to areas that significantly impede ecological flow, while pinch points are critical regions where ecological flow is concentrated [24,72]. Removing barriers and protecting pinch points can significantly improve landscape connectivity. Barriers and pinch points in different land-use types should be managed with specific protective measures tailored to the unique ecological characteristics of each area [71]. In corresponding cropland areas, measures such as limiting agricultural cultivation, promoting sustainable agricultural practices, and encouraging eco-friendly agricultural tourism can help minimize ecological resistance [77]. In water and wetland areas, it is necessary to create buffer zones, plant protective forests along the shoreline, control pollutant discharges, and improve water quality [78,79]. In construction land areas, efforts should focus on strengthening ecological greening and establishing wildlife corridors [80].

4.3. Restrictions and Future Exploration

A scientifically effective approach to balancing urbanization with ecological preservation is through the construction and optimization of an ESP. This study evaluated the ecological protection priority of ESP critical areas and provided important ideas for maximizing the benefits of ecological conservation under sustainable urban development. However, several limitations in this study warrant further investigation and refinement.

First, the research on ecosystem services and their changes is insufficient. Although ecosystem services were selected according to the actual situation in this study, the accuracy was still affected by the limitations of data, models, and technologies. The changes in ecosystem services are also closely related to population growth, climate change [81], economic development, and ecological protection [82,83], and have a close relationship with each other. Under the influence of urbanization and environmental impacts, the regional effects of ecosystem services will not remain static but rather respond in complex ways. For instance, there are multi-scale response characteristics of environmental factors to ES constraints [84], and the early growth of ecosystem services may be offset by the negative impacts of urbanization [85], among other issues. These are directions we can explore in greater depth in the future.

Secondly, threshold values are an issue in ESP construction. Many studies on ESPs have categorized the evaluation results and classified ecological security levels with corresponding thresholds, ignoring the internal differences and complexity of the region, as well as the significant differences across various assessment regions [86]. Ecological security assessments also involve many threshold issues, such as the thresholds for high-value areas of ecosystem services, distance thresholds when using landscape connectivity indices to identify ecological sources [87], and specific cost-weighted distance thresholds for identifying ecological corridors [88]. Although this study set thresholds based on previous findings, it did not account for inter-regional variability, and future studies should select more appropriate thresholds based on field conditions.

Finally, ESPs need to be explored in depth. Analyzing the drivers affecting ESP changes can reveal the effects of anthropogenic interventions and the natural environment on the ESP, providing a basis for developing ecological optimization strategies [89,90]. Future land-use changes are also an important direction in exploring the dynamic changes of ESPs. Simulating future land-use change under multiple scenarios and constructing ESPs can assist in formulating land-use optimization policies under different socio-economic conditions [91,92]. In the future, we can conduct a more in-depth analysis of ESP changes, especially the drivers of changes in ESP elements of different priorities, or predict ESP changes by simulating future land-use changes. This would help identify ecological corridors, pinch points, and barriers that may be threatened in the future, enabling us to propose targeted conservation strategies.

5. Conclusions

The construction and optimization of ESPs can effectively promote the balance between urbanization and regional ecological protection. Prioritizing key protection areas is essential for efficient resource allocation to enhance ecological benefits. However, most previous studies have only considered the ecological conservation prioritization of a single element of the ESP with a limited focus on evaluating ecological sources and corridors together. Based on the ESP pattern of WMA, this study conducted a comprehensive assessment of the ecological conservation priorities of both ecological sources and corridors. The results show that higher ecological source grades indicate greater landscape connectivity, higher ecosystem service values, and lower fragmentation levels. Similarly, higher corridor grades correspond to shorter corridor lengths and higher ecosystem service values. Primary ecological corridors connect fragmented and isolated ecological sources, playing a vital role in maintaining ecological connectivity. Secondary corridors link two independent ecological sources far apart and provide a basis for ecological protection in the intervening area. Based on the optimized ESP, we proposed hierarchical protection and management strategies for critical areas. This study incorporates ecological priority conservation into the ESP framework, broadening its applicability. The research conclusions provide a theoretical basis and reference for balancing urban economic development and ecological protection.

Supplementary Materials: The following supporting information can be downloaded at: https://www.mdpi.com/article/10.3390/land13111755/s1. References [93–96] are cited in the supplementary materials.

Author Contributions: Conceptualization, J.Z. and W.X.; Methodology, B.C. and H.L.; Validation, H.L. and L.Z.; Formal Analysis, B.C. and H.L.; Investigation, F.L.; Resources, L.Z. and Z.X.; Writing—Original Draft Preparation, B.C.; Writing—Review and Editing, J.Z.; Project Administration, J.Z.; Funding Acquisition, J.Z. All authors have read and agreed to the published version of the manuscript.

Funding: This research was supported by the Fundamental Research Foundation for the Chinese Academy of Forestry (No. CAFYBB2023ZA003) and the Fundamental Research Foundation for the Chinese Academy of Forestry (No. CAFYBB2019ZD001).

Data Availability Statement: The datasets generated during the current study are available from the corresponding author upon reasonable request.

Acknowledgments: The authors gratefully acknowledge the support of the funding. The authors are also deeply grateful to the editors and reviewers for their critical and constructive comments, which have significantly improved the quality of this manuscript.

Conflicts of Interest: The authors declare no conflicts of interest.

Appendix A

Table A1. Land-use transfer matrix for WMA from 2000 to 2020 (km^2).

Year	Land-Use Type	2020								
		Cropland	Forest	Shrubland	Grassland	Water	Wetland	Construction Land	Bare Land	Total
2000	Cropland	27,999.92	200.63	36.64	11.78	449.58	39.54	1173.82	7.15	29,919.06
	Forest	179.37	14,277.79	44.93	31.50	28.63	7.45	153.47	4.67	14,727.81
	Shrubland	37.78	54.27	2698.55	4.45	5.06	1.34	19.35	1.22	2822.02
	Grassland	11.23	36.40	4.84	1353.85	13.50	2.29	21.43	1.85	1445.38
	Water	91.67	16.72	3.06	5.82	4731.23	107.46	97.20	23.65	5076.81
	Wetland	9.63	1.01	0.23	0.24	189.29	426.16	16.39	1.08	644.02
	Construction land	113.03	8.02	1.11	0.65	10.09	6.99	2926.85	1.98	3068.72
	Bare land	3.60	0.36	0.01	1.57	62.92	3.35	7.73	163.67	243.20
	Total	28,446.23	14,595.19	2789.37	1409.87	5490.30	594.58	4416.23	205.27	57,947.03

Table A2. Land-use transfer matrix for WMA from 2000 to 2010 (km^2).

Year	Land-Use Type	2010								
		Cropland	Forest	Shrubland	Grassland	Water	Wetland	Construction Land	Bare Land	Total
2000	Cropland	27,826.37	274.21	51.97	22.46	724.23	139.80	862.86	17.17	29,919.06
	Forest	155.97	14,341.07	42.27	24.88	30.12	4.04	128.54	0.92	14,727.81
	Shrubland	28.95	48.51	2717.05	6.93	4.72	0.56	15.07	0.23	2822.02
	Grassland	10.31	38.30	13.28	1344.13	14.04	6.00	18.37	0.94	1445.38
	Water	182.32	15.35	3.10	5.73	4485.73	272.27	82.61	29.70	5076.81
	Wetland	86.01	2.42	0.40	0.93	289.56	237.93	16.77	10.02	644.02
	Construction land	84.77	10.11	2.21	1.59	29.82	3.29	2934.87	2.05	3068.72
	Bare land	11.01	1.28	0.02	1.75	51.85	51.56	8.02	117.71	243.20
	Total	28,385.70	14,731.26	2830.31	1408.39	5630.08	715.46	4067.12	178.73	57,947.03

Table A3. Land-use transfer matrix for WMA from 2010 to 2020 (km^2).

Year	Land-Use Type	2020								
		Cropland	Forest	Shrubland	Grassland	Water	Wetland	Construction Land	Bare Land	Total
2010	Cropland	26,737.00	265.64	51.15	16.01	293.50	87.81	920.55	14.04	28,385.70
	Forest	365.11	14,079.27	69.60	46.96	31.87	7.53	124.76	6.17	14,731.26
	Shrubland	73.86	72.67	2640.60	15.83	7.37	1.43	17.52	1.02	2830.31
	Grassland	26.45	39.54	10.12	1306.63	6.01	1.63	15.85	2.16	1408.39
	Water	476.01	33.37	5.73	7.46	4834.90	158.20	84.48	29.92	5630.08
	Wetland	112.85	3.23	0.59	3.94	251.40	314.72	12.87	15.85	715.46
	Construction land	638.10	100.51	11.50	12.16	50.00	13.74	3235.35	5.77	4067.12
	Barel and	17.12	1.08	0.01	0.79	15.20	9.42	4.82	130.28	178.73
	Total	28,446.51	14,595.31	2789.30	1409.78	5490.25	594.50	4416.18	205.20	57,947.03

References

1. Lin, L.; Wei, X.; Luo, P.; Wang, S.; Kong, D.; Yang, J. Ecological Security Patterns at Different Spatial Scales on the Loess Plateau. *Remote Sens.* **2023**, *15*, 1011. [CrossRef]
2. Gong, D.; Huang, M.; Lin, H. Construction of an Ecological Security Pattern in Rapidly Urbanizing Areas Based on Ecosystem Sustainability, Stability, and Integrity. *Remote Sens.* **2023**, *15*, 5728. [CrossRef]
3. Dai, L.; Wang, Z. Construction and optimization strategy of ecological security pattern based on ecosystem services and landscape connectivity: A case study of Guizhou Province, China. *Environ. Sci. Pollut. Res.* **2023**, *30*, 45123–45139. [CrossRef] [PubMed]
4. Tang, H.; Peng, J.; Jiang, H.; Lin, Y.; Dong, J.; Liu, M.; Meersmans, J. Spatial analysis enables priority selection in conservation practices for landscapes that need ecological security. *J. Environ. Manag.* **2023**, *345*, 118888. [CrossRef]
5. Tang, L.; Liang, G.; Gu, G.; Xu, J.; Duan, L.; Zhang, X.; Yang, X.; Lu, R. Study on the spatial-temporal evolution characteristics, patterns, and driving mechanisms of ecological environment of the Ecological Security Barriers on China's Land Borders. *Environ. Impact Assess. Rev.* **2023**, *103*, 107267. [CrossRef]
6. Yu, K. Security patterns and surface model in landscape ecological planning. *Landsc. Urban Plan.* **1996**, *36*, 1–17. [CrossRef]
7. Peng, J.; Yang, Y.; Liu, Y.; Hu, Y.; Du, Y.; Meersmans, J.; Qiu, S. Linking ecosystem services and circuit theory to identify ecological security patterns. *Sci. Total Environ.* **2018**, *644*, 781–790. [CrossRef]
8. Wang, Y.; Zhang, F.; Li, X.; Johnson, V.C.; Tan, M.L.; Kung, H.T.; Shi, J.C.; Bahtebay, J.; He, X. Methodology for Mapping the Ecological Security Pattern and Ecological Network in the Arid Region of Xinjiang, China. *Remote Sens.* **2023**, *15*, 2836. [CrossRef]
9. Jing, M.; Song, F.; Meng, K.; Su, F.; Wei, C. Optimization of landscape pattern in the main river basin of Liao River in China based on ecological network. *Environ. Sci. Pollut. Res.* **2023**, *30*, 65587–65601. [CrossRef]
10. Yang, J.; Xie, B.; Wang, T.; Mak-Mensah, E. Identification and optimization strategy of ecological security pattern in Maiji District of Gansu, China. *Ecol. Indic.* **2023**, *157*, 111309.
11. Yang, L.; Zhang, F.; Qin, L. Construction and stability evaluation of ecological networks in the Loess Plateau. *Ecol. Indic.* **2024**, *159*, 111697. [CrossRef]
12. Luo, J.; Fu, H. Construct the future wetland ecological security pattern with multi-scenario simulation. *Ecol. Indic.* **2023**, *153*, 110473. [CrossRef]
13. Li, J.; Dong, S.; Li, Y.; Wang, Y.; Li, Z.; Wang, M. Environmental governance of transnational regions based on ecological security: The China-Mongolia-Russia Economic Corridor. *J. Clean. Prod.* **2023**, *422*, 138625. [CrossRef]
14. Yang, J.; Deng, W.; Zhang, G.; Cui, X. Linking endangered species protection to construct and optimize ecological security patterns in the National ecological Civilization construction Demonstration Zone: A case study of Yichang, China. *Ecol. Indic.* **2024**, *158*, 111579. [CrossRef]
15. Tian, H.; Wang, H.; Lyu, X.; Li, X.; Yang, Y.; Zhang, Y.; Liu, J.; Lu, Y.; Zhao, X.; Qu, T.; et al. Construction and optimization of ecological security patterns in Dryland watersheds considering ecosystem services flows. *Ecol. Indic.* **2024**, *159*, 111664. [CrossRef]
16. Wang, N.; Zhao, Y. Construction of an ecological security pattern in Jiangnan water network area based on an integrated Approach: A case study of Gaochun, Nanjing. *Ecol. Indic.* **2024**, *158*, 111314. [CrossRef]
17. Chen, Z.; Lin, J.; Huang, J. Linking ecosystem service flow to water-related ecological security pattern: A methodological approach applied to a coastal province of China. *J. Environ. Manag.* **2023**, *345*, 118725. [CrossRef]
18. Sun, M.; Zhang, L.; Yang, R.; Li, X.; Zhang, Y.; Lu, Y. Construction of an integrated framework for assessing ecological security and its application in Southwest China. *Ecol. Indic.* **2023**, *148*, 110074. [CrossRef]
19. Xiang, H.; Zhang, J.; Mao, D.; Wang, M.; Yu, F.; Wang, Z.; Li, H. Optimizing ecological security patterns considering zonal vegetation distribution for regional sustainability. *Ecol. Eng.* **2023**, *194*, 107055. [CrossRef]
20. Gou, M.; Li, L.; Ouyang, S.; Shu, C.; Xiao, W.; Wang, N.; Hu, J.; Liu, C. Integrating ecosystem service trade-offs and rocky desertification into ecological security pattern construction in the Daning river basin of southwest China. *Ecol. Indic.* **2022**, *138*, 108845. [CrossRef]
21. Ji, Y.; Yang, L.; Dong, Q.; Zhou, S.; Jia, L.; Xun, B. Construction of eco-security model in the agro-pastoral interconnected zone in northern Shaanxi. *Ecol. Indic.* **2023**, *154*, 110832. [CrossRef]
22. Wang, J.; Bai, Y.; Huang, Z.; Ashraf, A.; Ali, M.; Fang, Z.; Lu, X. Identifying ecological security patterns to prioritize conservation and restoration: A case study in Xishuangbanna tropical region, China. *J. Clean. Prod.* **2024**, *444*, 141222. [CrossRef]
23. Wang, S.; Huang, Y.; Jiang, X.; Wang, T.; Jin, Y. Identification and Optimization of Ecological Security Patterns in the Xiangyang Metropolitan Area. *IEEE J. Sel. Top. Appl. Earth Observ. Remote Sens.* **2023**, *16*, 8671–8679. [CrossRef]
24. Lai, X.; Yu, H.; Liu, G.; Zhang, X.; Feng, Y.; Ji, Y.; Zhao, Q.; Jiang, J.; Gu, X. Construction and Analysis of Ecological Security Patterns in the Southern Anhui Region of China from a Circuit Theory Perspective. *Remote Sens.* **2023**, *15*, 1385. [CrossRef]
25. Gao, Y.; Ma, L.; Liu, J.; Zhuang, Z.; Huang, Q.; Li, M. Constructing Ecological Networks Based on Habitat Quality Assessment: A Case Study of Changzhou, China. *Sci. Rep.* **2017**, *7*, 46073. [CrossRef]
26. Zhang, Y.; Yang, R.; Sun, M.; Lu, Y.; Zhang, L.; Yin, Y.; Li, X. Identification of spatial protection and restoration priorities for ecological security pattern in a rapidly urbanized region: A case study in the Chengdu-Chongqing economic Circle, China. *J. Environ. Manag.* **2024**, *366*, 121789. [CrossRef] [PubMed]
27. Zhao, Y.; He, L.; Bai, W.; He, Z.; Luo, F.; Wang, Z. Prediction of ecological security patterns based on urban expansion: A case study of Chengdu. *Ecol. Indic.* **2024**, *158*, 111467. [CrossRef]

28. Jia, Q.; Jiao, L.; Lian, X.; Wang, W. Linking supply-demand balance of ecosystem services to identify ecological security patterns in urban agglomerations. *Sust. Cities Soc.* **2023**, *92*, 104497. [CrossRef]
29. Liu, H.; Wang, Z.; Zhang, L.; Tang, F.; Wang, G.; Li, M. Construction of an ecological security network in the Fenhe River Basin and its temporal and spatial evolution characteristics. *J. Clean. Prod.* **2023**, *417*, 137961. [CrossRef]
30. Wei, B.; Kasimu, A.; Fang, C.; Reheman, R.; Zhang, X.; Han, F.; Zhao, Y.; Aizizi, Y. Establishing and optimizing the ecological security pattern of the urban agglomeration in arid regions of China. *J. Clean. Prod.* **2023**, *427*, 139301. [CrossRef]
31. Wang, W.; Li, B.; Su, F.; Jiang, Z.; Chen, S. Identifying Ecological Security Patterns Meeting Future Urban Expansion in Changsha-Zhuzhou-Xiangtan Urban Agglomeration, China. *Remote Sens.* **2023**, *15*, 3141. [CrossRef]
32. Liu, H.; Zhu, J.; Zeng, L.; Gou, M.; Chen, B.; Lv, J.; Xiao, W. Identification of Urban Ecological Security Pattern Based on Ecosystem Services Supply-Demand. *Ecosyst. Health Sustain.* **2024**, *10*, 0146. [CrossRef]
33. Peng, Y.; Cheng, W.; Xu, X.; Song, H. Analysis and prediction of the spatiotemporal characteristics of land-use ecological risk and carbon storage in Wuhan metropolitan area. *Ecol. Indic.* **2024**, *158*, 111432. [CrossRef]
34. Deng, Y.; Jiang, W.; Tang, Z.; Li, J.; Lv, J.; Chen, Z.; Jia, K. Spatio-Temporal Change of Lake Water Extent in Wuhan Urban Agglomeration Based on Landsat Images from 1987 to 2015. *Remote Sens.* **2017**, *9*, 270. [CrossRef]
35. Zeng, W.; Tang, H.; Liang, X.; Hu, Z.; Yang, Z.; Guan, Q. Using ecological security pattern to identify priority protected areas: A case study in the Wuhan Metropolitan Area, China. *Ecol. Indic.* **2023**, *148*, 110121. [CrossRef]
36. Peng, K.; Jiang, W.; Deng, Y.; Liu, Y.; Wu, Z.; Chen, Z. Simulating wetland changes under different scenarios based on integrating the random forest and CLUE-S models: A case study of Wuhan Urban Agglomeration. *Ecol. Indic.* **2020**, *117*, 106671. [CrossRef]
37. Wang, Z.; Zhang, J.; Chen, J.; Gao, H.; Li, J.; Li, M. Determining the ecological security pattern and important ecological regions based on the supply-demand of ecosystem services: A case study of Xuzhou City, China. *Front. Public Health* **2023**, *11*, 1087588. [CrossRef] [PubMed]
38. Chen, J.; Xue, J.; Gu, K.; Wang, Y. Balancing urban expansion with ecological integrity: An ESP framework for rapidly urbanizing small and medium-sized cities, with insights from Suizhou, China. *Ecol. Inform.* **2024**, *80*, 102508.
39. Lu, Y.; Liu, Y.; Huang, D.; Liu, Y. Evolution Analysis of Ecological Networks Based on Spatial Distribution Data of Land Use Types Monitored by Remote Sensing in Wuhan Urban Agglomeration, China, from 2000 to 2020. *Remote Sens.* **2022**, *14*, 2618. [CrossRef]
40. Wen, L.; Chatalova, L.; Gao, X.; Zhang, A. Reduction of carbon emissions through resource-saving and environment-friendly regional economic integration? Evidence from Wuhan metropolitan area, China. *Technol. Forecast. Soc. Chang.* **2021**, *166*, 120590. [CrossRef]
41. Wang, T.; Li, H.; Huang, Y. The complex ecological network's resilience of the Wuhan metropolitan area. *Ecol. Indic.* **2021**, *130*, 108101. [CrossRef]
42. Wu, Y.; Shi, K.; Chen, Z.; Liu, S.; Chang, Z. Developing Improved Time-Series DMSP-OLS-Like Data (1992–2019) in China by Integrating DMSP-OLS and SNPP-VIIRS. *IEEE Trans. Geosci. Remote Sens.* **2022**, *60*, 4407714. [CrossRef]
43. Das, M.; Das, A. Dynamics of Urbanization and its impact on Urban Ecosystem Services (UESs): A study of a medium size town of West Bengal, Eastern India. *J. Urban Manag.* **2019**, *8*, 420–434. [CrossRef]
44. Sharma, R.; Malaviya, P. Ecosystem services and climate action from a circular bioeconomy perspective. *Renew. Sust. Energy Rev.* **2023**, *175*, 113164. [CrossRef]
45. Peng, K.; Jiang, W.; Ling, Z.; Hou, P.; Deng, Y. Evaluating the potential impacts of land use changes on ecosystem service value under multiple scenarios in support of SDG reporting: A case study of the Wuhan urban agglomeration. *J. Clean. Prod.* **2021**, *307*, 127321. [CrossRef]
46. Sahle, M.; Saito, O.; Fürst, C.; Yeshitela, K. Quantifying and mapping of water-related ecosystem services for enhancing the security of the food-water-energy nexus in tropical data-sparse catchment. *Sci. Total Environ.* **2019**, *646*, 573–586. [CrossRef]
47. Liu, H.; Xiao, W.; Zhu, J.; Zeng, L.; Li, Q. Urbanization Intensifies the Mismatch between the Supply and Demand of Regional Ecosystem Services: A Large-Scale Case of the Yangtze River Economic Belt in China. *Remote Sens.* **2022**, *14*, 5147. [CrossRef]
48. Li, M.; Zhou, Y.; Xiao, P.; Tian, Y.; Huang, H.; Xiao, L. Evolution of Habitat Quality and Its Topographic Gradient Effect in Northwest Hubei Province from 2000 to 2020 Based on the InVEST Model. *Land* **2021**, *10*, 857. [CrossRef]
49. Zhang, Z.; Peng, J.; Xu, Z.; Wang, X.; Meersmans, J. Ecosystem services supply and demand response to urbanization: A case study of the Pearl River Delta, China. *Ecosyst. Serv.* **2021**, *49*, 101274. [CrossRef]
50. Schirpke, U.; Meisch, C.; Marsoner, T.; Tappeiner, U. Revealing spatial and temporal patterns of outdoor recreation in the European Alps and their surroundings. *Ecosyst. Serv.* **2018**, *31*, 336–350. [CrossRef]
51. Peng, J.; Zhao, S.; Dong, J.; Liu, Y.; Meersmans, J.; Li, H.; Wu, J. Applying ant colony algorithm to identify ecological security patterns in megacities. *Environ. Modell. Softw.* **2019**, *117*, 214–222. [CrossRef]
52. He, G.; Ruan, J. Study on ecological security evaluation of Anhui Province based on normal cloud model. *Environ. Sci. Pollut. Res.* **2022**, *29*, 16549–16562. [CrossRef] [PubMed]
53. Qian, W.; Zhao, Y.; Li, X. Construction of ecological security pattern in coastal urban areas: A case study in Qingdao, China. *Ecol. Indic.* **2023**, *154*, 110754. [CrossRef]
54. Li, C.; Huang, L.; Xu, Q.; Cao, Z. Synergistic ecological network approach for sustainable development of highly urbanized area in the Bay Bottom region: A study in Chengyang District, Qingdao. *Ecol. Indic.* **2024**, *158*, 111443. [CrossRef]

55. Zhou, G.; Huan, Y.; Wang, L.; Zhang, R.; Liang, T.; Han, X.; Feng, Z. Constructing a multi-leveled ecological security pattern for improving ecosystem connectivity in the Asian water Tower region. *Ecol. Indic.* **2023**, *154*, 110597. [CrossRef]
56. Fan, F.; Wen, X.; Feng, Z.; Gao, Y.; Li, W. Optimizing urban ecological space based on the scenario of ecological security patterns: The case of central Wuhan, China. *Appl. Geogr.* **2022**, *138*, 102619. [CrossRef]
57. Liu, X.; Su, Y.; Li, Z.; Zhang, S. Constructing ecological security patterns based on ecosystem services trade-offs and ecological sensitivity: A case study of Shenzhen metropolitan area, China. *Ecol. Indic.* **2023**, *154*, 110626. [CrossRef]
58. Chen, X.; Kang, B.; Li, M.; Du, Z.; Zhang, L.; Li, H. Identification of priority areas for territorial ecological conservation and restoration based on ecological networks: A case study of Tianjin City, China. *Ecol. Indic.* **2023**, *146*, 109809. [CrossRef]
59. Li, Z.; Chang, J.; Li, C.; Gu, S. Ecological Restoration and Protection of National Land Space in Coal Resource-Based Cities from the Perspective of Ecological Security Pattern: A Case Study in Huaibei City, China. *Land* **2023**, *12*, 442. [CrossRef]
60. McRae, B.H.; Dickson, B.G.; Keitt, T.H.; Shah, V.B. Using circuit theory to model connectivity in ecology, evolution, and conservation. *Ecology* **2008**, *89*, 2712–2724. [CrossRef]
61. Tu, W.; Du, Y.; Yi, J.; Liang, F.; Wang, N.; Qian, J.; Huang, S.; Luo, P.; Wang, X. Assessment of the dynamic ecological networks on the Qinghai-Tibet Plateau using human's digital footprints. *Ecol. Indic.* **2023**, *147*, 109954. [CrossRef]
62. Yu, H.; Gu, X.; Liu, G.; Fan, X.; Zhao, Q.; Zhang, Q. Construction of Regional Ecological Security Patterns Based on Multi-Criteria Decision Making and Circuit Theory. *Remote Sens.* **2022**, *14*, 527. [CrossRef]
63. Zhang, L.; Qiang, Z.; Xu, E. Improving the ecological network optimization with landscape connectivity: A case study of Neijiang City, Sichuan Province. *Environ. Sci. Pollut. Res.* **2023**, *30*, 54753–54769. [CrossRef]
64. Li, S.; Zhao, Y.; Xiao, W.; Yue, W.; Wu, T. Optimizing ecological security pattern in the coal resource-based city: A case study in Shuozhou City, China. *Ecol. Indic.* **2021**, *130*, 108026. [CrossRef]
65. Qiao, Q.; Zhen, Z.; Liu, L.; Luo, P. The Construction of Ecological Security Pattern under Rapid Urbanization in the Loess Plateau: A Case Study of Taiyuan City. *Remote Sens.* **2023**, *15*, 1523. [CrossRef]
66. Teng, M.; Wu, C.; Zhou, Z.; Lord, E.; Zheng, Z. Multipurpose greenway planning for changing cities: A framework integrating priorities and a least-cost path model. *Landsc. Urban. Plan.* **2011**, *103*, 1–14. [CrossRef]
67. Gao, J.; Du, F.; Zuo, L.; Jiang, Y. Integrating ecosystem services and rocky desertification into identification of karst ecological security pattern. *Landsc. Ecol.* **2021**, *36*, 2113–2133. [CrossRef]
68. Liu, Z.; Gan, X.; Dai, W.; Huang, Y. Construction of an Ecological Security Pattern and the Evaluation of Corridor Priority Based on ESV and the "Importance-Connectivity" Index: A Case Study of Sichuan Province, China. *Sustainability* **2022**, *14*, 3985. [CrossRef]
69. Huang, L.; Wang, D.; He, C. Ecological security assessment and ecological pattern optimization for Lhasa city (Tibet) based on the minimum cumulative resistance model. *Environ. Sci. Pollut. Res.* **2022**, *29*, 83437–83451. [CrossRef]
70. Ding, M.; Liu, W.; Xiao, L.; Zhong, F.; Lu, N.; Zhang, J.; Zhang, Z.; Xu, X.; Wang, K. Construction and optimization strategy of ecological security pattern in a rapidly urbanizing region: A case study in central-south China. *Ecol. Indic.* **2022**, *136*, 108604. [CrossRef]
71. Lan, Y.; Wang, J.; Liu, Q.; Liu, F.; Liu, L.; Li, J.; Luo, M. Identification of critical ecological restoration and early warning regions in the five-lakes basin of central Yunnan. *Ecol. Indic.* **2024**, *158*, 111337. [CrossRef]
72. Wei, W.; Liu, C.; Ma, L.; Xie, B.; Zhou, J.; Nan, S. Optimization strategies of ecological security patterns through importance of ecosystem services and ecological sensitivity-A case study in the Yellow River Basin. *Land Degrad. Dev.* **2024**, *35*, 985–1001. [CrossRef]
73. Yang, Z.; Ma, C.; Liu, Y.; Zhao, H.; Hua, Y.; Ou, S.; Fan, X. Provincial-Scale Research on the Eco-Security Structure in the Form of an Ecological Network of the Upper Yellow River: A Case Study of the Ningxia Hui Autonomous Region. *Land* **2023**, *12*, 1341. [CrossRef]
74. Kang, J.; Qing, Y.; Lu, W. Construction and optimization of the Saihanba ecological network. *Ecol. Indic.* **2023**, *153*, 110401. [CrossRef]
75. Liu, Q.; Sun, Y.; Mei, Y.; Jian, Z.; Pan, F.; Zhang, L. Construction and Analysis of Ecological Security Pattern of Qingdao Based on MSPA and MCR Models. *Pol. J. Environ. Stud.* **2023**, *32*, 155–169. [CrossRef]
76. Ding, G.; Yi, D.; Yi, J.; Guo, J.; Ou, M.; Ou, W.; Tao, Y.; Pueppke, S.G. Protecting and constructing ecological corridors for biodiversity conservation: A framework that integrates landscape similarity assessment. *Appl. Geogr.* **2023**, *160*, 103098. [CrossRef]
77. Chen, H.; Yan, W.; Li, Z.; Wende, W.; Xiao, S. A framework for integrating ecosystem service provision and connectivity in ecological spatial networks: A case study of the Shanghai metropolitan area. *Sust. Cities Soc.* **2024**, *100*, 105018. [CrossRef]
78. Li, Q.; Zhou, Y.; Yi, S. An integrated approach to constructing ecological security patterns and identifying ecological restoration and protection areas: A case study of Jingmen, China. *Ecol. Indic.* **2022**, *137*, 108723. [CrossRef]
79. Li, S.; He, W.; Wang, L.; Zhang, Z.; Chen, X.; Lei, T.; Wang, S.; Wang, Z. Optimization of landscape pattern in China Luojiang Xiaoxi basin based on landscape ecological risk assessment. *Ecol. Indic.* **2023**, *146*, 109887. [CrossRef]
80. Wang, Y.; Zhang, L.; Song, Y. Study on the Construction of the Ecological Security Pattern of the Lancang River Basin (Yunnan Section) Based on InVEST-MSPA-Circuit Theory. *Sustainability* **2023**, *15*, 477. [CrossRef]
81. Chen, S.; Liu, X.; Yang, L.; Zhu, Z. Variations in Ecosystem Service Value and Its Driving Factors in the Nanjing Metropolitan Area of China. *Forests* **2023**, *14*, 113. [CrossRef]

82. Huang, C.; Zhao, D.; Liu, C.; Liao, Q. Integrating territorial pattern and socioeconomic development into ecosystem service value assessment. *Environ. Impact Assess. Rev.* **2023**, *100*, 107088. [CrossRef]
83. Guo, X.; Zhang, Y.; Guo, D.; Lu, W.; Xu, H. How does ecological protection redline policy affect regional land use and ecosystem services? *Environ. Impact Assess. Rev.* **2023**, *100*, 107062. [CrossRef]
84. Liu, Y.; Zhang, C.; Zeng, H. Constraint effects among several key ecosystem service types and their influencing factors: A case study of the Pearl River Delta, China. *Ecol. Indic.* **2023**, *146*, 109883. [CrossRef]
85. Qiu, Z.; Guan, Y.; Zhou, K.; Kou, Y.; Zhou, X.; Zhang, Q. Spatiotemporal Analysis of the Interactions between Ecosystem Services in Arid Areas and Their Responses to Urbanization and Various Driving Factors. *Remote Sens.* **2024**, *16*, 520. [CrossRef]
86. Wen, J.; Hou, K. Research on the progress of regional ecological security evaluation and optimization of its common limitations. *Ecol. Indic.* **2021**, *127*, 107797. [CrossRef]
87. Xie, J.; Xie, B.; Zhou, K.; Li, J.; Xiao, J.; Liu, C. Impacts of landscape pattern on ecological network evolution in Changsha-Zhuzhou-Xiangtan Urban Agglomeration, China. *Ecol. Indic.* **2022**, *145*, 109716. [CrossRef]
88. Zhang, Y.; Zhao, Z.; Yang, Y.; Fu, B.; Ma, R.; Lue, Y.; Wu, X. Identifying ecological security patterns based on the supply, demand and sensitivity of ecosystem service: A case study in the Yellow River Basin, China. *J. Environ. Manag.* **2022**, *315*, 115158. [CrossRef]
89. Wang, C.; Wang, Q.; Liu, N.; Sun, Y.; Guo, H.; Song, X. The impact of LUCC on the spatial pattern of ecological network during urbanization: A case study of Jinan City. *Ecol. Indic.* **2023**, *155*, 111004. [CrossRef]
90. Cao, W.; Jia, G.; Yang, Q.; Sun, H.; Wang, L.; Svenning, J.C.; Wen, L. Construction of ecological network and its temporal and spatial evolution characteristics: A case study of Ulanqab. *Ecol. Indic.* **2024**, *166*, 112344. [CrossRef]
91. Li, L.; Huang, X.; Wu, D.; Yang, H. Construction of ecological security pattern adapting to future land use change in Pearl River Delta, China. *Appl. Geogr.* **2023**, *154*, 102946. [CrossRef]
92. Cai, G.; Xiong, J.; Wen, L.; Weng, A.; Lin, Y.; Li, B. Predicting the ecosystem service values and constructing ecological security patterns in future changing land use patterns. *Ecol. Indic.* **2023**, *154*, 110787. [CrossRef]
93. Yang, S.; Zhao, W.; Liu, Y.; Wang, S.; Wang, J.; Zhai, R. Influence of land use change on the ecosystem service trade-offs in the ecological restoration area: Dynamics and scenarios in the Yanhe watershed, China. *Sci. Total Environ.* **2018**, *644*, 556–566. [CrossRef] [PubMed]
94. Hamel, P.; Valencia, J.; Schmitt, R.; Shrestha, M.; Piman, T.; Sharp, R.P.; Francesconi, W.; Guswa, A.J. Modeling seasonal water yield for landscape management: Applications in Peru and Myanmar. *J. Environ. Manag.* **2020**, *270*, 110792. [CrossRef] [PubMed]
95. Jiang, C.; Wang, F.; Zhang, H.; Dong, X. Quantifying changes in multiple ecosystem services during 2000-2012 on the Loess Plateau, China, as a result of climate variability and ecological restoration. *Ecol. Eng.* **2016**, *97*, 258–271. [CrossRef]
96. Paracchini, M.L.; Zulian, G.; Kopperoinen, L.; Maes, J.; Schägner, J.P.; Termansen, M.; Zandersen, M.; Perez-Soba, M.; Scholefield, P.A.; Bidoglio, G. Mapping cultural ecosystem services: A framework to assess the potential for outdoor recreation across the EU. *Ecol. Indic.* **2014**, *45*, 371–385. [CrossRef]

land

MDPI

Article

Interplay of Urbanization and Ecological Environment: Coordinated Development and Drivers

Ruixu Chen [1], Yang Chen [2], Oleksii Lyulyov [3,4] and Tetyana Pimonenko [3,4,*]

1 School of Marxism, Fujian Medical University, Fuzhou 350122, China; 13765175707@163.com
2 School of Economics, Fujian Normal University, Fuzhou 350108, China; cheny3598@gmail.com
3 Department of Management, Faculty of Applied Sciences, WSB University, 41-300 Dabrowa Gornicza, Poland; alex_lyulev@econ.sumdu.edu.ua
4 Department of Marketing, Sumy State University, 2, Rymskogo-Korsakova St., 40007 Sumy, Ukraine
* Correspondence: tetyana_pimonenko@econ.sumdu.edu.ua

Abstract: The interplay between urbanization and ecological environmental efficiency has gained increasing significance in the context of sustainable development, as rapid urban growth poses challenges to resource consumption, greenhouse gas emissions, and the overall ecological well-being of urban areas. Understanding and analyzing the coordinated development of urbanization and ecological environmental efficiency, as well as assessing the influence of drivers on this relationship, is crucial for developing effective policies and strategies that promote environmentally sustainable urban development. This study establishes an urbanization index based on four key aspects: economy, society, population, and ecology. This investigation focuses on 30 provinces in China spanning from 2011 to 2020. The following methods are applied: global Malmquist–Luenberger productivity index, entropy method, TOPSIS model, coupled coordination degree model, panel-corrected standard error (PCSE), and feasible generalized least squares (FGLS) models. The empirical results demonstrate a favorable level of coordinated development between urbanization and the ecological environment overall, with more pronounced regional evolution trends. The trade openness, energy structure, and digitalization level play significant roles in effectively promoting the coordinated development of urbanization and the ecological environment to varying extents. The growth of trade openness and digitalization level promote coordinated development between urbanization and the ecological environment by 0.125 and 0.049, respectively. However, the increase in the energy structure decreases it by 0.509. These results have significant implications for policymakers, urban planners, and stakeholders, emphasizing the need for a balanced approach that prioritizes ecological environmental protection in urbanization efforts. This study underscores the importance of sustainable urban development strategies to ensure long-term ecological and environmental sustainability.

Keywords: sustainable development; urbanization; governance; infrastructure; quality of life

Citation: Chen, R.; Chen, Y.; Lyulyov, O.; Pimonenko, T. Interplay of Urbanization and Ecological Environment: Coordinated Development and Drivers. *Land* **2023**, *12*, 1459. https://doi.org/10.3390/land12071459

Academic Editors: Luca Congedo, Francesca Assennato and Michele Munafò

Received: 12 June 2023
Revised: 10 July 2023
Accepted: 20 July 2023
Published: 21 July 2023

1. Introduction

China has undergone a remarkable surge in urbanization, with the urbanization rate increasing from 17.92% to 65.22% between 1978 and 2022. However, scholars [1,2] outline that this rapid urbanization has led to various challenges, including resource depletion, environmental degradation, imbalanced spatial expansion of cities, and a severe urban–rural divide. Past studies [3–5] show that to address these issues, there is a need to modernize the urbanization strategy by integrating the concept of ecological civilization into the process. Considering the findings [6–8], this strategy should promote green, circular, low-carbon development, emphasize the efficient and sustainable utilization of land, water, energy, and other resources, and strengthen environmental protection and ecological restoration. This approach emphasizes the efficient and sustainable utilization of resources, environmental protection, and ecological restoration. However, it is important to assess whether the new urbanization strategy effectively contributes to improving the ecological environment and to understand the overall characteristics and spatial patterns of its impact on ecological environmental efficiency.

This research aims to comprehensively understand and analyze the coordinated development of urbanization and ecological environmental efficiency while also assessing the influence of drivers on this relationship. This understanding is crucial for the development of effective policies and strategies that promote environmentally sustainable urban development. This study employs a diverse set of methodologies to accomplish its objectives: the global Malmquist–Luenberger productivity index to evaluate ecological environmental efficiency; the entropy method and TOPSIS model to assess the urbanization index; and panel-corrected standard error (PCSE) and feasible generalized least squares (FGLS) models to measure the influence of drivers on the coordination between urbanization and ecological environmental efficiency. Additionally, this study utilizes nuclear density methods to analyze the evolutionary trend and coordination between urbanization and the ecological environment over time. By adopting these approaches, the research aims to provide comprehensive insights into the interplay of urbanization and ecological environmental efficiency. The findings of this study hold significant implications for policymakers, urban planners, and stakeholders, emphasizing the necessity of a balanced approach that prioritizes ecological environmental protection in urbanization efforts. Ultimately, this research aims to foster environmentally sustainable urban development by integrating these insights and recommendations.

This paper has the following structure: literature review—analysis of the theoretical background on urbanization effect on ecological environmental efficiency; materials and methods—describing variables and sources for analysis and the methods and instruments to check the research hypothesis; results—explaining the results of the analysis; discussion and conclusions—exploring the core findings, outlining the policy implication, limitations, and further directions for investigations.

2. Literature Review

Urban planners and sociologists have expressed concerns about the relationship between ecological environmental efficiency and urbanization. They have introduced concepts such as "pastoral cities" [9], "satellite towns" [10], "eutopia" [11], "organic planning" [12], and "organic evacuation" [13]. Scholars [14–18] acknowledge three coupling states between urbanization and the ecological environment. First, there could be a positive coupling where urbanization promotes improvements in ecological environmental quality through scale effects and technological progress resulting from population and industry agglomeration and distribution [19–22]. Second, a negative coupling could occur where urbanization poses challenges to sustainable development, as it results in heightened resource and energy consumption, elevated greenhouse gas emissions, and degradation of the ecological environment. These factors hinder the long-term sustainability of cities [23–27]. Third, there exists a dynamic coupling between urbanization and the ecological environment. Scholars [28–31] outline that this relationship is not a simple linear one but rather follows patterns such as double exponentials, inverted U-shaped curves (environmental Kuznets curve), or S-shaped curves, illustrating the interaction and mutual influence between urbanization and the ecological environment as they progress from low to high levels.

It should be noted that different disciplines have examined the impact of urbanization on the ecological environment from various perspectives. Environmental science [32–35] focuses on studying pollution, destruction, and protection of groundwater, climate change, air quality, and soil during the urbanization process to understand the ecological and environmental effects. Ecology [20,36,37] measures the impact of urbanization by assessing changes in biodiversity resulting from urban development. Systematics comprehensively analyzes the ecological and environmental effects of urbanization, considering aspects such as resources, environment, system, economy, and society [38–41].

Scholars [42–44] examine the impact of environmental regulation on urbanization from two angles: technology and industry. On the one hand, the effects of environmental regulation on urbanization have been studied in terms of technological innovation, with many researchers suggesting that environmental regulations can effectively drive technological advancements and progress [45–49]. On the other hand, scholars have investigated the impact of environmental regulation on industrial transfers, transformations in industrial structure, upgrades, and agglomeration from an industrial perspective [50–54].

The scientific community [55–58] conducts extensive research on the relationship between urbanization and ecological environmental efficiency, producing valuable insights that serve as a reference for further studies in this field. However, most of the existing research focuses on exploring the coupling relationship between urbanization and the ecological environment and analyzing the ecological and environmental effects of urbanization from individual disciplinary perspectives. This study aims to address the lack of in-depth discussion on whether the urbanization process affects the efficiency of the ecological environment and whether it hampers ecological environment improvement, as well as to investigate the underlying mechanisms involved. Over the past few decades, China has rapidly urbanized and has also prioritized the construction of ecological civilization.

Considering the above, the analysis of the coordination between urbanization and the ecological environment over time requires further exploration, particularly in terms of understanding the specific mechanisms and processes involved. In addition, there is a lack of in-depth discussion on whether the urbanization process affects the efficiency of the ecological environment and hampers ecological environment improvement. In addition, the multidimensional aspects of ecological environmental efficiency in relation to urbanization require comprehensive analysis. Existing studies have often focused on limited aspects, neglecting the multidimensional nature of the issue. There is a need to bridge the knowledge gap by considering the evolutionary trend and coupling between urbanization and the ecological environment.

3. Materials and Methods

3.1. Research Model

To analyze the coupling and coordination level between urbanization and ecological environmental efficiency, this study employs the coupling and coordination degree model. The model is expressed as follows:

(1) The comprehensive development model is utilized to assess the development level of both urbanization (U_1) and the ecological environment (U_2).

(2) The coupled coordination degree model (1) consists of two components: the coupling degree model and the coordination degree model. The coupling degree model (2) is employed to quantify the level of interaction between multiple systems, while the coordination degree model (3) is used to assess the degree of coordinated development between these systems.

$$D = \sqrt{C \times T}, \quad (1)$$

where D is the coupling and coordination degree between new urbanization and the ecological environment. If D is higher, the relationship between the new urbanization and ecological environment is better; T is the comprehension coordination evaluation index:

$$T = \alpha \times U_1 + \beta \times U_2; \quad (2)$$

where α and β are weighting coefficients (0.5).

C is the coupling degree of urbanization and ecological environment, and the value range is [0, 1]:

$$C = \frac{2\sqrt{U_1 U_2}}{U_1 + U_2}. \quad (3)$$

The coupling and coordination degree between urbanization and the ecological environment, as identified in study [59], is classified into seven phases and three intervals (Table 1).

Table 1. Coupled coordination degree stage division.

D	Coupling Coordination Phase	Coupling Coordination Interval
0.900~1.000	Quality coordination	
0.800~0.899	Good coordination	Accept
0.700~0.799	Intermediate coordination	
0.650~0.699	Primary coordination II	
0.600~0.649	Primary coordination I	Forced to accept
0.500~0.599	Forced coordination	
0.000~0.499	Disorder coordination	Not accept

The kernel density estimation method was employed to analyze the distribution of the comprehensive index measuring the efficiency of the coupled coordination degree between urbanization and the ecological environment in China:

$$f(x) = \frac{1}{Nh}\sum_{i=1}^{N} K\left(\frac{X_i - x}{h}\right) \quad (4)$$

where N is the numeral of observations; X_i is the independently and identically distributed annotations; x is the mean of the observations; $K(\cdot)$ is the kernel density function; and h is the bandwidth. The lesser the bandwidth is, the more accurate the estimation is.

$$K(x) = \frac{1}{\sqrt{2\pi}}\exp\left(-\frac{x^2}{2}\right) \quad (5)$$

The panel data model is employed to analyze the driving factors that influence the coupling and coordination between new urbanization and the ecological environment:

$$D_{i,t} = \alpha + \beta \sum Z_{i,t} + \varepsilon_{i,t} \quad (6)$$

where $D_{i,t}$ is the coupled coordination degree between urbanization and the ecological environment in China; $Z_{i,t}$ are the set of driving factors; and $\varepsilon_{i,t}$ are random error items.

To analyze panel data that include both time series and cross-sectional observations, it is crucial to consider the unique characteristics of this data type. Tests, such as the Wooldridge test for autocorrelation in panel data, the Modified Wald test for groupwise heteroskedasticity, and Pesaran's test of cross-sectional independence, were conducted to assess the data's characteristics and determine the appropriate specification and econometric method. Heteroskedasticity refers to situations where the variance in errors varies across different observations, while autocorrelation occurs when errors within a time series are correlated over time. By adjusting the standard errors, the PCSE model effectively accounts for these issues, resulting in more reliable and efficient coefficient estimation. The feasible generalized least squares (FGLS) model proves particularly valuable when correlations exist not only over time but also across different cross-sectional units. By simultaneously incorporating these correlations, the FGLS model provides a more comprehensive understanding of the data structure and enables more robust parameter estimation. This method takes into account the potential dependencies between observations, leading to more accurate and reliable statistical inference.

3.2. Data and Variable Description

Considering past studies [60–65], the urbanization index (U_1) incorporates 17 second-level level indicators from four key aspects: population urbanization, economic urbanization, social urbanization, and ecological environment urbanization (Table 2).

Table 2. Construction of the urbanization index.

Variable	Calculation Method	Unit
	Population urbanization	
Urbanization rate	The urbanization rate of permanent residents	%
Density of population	Urban population density	square kilometer
Employment status	Registered urban unemployment rate	%
Employment structure	The share of employment in secondary and tertiary industries	%
Population education	The average number of students per 100,000 institutions of higher learning	Person
	Economic urbanization	
Economic development level	GDP per capita	Yuan/person
Economic structure	The added value of the tertiary industry accounted for GDP	%
Government receipts	General public budget revenue	100 million
Investment level	Investment in the fixed assets	100 million
Residents' income	Disposable income of urban residents per capita	Yuan/person
	Social urbanization	
Public service	The share of education expenditure in government expenditure	%
	Number of health technicians per thousand people	1000 people
Infrastructure	Public transport vehicles per 10,000 people	vehicle
	Urban road area per capita	square meter
Quality of life	Public library collections per capita	volume
	Telephone penetration	%
	Ecological environment urbanization	
Garbage disposal	The harmless treatment rate of household garbage	%
Ecological foundation	Green coverage rate of the built-up area	%
Sewage treatment	Daily urban sewage treatment capacity	10,000 m^3
Air quality	Total industrial sulfur dioxide emissions	Ten thousand tons

This study applies the entropy-based TOPSIS method to assess the urbanization index [53]. The entropy method calculates the weight based on the variability among indicators, meaning that a higher entropy weight indicates a greater dispersion of data within the index. The advantage of the entropy method is that it objectively determines the weight of the index based on the information reflected by the indicators [54]. The specific operational steps are outlined below:

Step 1: Assessment of the entropy weight of each index.

Normalize the initial index:

$$r_{ij} = \frac{x_{ij}}{\sum_{j=1}^{n} x_{ij}} \tag{7}$$

where x_{ij} are the individual index values, $i = 1, 2, 3, \ldots, m; j = 1, 2, 3, \ldots, n$.

Assessment of the index entropy value:

$$e_i = -\frac{\sum_{j=1}^{n} r_{ij} * Inr_{ij}}{Inn} \tag{8}$$

where e_i is the $i-$th index entropy value.

Assessment of the index weight:

$$\omega_i = \frac{1 - e_i}{\sum_{i=1}^{m} (1 - e_i)} \tag{9}$$

Step 2: Establish the TOPSIS comprehensive evaluation model.

Using the normalized matrix obtained when calculating the entropy weight:

$$Z = \begin{bmatrix} z_{11} & \cdots & z_{1m} \\ \vdots & \ddots & \vdots \\ z_{n1} & \cdots & z_{nm} \end{bmatrix} \quad (10)$$

$$z_i^+ = max\{z_{1i}^+, z_{2i}^+, \ldots\ldots z_{mi}^+\} z_i^- = min\{z_{1i}^-, z_{2i}^-, \ldots\ldots z_{mi}^-\} \quad (11)$$

Calculate the distance between each index z_i^+ and z_i^- on different evaluation objects. The formula is as follows:

$$Dist_i^+ = \sqrt{\sum_{j=1}^{m} \omega_j (z_j^+ - z_{ij})^2} \quad (12)$$

$$Dist_i^- = \sqrt{\sum_{j=1}^{m} \omega_j (z_j^- - z_{ij})^2} \quad (13)$$

where ω_j is the entropy weight of the jth index

Step 3: Calculate the score of each indicator.

$$s_i = \frac{Dist_i^-}{Dist_i^- + Dist_i^+} \quad (14)$$

The value range of s_i is 0 to 1, and the closer the score is to 1, the higher the urbanization development level in the region; otherwise, the development level is low.

This study utilized the global Malmquist–Luenberger productivity index to assess ecological efficiency in China. This index combines the concepts of the Malmquist and Luenberger productivity indexes to provide a comprehensive assessment of productivity changes. The index captures both efficiency change, which refers to the ability to use resources effectively, and technological change, which represents changes in the production frontier or best practices. By considering both factors, it offers a more holistic measure of productivity changes. Additionally, the Malmquist–Luenberger productivity index is capable of handling multiple inputs and outputs simultaneously, enabling a comprehensive analysis of productivity changes across various dimensions. This makes it well suited for evaluating the performance of complex systems, such as China, with diverse production processes. This index measures the efficiency of the ecological environment while considering undesired outputs:

$$U_{2t,t+1} = \frac{1 + R_G(x_{it}, y_{it}, z_{it}; g_x, g_y, g_z)}{1 + R_G\left(x_{i(t+1)}, y_{i(t+1)}, z_{i(t+1)}; g_x, g_y, g_z\right)} \quad (15)$$

where U_2 is the total efficiency index of the ecological environment; x, y, and z are the input, expected output, and unexpected output, respectively; t is the period; and i is the region.

Based on previous studies [66–68], in model (15), x_{it} (input variables) are the following: coal consumption, total water supply, built-up area, urban employment, and fixed asset investment. The output variable (y_{it}) is the index comprising GDP, while z_{it} (unexpected output) is carbon dioxide discharge, sulfur dioxide discharge, and industrial wastewater discharge. The Malmquist–Luenberger productivity index allows the incorporation of all selected variables simultaneously. Including environmental variables as unexpected output into model (15) enables a more environmentally conscious assessment of productivity changes. This helps to identify areas where improvements can be made in terms of resource usage and environmental impact.

The industrial structure, energy structure, environmental regulation, and digitalization level are identified as crucial driving factors that impact the coupling and coordination between new urbanization and the ecological environment. The industrial structure (Instr—

the share of the secondary industry in regional GDP, %) refers to the composition and characteristics of industries within urban areas, with a focus on their resource consumption and environmental impact [69]. The energy structure (Ens—the proportion of coal consumption in total energy consumption) pertains to the sources of energy used in urban areas and their efficiency, emphasizing the transition to cleaner and more sustainable energy sources [70]. Trade openness (Open—the share of foreign direct investment above a designated size in regional GDP) facilitates the exchange of goods, services, and knowledge between regions, which can lead to the transfer of environmentally friendly technologies and practices. Regions can access cleaner and more sustainable production methods, reducing their environmental impact and promoting ecological efficiency. Environmental regulation (Er) involves the policies and regulations in place to mitigate the negative environmental impacts of urbanization and promote sustainable practices. This study constructs a comprehensive index system of environmental regulation (Er) based on the entropy-based TOPSIS method (Formulas (7)–(14)) to reflect the intensity of environmental regulation more accurately in each province. Based on previous studies [50–54], common indicators were chosen to describe environmental regulation: industrial wastewater discharge volume (10 thousand tons) [71], industrial sulfur dioxide emissions (10 thousand tons) [71], industrial smoke (powder) dust emissions (10 thousand tons) [72], and the amount of pollutant discharge fee (10 thousand tons) [71]. Last, the digitalization level (Dig—internet penetration rate) signifies the integration of digital technologies in urban systems, enabling data-driven decision making and smart solutions for efficient resource management and environmental monitoring. By considering these driving factors, policymakers and urban planners can develop strategies and interventions that foster coordinated and sustainable development between new urbanization and the ecological environment.

This paper utilizes panel data from 30 Chinese provinces spanning the period from 2010 to 2020 (excluding Xizang, Hong Kong, Macao, and Taiwan). The data for each index variable are collected from various sources, including the China Statistical Yearbook [71], China Environmental Statistical Yearbook [72], and the statistical yearbooks of each province [73].

4. Results

As shown in Table 3, the environmental ecological efficiency exhibits a range of 3.885 among the 300 observed values, with the highest efficiency reaching 0.033. On the other hand, the new urbanization index ranges from 0.157 to 0.639, indicating a relatively small difference and implying the rapid development of China's urbanization. Moreover, noticeable differences can be observed in the economic development level, industrial structure, trade openness, environmental regulation, and energy structure.

Table 3. Descriptive statistics.

Variable	Symbol	Mean	Std.	Min	Max
Ecological environmental efficiency	GML	1.048	0.270	0.033	3.885
Urbanization index	Nurb	0.325	0.101	0.157	0.639
Industrial structure	Instr	1.219	0.696	0.518	5.297
Trade openness	Open	0.274	0.290	0.008	1.464
Environmental regulation	Er	0.004	0.004	0.000	0.031
Energy structure	Ens	0.033	0.023	0.004	0.095
Digital infrastructure	Dig	6.534	0.916	3.728	8.266

The calculation of the coupling and coordination level between urbanization and ecological environmental efficiency is presented in Table 4. The variation range of the coupling coordination is 0.754, with the highest observation efficiency reaching 1.099 among the 300 observations.

Table 4. Results of the coupling and coordination degree of urbanization and ecological environment efficiency.

Province	2011	2012	2013	2014	2015	2016	2017	2018	2019	2020
Beijing	0.76	0.84	0.86	0.85	0.87	1.00	0.89	0.88	0.92	0.91
Tianjin	0.76	0.76	0.73	0.75	0.61	1.00	0.74	0.83	0.75	0.87
Hebei	0.69	0.71	0.71	0.71	0.73	0.75	0.77	0.79	0.78	0.80
Shanghai	0.82	0.81	0.85	0.83	0.85	0.87	1.00	0.88	0.95	0.91
Jiangsu	0.77	0.80	0.82	0.80	0.84	0.86	0.88	0.88	0.89	0.92
Zhejiang	0.77	0.78	0.79	0.80	0.82	0.84	0.85	0.86	0.87	0.87
Fujian	0.70	0.73	0.75	0.75	0.78	0.78	0.80	0.81	0.83	0.83
Shandong	0.73	0.75	0.78	0.78	0.79	0.83	0.83	0.84	0.83	0.85
Guangdong	0.77	0.79	0.81	0.80	0.83	0.85	0.87	0.88	0.90	0.90
Hainan	0.50	0.69	0.69	0.71	0.66	0.94	0.78	0.72	0.75	0.74
Shanxi	0.65	0.69	0.68	0.69	0.70	0.72	0.74	0.78	0.74	0.76
Anhui	0.68	0.71	0.71	0.72	0.74	0.75	0.77	0.79	0.82	0.81
Jiangxi	0.69	0.70	0.71	0.71	0.72	0.74	0.76	0.77	0.33	0.80
Henan	0.70	0.71	0.72	0.73	0.75	0.78	0.80	0.80	0.83	0.82
Hubei	0.69	0.71	0.73	0.74	0.76	0.79	0.79	0.81	0.82	0.78
Hunan	0.68	0.69	0.71	0.72	0.73	0.75	0.77	0.78	0.81	0.82
Guangxi	0.67	0.69	0.69	0.70	0.73	0.74	0.76	0.77	0.77	0.78
Nei Monggol	0.67	0.68	0.69	0.70	0.70	0.74	0.71	0.75	0.74	0.75
Chongqing	0.67	0.70	0.72	0.70	0.74	0.75	0.75	0.77	0.80	0.77
Sichuan	0.69	0.71	0.71	0.72	0.73	0.76	0.78	0.80	0.81	0.81
Guizhou	0.61	0.64	0.66	0.67	0.68	0.72	0.71	0.73	0.76	0.75
Yunnan	0.65	0.68	0.67	0.68	0.69	0.72	0.74	0.74	0.80	0.77
Xizang	0.73	0.74	0.74	0.74	0.73	0.76	0.78	0.79	0.79	0.80
Shaanxi	0.62	0.66	0.66	0.68	0.69	0.71	0.72	0.76	0.74	0.75
Gansu	0.63	0.56	0.80	0.58	0.78	0.68	0.70	0.72	0.73	0.74
Qinghai	0.59	0.65	0.63	0.66	0.69	0.71	0.74	0.76	0.77	0.82
Ningxia	0.68	0.66	0.69	0.70	0.70	0.72	0.76	0.74	0.77	0.75
Xinjiang	0.67	0.69	0.69	0.70	0.73	0.74	0.76	0.77	0.77	0.78
Liaoning	0.52	0.53	0.50	0.50	0.67	0.65	0.60	0.59	0.58	0.70
Jilin	0.37	0.42	0.41	0.44	0.46	0.56	0.51	0.45	0.50	0.63
Heilongjiang	0.44	0.42	0.42	0.42	0.53	0.57	0.51	0.48	0.48	0.64

The results presented in Table 4 indicate that from 2011 to 2020, the average level of coupling and coordination between China's urbanization development and the ecological environment has remained consistently high. However, notable differences are observed among the eastern, western, and northeast regions, indicating varying degrees of coupling and coordination in these areas (Figure 1).

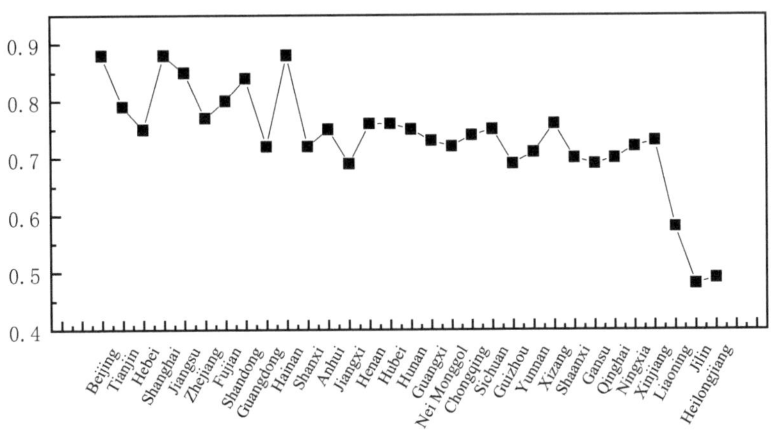

Figure 1. The average value of the coupling and coordination degree of urbanization and ecological environment efficiency in 30 provinces in China.

According to Figure 1, the eastern region demonstrates a high level of coupling and coordination between urbanization and ecological environment efficiency. In contrast, the central region maintains an intermediate level, while the northeast region shows a primary level.

The results obtained through the kernel density estimation method (Figure 2a) demonstrate an increasing trend in the coupling and coordination degree between urbanization and ecological environment efficiency over time. The overall pattern reveals a prominent peak, indicating that China is actively prioritizing ecological and environmental protection in its urbanization efforts, aligning with national policies. However, upon closer examination of the four regions individually (Figure 2b–e), it becomes evident that there are substantial variations in the coordinated development of urbanization and the ecological environment.

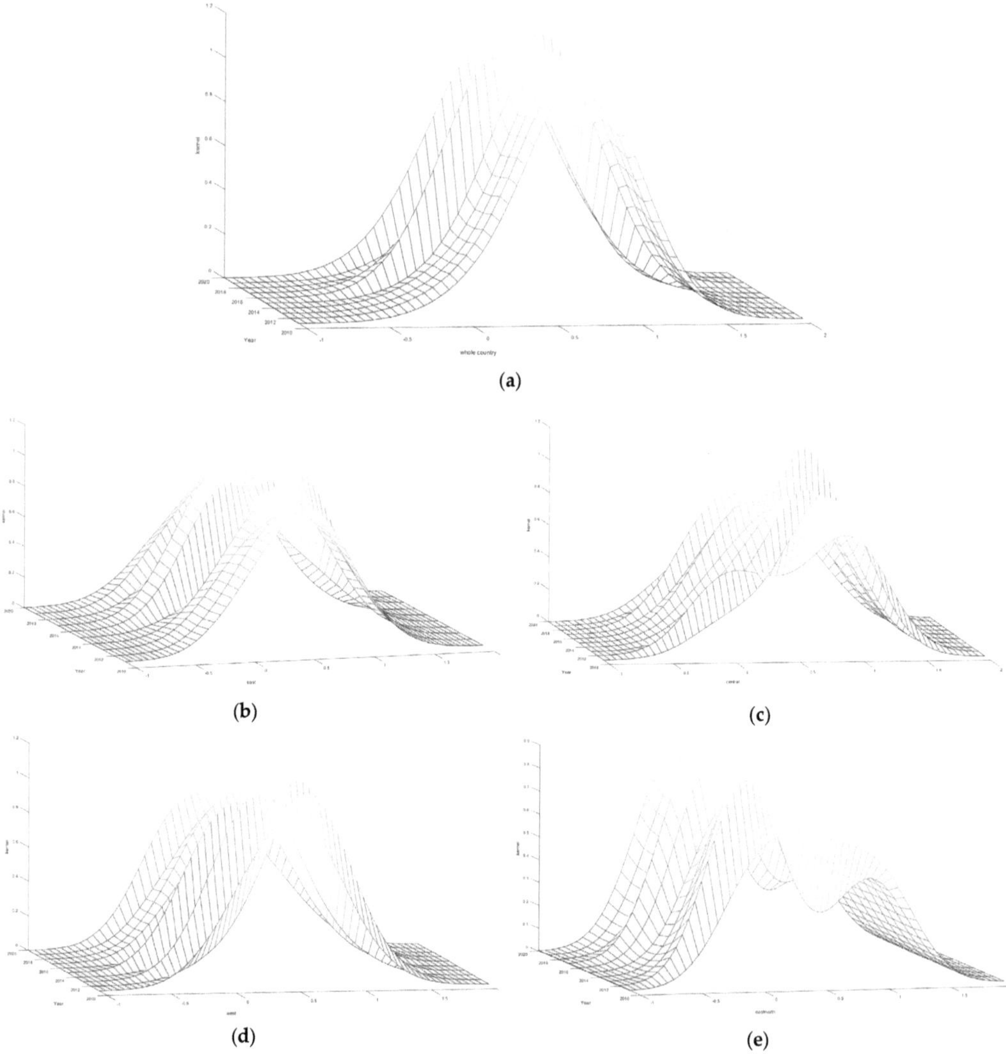

Figure 2. National and four major regional evolution trends. (**a**) Whole country; (**b**) East; (**c**) Central; (**d**) West; (**e**) East–north.

In the eastern region, the strong coupling between urbanization and ecological environment efficiency indicates that urban development in this area has been carefully planned with ecological considerations in mind. Strategies such as green infrastructure, sustainable urban planning, and the adoption of eco-friendly technologies have likely been implemented. As urbanization progresses, there is a simultaneous emphasis on protecting the environment, resulting in a positive influence on ecological environmental protection. The central region, although not as well integrated as the eastern region, still exhibits some degree of coordination between urban development and ecological protection. Efforts to protect the environment exist but may not be as comprehensive or effectively implemented due to competing development priorities. Despite this, the positive relationship indicates progress in balancing urbanization and environmental protection. In contrast, the primary level of coupling between urbanization and ecological environment efficiency in the northeast region suggests minimal coordination between urban development and environmental protection efforts. Urbanization in this area appears to have been carried out without sufficient consideration for ecological concerns, potentially leading to negative environmental impacts. This lack of coordination may stem from factors such as rapid urbanization, inadequate environmental regulations, or a focus on short-term economic gains without long-term sustainability considerations. Overall, the findings highlight the importance of well-planned urbanization that incorporates ecological environment protection measures. When urban development considers ecological factors, it positively influences environmental protection. Conversely, unplanned or poorly coordinated urbanization can have adverse effects on the ecological environment. Therefore, policymakers and urban planners in regions with lower levels of coupling should prioritize integrating ecological considerations into urban development, aiming for a sustainable and balanced approach that benefits both urbanization and environmental protection.

The results of Pesaran's test indicate a test statistic of 5.332 and a probability value of 0.000 (Table 5), suggesting strong evidence of cross-sectional dependence in the panel data. The Modified Wald test reveals a test statistic of 47,800.90, with a probability value of 0.0000, indicating the presence of significant groupwise heteroskedasticity in the panel data. Furthermore, the Wooldridge test demonstrates a test statistic of 9.258 and a probability value of 0.0049, indicating the presence of autocorrelation in the panel data, although to a lesser extent compared to the other two tests.

Table 5. Panel data tests: Wooldridge test for autocorrelation, Modified Wald test for groupwise heteroskedasticity, and Pesaran's test of cross-sectional independence.

Test	Statistic	Probability
Pesaran's test of cross-sectional independence	5.332	0.000
Modified Wald test for groupwise heteroskedasticity	47,800.90	0.0000
Wooldridge test for autocorrelation	9.258	0.0049

Table 6 provides the findings from the PCSE (panel-corrected standard error) and FGLS (feasible generalized least squares) models, which examine the influence of various drivers on the coordination between urbanization and ecological environmental efficiency.

The results suggest that certain factors significantly impact the coupling and coordination between new urbanization and the ecological environment. Specifically, the variables of trade openness (Open), energy structure (Ens), and digitalization level (Dig) demonstrate statistically significant effects. Trade openness, as indicated by the coefficients of 0.125 (PCSE) and 0.108 (FGLS), shows a positive relationship with the coupling and coordination between urbanization and ecological environmental efficiency. This implies that greater openness to trade contributes to enhanced coordination between urbanization and the ecological environment. Similarly, the energy structure variable (Ens) displays a negative relationship with a statistically significant impact. The coefficients of −0.509 (PCSE) and −0.595 (FGLS) suggest that a lower proportion of coal consumption in total energy consumption positively affects the coordination between urbanization and ecological environmental efficiency. Additionally, the digitalization level (Dig) exhibits a positive and statistically significant relationship with a coefficient of 0.049 in both the PCSE and FGLS models. This indicates that a higher internet penetration rate contributes to the coordination and coupling between urbanization and the ecological environment. However, the variables of industrial structure (Instr) and environmental regulation (Er) do not show statistically significant effects on the coordination between urbanization and ecological environmental efficiency.

Table 6. The results of the PCSE and FGLS models.

Variable	PCSE		FGLS	
	Coefficient	Probability	Coefficient	Probability
Instr	−0.007	0.313	0.002	0.623
Open	0.125	0.000	0.108	0.000
Er	1.349	0.394	0.435	0.554
Ens	−0.509	0.029	−0.595	0.000
Dig	0.049	0.000	0.049	0.000
const	0.422	0.000	0.409	0.000
Wald chi2	413.96	0.000	876.69	0.000
R-squared	0.479		–	

5. Discussion

The findings of this study regarding the coupling and coordination level between urbanization and ecological environmental efficiency align with previous investigations in the field. The average level of coupling and coordination between China's urbanization development and the ecological environment remained consistently high from 2011 to 2020, which is in line with the conclusions reached by previous studies [74–76].

Notable differences among the eastern, western, and northeast regions, indicating varying degrees of coupling and coordination, are also consistent with the research conducted by [74–76]. The comparative analysis with previous studies reinforces the importance of well-planned urbanization that incorporates ecological environment protection measures. The positive relationship between coupling and ecological environment efficiency highlights the need for policymakers and urban planners to integrate ecological considerations into urban development, as emphasized in [49,50].

The findings outline the importance of considering ecological environmental protection and efficiency as essential components of China's long-term urbanization strategy. Such conclusions were also attained by scholars [3,5]. Considering previous studies [3,5], the realization of new urbanization is crucial for the country's development, but it should be accompanied by strengthened measures to protect and enhance the ecological environment. This implies that policymakers and urban planners should prioritize sustainable practices and incorporate ecological considerations into urban development plans.

The empirical results show that driving factors (trade openness, energy structure, digitalization level) positively affect the coupling coordination degree between urbanization and ecological environmental efficiency. The obtained results are consistent with previous studies [77–79]. Thus, trade openness is an important aspect to consider, as it reflects the degree of integration of a region's economy with the global market. Increased trade openness led to a higher volume of trade activities, which has implications for environmental sustainability. However, considering the studies [70,77], a higher level of trade openness results in greater environmental pressures if environmental regulations and monitoring systems are not effectively implemented. Scholars [20,51,80] outline that urbanization is often accompanied by an increased demand for energy, particularly in rapidly developing regions. Higher electricity consumption indicates greater energy requirements for urban infrastructure, industries, and households. Managing electricity consumption and transitioning to cleaner and more sustainable energy sources are critical aspects of promoting ecological environmental efficiency in the context of urbanization. A higher level of internet penetration promotes coordination between urbanization and ecological environmental efficiency by facilitating the exchange of information, enhancing collaboration among stakeholders, and fostering citizen engagement. By leveraging the power of the internet, policymakers and urban planners can effectively integrate ecological considerations into urban development, leading to a more sustainable and balanced approach that benefits both urbanization and environmental protection [81,82].

The results of this study contribute to the existing knowledge on the coordinated development of urbanization and ecological environmental efficiency. The insights gained from this study, in conjunction with the findings of past research, provide a robust foundation for policymakers and practitioners involved in urban planning, as well as environmental protection and conservation efforts [3,5]. By building upon the knowledge of coupling and coordination between urbanization and ecological environment efficiency, stakeholders could make informed decisions and develop strategies that promote sustainable urbanization and effectively preserve the environment. These findings have implications for policymakers, urban planners, and environmentalists seeking to promote sustainable urban development and preserve the ecological balance.

6. Conclusions

This study utilized panel data from 30 provinces in China spanning the years 2011 to 2020. This research employed the entropy right TOPSIS method, Malmquist–Luenberger productivity index, PCSE, FGLS, and kernel density to measure the coordinated development of urbanization and ecological environmental efficiency and to analyze the core drivers of their link. This study demonstrates that the average level of coupling and coordination between China's urbanization development and the ecological environment has remained consistently high from 2011 to 2020. However, notable differences are observed among the eastern, western, and northeast regions, indicating varying degrees of coupling and coordination in these areas. These findings highlight the importance of understanding and addressing regional disparities when considering the relationship between urbanization and ecological environmental efficiency. Future efforts should focus on promoting sustainable and balanced urban development strategies that prioritize ecological considerations, particularly in regions with lower levels of coupling. By integrating ecological factors into urban planning and implementation, policymakers and urban planners can strive for a harmonious coexistence between urbanization and environmental protection.

Several variables were also identified as significant factors influencing the coordinated development of urbanization and ecological environmental efficiency. The realization of new urbanization is a crucial long-term strategy for China, necessitating strengthened protection and efficiency of the ecological environment. To comprehensively promote new urbanization and enhance ecological environmental efficiency, the following recommendations are proposed:

1. Given the observed variations in coupling and coordination levels between urbanization and the ecological environment across different regions, policymakers should focus on regional planning and coordination strategies. This involves tailoring policies and approaches to the specific needs and challenges of each region, considering their unique characteristics and development priorities.
2. Urban planning should prioritize ecological and environmental protection, incorporating designs that emphasize small fragments, low density, and organic arrangement. The creation of green spaces and ecological parks should be emphasized to improve urban air and water quality. Energy conservation and environmental protection should be regarded as essential elements of urban planning, with comprehensive promotion of energy-saving and emission reduction technologies [80,83], facilitating the coordinated development of new urbanization and ecological/environmental protection. Thus, Curitiba (the capital city of the state of Paraná in Brazil) is often hailed as a model for sustainable urban planning. The city has implemented innovative

strategies to address urbanization and environmental challenges [84,85]. For instance, it has prioritized the development of an efficient public transportation system, including a well-integrated bus rapid transit (BRT) network. This emphasis on public transport has reduced congestion and air pollution, leading to improved ecological environmental efficiency in the city. Contrasting the Chinese model, Curitiba's approach highlights the importance of sustainable transportation solutions in achieving ecological balance in urban areas.

3. Urban green spaces are vital components of the urban ecological environment. It is recommended that new urbanization efforts prioritize strengthening urban greening coverage by increasing the area of urban green spaces, establishing new green spaces and ecological parks, developing water systems and wetlands, improving the urban ecological environment, and enhancing the overall image and appeal of cities. Portland (the United States of America) is renowned for its sustainable urban development policies [86,87]. The city has implemented land-use planning that encourages mixed-use neighborhoods, preserves green spaces, and promotes public transportation and cycling infrastructure. Portland's emphasis on compact urban growth and preservation of natural areas has contributed to improved ecological environmental efficiency.
4. Given the negative impact of the energy structure variable on the coordination between urbanization and ecological environmental efficiency, policymakers should prioritize energy transition and diversification efforts. This involves reducing the reliance on coal consumption and promoting the use of cleaner and renewable energy sources [88–90]. Implementing policies that incentivize the adoption of sustainable energy practices and technologies will contribute to a more coordinated and environmentally friendly urbanization process.
5. The positive correlation between digitalization level and the coordination of urbanization with the ecological environment emphasizes the significance of embracing digital technologies and fostering innovation. Policymakers should create an enabling environment for the digital transformation of urban areas, including enhancing internet infrastructure and promoting digital solutions for environmental monitoring, resource management, and sustainable urban development [91]. This includes encouraging the use of digital technologies in urban infrastructure planning, construction, and management to enhance efficiency and reduce environmental impacts. Additionally, implementing comprehensive data collection and analysis systems can effectively monitor urban environmental parameters and inform decision-making processes for urban planning, resource allocation, and environmental protection. Furthermore, ensuring widespread access to digital connectivity and utilizing digital platforms for public engagement, information sharing, and environmental education will raise awareness and empower individuals to actively contribute to the advancement of sustainable urbanization.

This research holds significance in the international context by offering insights, recommendations, and empirical evidence on the coordinated development of urbanization and ecological environmental efficiency and the core drivers of their link. Its findings have the potential to inform policy decisions, urban planning practices, and conservation efforts globally, ultimately contributing to a more sustainable and environmentally conscious approach to urban development.

Despite the valuable results, this study has a few limitations. The findings of this study are specific to the context of China and may not be directly applicable to other countries or regions. Different socioeconomic and environmental factors could influence the coordination between urbanization and ecological environmental efficiency in different contexts. This study primarily focuses on examining the relation between new urbanization and ecological environment efficiency. However, establishing a causal relationship requires further investigation using experimental or quasi-experimental research designs. Additionally, the possibility of reverse causality, where improvements in ecological environment efficiency may also influence new urbanization, should be considered.

Author Contributions: Conceptualization, R.C., Y.C., O.L. and T.P.; methodology R.C., Y.C., O.L. and T.P.; validation, R.C., Y.C., O.L. and T.P.; formal analysis, R.C., Y.C., O.L. and T.P.; investigation, R.C., Y.C., O.L. and T.P.; resources, R.C., Y.C., O.L. and T.P.; data curation, R.C., Y.C., O.L. and T.P.; writing—original draft preparation, R.C., Y.C., O.L. and T.P.; writing—review and editing, R.C., Y.C., O.L. and T.P.; visualization, R.C., Y.C., O.L. and T.P. All authors have read and agreed to the published version of the manuscript.

Funding: This research received no external funding.

Data Availability Statement: Not applicable.

Conflicts of Interest: The authors declare no conflict of interest.

References

1. Lu, A.; Chen, M. Some understandings on the compilation background of the "National New Urbanization Plan (2014–2020)". *J. Geogr.* **2015**, *2*, 179–185.
2. Yu, Y. Performance Analysis of Public Investment in Chinese University Education Based on Regional Differences and Influencing Factors. *Bus. Ethics Leadersh.* **2023**, *7*, 37–49. [CrossRef]
3. Zhang, Y. China's green urbanization in the perspective of ecological civilization. *Chin. J. Urban Environ. Stud.* **2021**, *9*, 2150001. [CrossRef]
4. Dzwigol, H.; Trushkina, N.; Kwilinski, A. The organizational and economic mechanism of implementing the concept of green logistics. *Virtual Econ.* **2021**, *4*, 41–75. [CrossRef]
5. Lv, T.; Hu, H.; Xie, H.; Zhang, X.; Wang, L.; Shen, X. An empirical relationship between urbanization and carbon emissions in an ecological civilization demonstration area of China based on the STIRPAT model. *Environ. Dev. Sustain.* **2023**, *25*, 2465–2486. [CrossRef]
6. Chygryn, O.; Shevchenko, K. Energy industry development: Key trends and the core determinants. *SocioEconomic Chall.* **2023**, *7*, 115–128. [CrossRef]
7. Lahouirich, M.W.; El Amri, A.; Oulfarsi, S.; Sahib Eddine, A.; El Bayed Sakalli, H.; Boutti, R. From financial performance to sustainable development: A great evolution and an endless debate. *Financ. Mark. Inst. Risks* **2022**, *6*, 68–79. [CrossRef]
8. Kolomiiets, S.; Jakubowska, A.; Goreva, E. Management of Urban Amenities in the Context of Public Health Care for Local Communities. *Health Econ. Manag. Rev.* **2021**, *2*, 96–102. [CrossRef]
9. Howard, E. *Garden Cities of Tomorrow*; Faber: London, UK, 1946; pp. 9–28.
10. Unwin, T. *Rural Urban Interaction in Developing Countries: A Theoretical Perspective*; Potter, R.B., Ed.; Routledge: Abingdon, UK, 1989; pp. 11–32.
11. Geddes, P. *Cities in Evolution: An Introduction to the Town Planning Movement and to the Study of Civics*; Williams: London, UK, 1915.
12. Turley, A.C. *Urban Culture: Exploring Cities and Cultures*; Routledge: Oxfordshire, UK, 2015.
13. Illir, S. *City: Its Development, Decline, and Future*; Construction Industry in China Press: Beijing, China, 1986.
14. Fang, C.; Liu, H.; Wang, S. The coupling curve between urbanization and the eco-environment: China's urban agglomeration as a case study. *Ecol. Indic.* **2021**, *130*, 108107. [CrossRef]
15. Šulyová, D.; Kubina, M. Quality of life in the concept of strategic management for smart cities. *Forum Sci. Oeconomia* **2022**, *10*, 9–24. [CrossRef]
16. Meyer, D.F.; Neethling, J.R. An assessment of the financial health of the south african metropolitan municipal regions. *Forum Sci. Oeconomia* **2021**, *9*, 59–77. [CrossRef]
17. Liu, N.; Liu, C.; Xia, Y.; Da, B. Examining the coordination between urbanization and eco-environment using coupling and spatial analyses: A case study in China. *Ecol. Indic.* **2018**, *93*, 1163–1175. [CrossRef]
18. Zou, C.; Zhu, J.; Lou, K.; Yang, L. Coupling coordination and spatiotemporal heterogeneity between urbanization and ecological environment in Shaanxi Province, China. *Ecol. Indic.* **2022**, *141*, 109152. [CrossRef]
19. Yan, X.; Cheng, C.; Yi, G.; Bai, J. The economic threshold effect of urbanization in the Yangtze River Economic Belt on energy consumption. *Econ. Geogr.* **2019**, *1*, 73–81.
20. Kuzior, A.; Sira, M.; Brozek, P. Using blockchain and artificial intelligence in energy management as a tool to achieve energy efficiency. *Virtual Econ.* **2022**, *5*, 69–90. [CrossRef]
21. Wang, W.; Mao, W. Measurement and analysis of urbanization and ecological environment response in Wuling Mountain area. Take Xiangxi Autonomous Prefecture as an example. *Econ. Geogr.* **2016**, *6*, 148–154.
22. Yu, B.; Chen, L. Can the new type of urbanization reduce overcapacity? *Quant. Econ. Tech. Econ. Res.* **2019**, *1*, 22–41.
23. Bacha, E.L.; Bacha, E.L.; Klein, H.S.; Klein, J. (Eds.) *Social Change in Brazil, 1945–1985: The Incomplete Transition*; University of New Mexico Press: Albuquerque, NM, USA, 1989.

24. Ramadania, R.; Ratnawati, R.; Juniwati, J.; Afifah, N.; Heriyadi, H.; Darma, D.C. Impulse buying and hedonic behavior: A mediation effect of positive emotions. *Virtual Econ.* **2022**, *5*, 43–64. [CrossRef]
25. Wyrwa, J.; Zaraś, M.; Wolak, K. Smart solutions in cities during the COVID-19 pandemic. *Virtual Econ.* **2021**, *4*, 88–103. [CrossRef]
26. Poumanyvong, P.; Kaneko, S. Does urbanization lead to less energy use and lower CO_2 emissions? A cross-country analysis. *Ecol. Econ.* **2010**, *70*, 434–444. [CrossRef]
27. Sadorsky, P. The effect of urbanization on CO_2, emissions in emerging economies. *Energy Econ.* **2014**, *1*, 147–153. [CrossRef]
28. Duan, W. Analysis of the interactive coupling mechanism and regularity between urbanization and ecological environment. *Environ. Dev.* **2017**, *10*, 185.
29. Sun, H.; Huang, Z.; Xu, D.; Shi, X.; Liu, H.; Tan, L.; Ge, J. Spatial characteristics and driving mechanism of the coupling between urbanization and ecological environment in the Pan-Yangtze River Delta urban agglomeration. *Econ. Geogr.* **2017**, *2*, 163–170.
30. Beckerman, W. Economic growth and the environment: Whose growth? Whose environment? *World Dev.* **1992**, *20*, 481–496. [CrossRef]
31. Zhang, J.; Jiao, W.; Han, B. The coordinated evolution characteristics and spatial coupling relationship between urbanization and ecosystem service. *J. Ecol.* **2020**, *10*, 3271–3282.
32. Gandy, M. Rethinking urban metabolism: Water, space and the modern city. *City* **2004**, *8*, 363–379. [CrossRef]
33. Peterson, T.C.; Gallo, K.P.; Lawrimore, J.; Owen, T.W.; Huang, A.; McKittrick, D.A. Global rural temperature trends. *Geophys. Res. Lett.* **1999**, *26*, 329–332. [CrossRef]
34. Rinehart, L.R.; Fujita, E.M.; Chow, J.C.; Magliano, K.; Zielinska, B. Spatial distribution of PM2.5 associated organic compounds in central California. *Atmos. Environ.* **2006**, *40*, 290–303. [CrossRef]
35. Cui, L.; Shi, J. Urbanization and its environmental effects in Shanghai, China. *Urban Clim.* **2012**, *2*, 1–15. [CrossRef]
36. Cam, E.; Nichols, J.D.; Sauer, J.R.; Hines, J.E.; Flather, C.H. Relative species richness and community completeness: Birds and urbanization in the Mid-Atlantic States. *Ecol. Appl.* **2000**, *10*, 1196–1210. [CrossRef]
37. Zanette, L.R.S.; Martins, R.P.; Ribeiro, S.P. Effects of urbanization on Neotropical wasp and bee assemblages in a Brazilian metropolis. *Landsc. Urban Plan.* **2005**, *71*, 105–121. [CrossRef]
38. Grimm, N.B.; Faeth, S.H.; Golubiewski, N.E.; Redman, C.L.; Wu, J.; Bai, X.; Briggs, J.M. Global change and the ecology of cities. *Science* **2008**, *319*, 756–760. [CrossRef]
39. Skvarciany, V.; Vidžiūnaitė, S. Decent work and economic growth: The case study of the BRICS countries. *Forum Sci. Oeconomia* **2022**, *10*, 73–89. [CrossRef]
40. Ujwary-Gil, A.; Re, M.L.; Parente, F. The actor-network model of economic networks in a geo-economic context: The conceptual considerations. *Forum Sci. Oeconomia* **2022**, *10*, 9–28. [CrossRef]
41. Alberti, M. The effects of urban patterns on ecosystem function. *Int. Reg. Sci. Rev.* **2005**, *28*, 168–192. [CrossRef]
42. Tang, M.; Li, Z.; Hu, F.; Wu, B. How does land urbanization promote urban eco-efficiency? The mediating effect of industrial structure advancement. *J. Clean. Prod.* **2020**, *272*, 122798. [CrossRef]
43. Wang, Y.; Li, X.; Kang, Y.; Chen, W.; Zhao, M.; Li, W. Analyzing the impact of urbanization quality on CO_2 emissions: What can geographically weighted regression tell us? *Renew. Sustain. Energy Rev.* **2019**, *104*, 127–136. [CrossRef]
44. Liu, K.; Qiao, Y.; Zhou, Q. Analysis of China's industrial green development efficiency and driving factors: Research based on MGWR. *Int. J. Environ. Res. Public Health* **2021**, *18*, 3960. [CrossRef]
45. Brunnermeier, S.B.; Cohen, M.A. Determinants of environmental innovation in US manufacturing industries. *J. Environ. Econ. Manag.* **2003**, *45*, 278–293. [CrossRef]
46. Kharazishvili, Y.; Kwilinski, A.; Dzwigol, H.; Liashenko, V. Strategic European integration scenarios of Ukrainian and polish research, education and innovation spaces. *Virtual Econ.* **2021**, *4*, 7–40. [CrossRef] [PubMed]
47. Sadigov, R. Impact of Digitalization on Entrepreneurship Development in the Context of Business Innovation Management. *Mark. Manag. Innov.* **2022**, *1*, 167–175. [CrossRef]
48. Trzeciak, M.; Kopec, T.P.; Kwilinski, A. Constructs of project programme management supporting open innovation at the strategic level of the organization. *J. Open Innov. Technol. Mark. Complex.* **2022**, *8*, 58. [CrossRef]
49. Li, Q.; Nie, R. Environmental regulation and regional technology innovation: Empirical analysis based on interprovincial panel data in China. *J. Zhongnan Univ. Econ. Law* **2009**, *4*, 18–23.
50. Fu, J.; Li, L. Empirical study on environmental regulation, factor endowment and industrial international competitiveness—Based on the panel data of China's manufacturing industry. *Manag. World* **2010**, *10*, 87–98.
51. Ziabina, Y.; Navickas, V. Innovations in Energy Efficiency Management: Role of Public Governance. *Mark. Manag. Innov.* **2022**, *4*, 218–227. [CrossRef]
52. Fan, Q.; Chu, C.; Gao, J. The impact of environmental regulation and industrial structure upgrading on high-quality economic development. *China's Popul. Resour. Environ.* **2020**, *6*, 84–94.
53. Chen, M.; Xu, Y.; Zhou, Y. Statistical evaluation and regional comparison of high-quality economic development under the new development concept. Based on the improved TOPSIS comprehensive evaluation model. *Stat. J.* **2020**, *2*, 1–14.
54. Yuan, C. Evaluation and analysis of high quality economic development of cities in Jiangsu province based on entropy weight method. *Bus. Econ.* **2022**, *4*, 19–20.
55. Olawumi, T.O.; Chan, D.W. A scientometric review of global research on sustainability and sustainable development. *J. Clean. Prod.* **2018**, *183*, 231–250. [CrossRef]

56. Swyngedouw, E.; Heynen, N.C. Urban political ecology, justice and the politics of scale. *Antipode* **2003**, *35*, 898–918. [CrossRef]
57. Bai, X.; Chen, J.; Shi, P. Landscape urbanization and economic growth in China: Positive feedbacks and sustainability dilemmas. *Environ. Sci. Technol.* **2012**, *46*, 132–139. [CrossRef]
58. Wang, S.K. Types and Selection of Weight Matrix in Spatial Econometric Model. *Econ. Math.* **2013**, *30*, 57–63.
59. Naikoo, M.W.; Shahfahad Talukdar, S.; Ishtiaq, M.; Rahman, A. Modeling built-up land expansion probability using the integrated fuzzy logic and coupling coordination degree model. *J. Environ. Manag.* **2023**, *325 Pt A*, 116441. [CrossRef]
60. Chu, Y.W. China's new urbanization plan: Progress and structural constraints. *Cities* **2020**, *103*, 102736. [CrossRef]
61. Tovmasyan, G.; Gevorgyan, M. The History, Culture and Architecture as a Potential of Urban Tourism Development: Evidence from Armenia. *SocioEconomic Chall.* **2022**, *6*, 42–49. [CrossRef]
62. Acheampong, S.; Pimonenko, T.; Lyulyov, O. Sustainable Marketing Performance of Banks in the Digital Economy: The Role of Customer Relationship Management. *Virtual Econ.* **2023**, *6*, 19–37. [CrossRef]
63. Salman, R.T.; Sanni, P.; Olaniyi, T.A.; Yahaya, K.A. Governance Transparency of Tax Revenue Performance in West Africa. *Bus. Ethics Leadersh.* **2022**, *6*, 14–24. [CrossRef]
64. Chen, M.; Liu, W.; Lu, D.; Chen, H.; Ye, C. Progress of China's new-type urbanization construction since 2014: A preliminary assessment. *Cities* **2018**, *78*, 180–193. [CrossRef]
65. Kim, J.U. A bumpy road to cities: Analysis of the obstacles and limits of China's new urbanization. *Pac. Focus* **2015**, *30*, 372–388. [CrossRef]
66. Yage, J.; Xiaohe, X.; Hong, W. Research on the changes of environmental regulation and TFP under the urban agglomeration strategy in China. *Sci. Technol.* **2020**, *1*, 46–55.
67. Zhou, X.; Xue, Z.; Seydehmet, J. An empirical study on industrial eco-efficiency in arid resource exploitation region of northwest China. *Environ. Sci. Pollut. Res.* **2021**, *28*, 53394–53411. [CrossRef]
68. Arefieva, O.; Polous, O.; Arefiev, S.; Tytykalo, V.; Kwilinski, A. Managing sustainable development by human capital reproduction in the system of company's organizational behavior. *IOP Conf. Ser. Earth Environ. Sci.* **2021**, *628*, 012039. [CrossRef]
69. Ginevičius, R.; Nazarko, J.; Gedvilaitė, D.; Dacko-Pikiewicz, Z. Quantifying the economic development dynamics of a country based on the Lorenz curve. *Econ. Manag.* **2021**, *24*, 55–65. [CrossRef]
70. Kurowska-Pysz, J.; Wróblewski, Ł.; Szczepańska-Woszczyna, K. Identification and assessment of barriers to the development of cross-border cooperation. In Proceedings of the 31st International Business Information Management Association Conference, IBIMA 2018: Innovation Management and Education Excellence through Vision 2020, Milan, Italy, 28–26 April 2018; pp. 3317–3327.
71. China Statistical Yearbook. Available online: http://www.stats.gov.cn/sj/ndsj/2022/indexeh.htm (accessed on 10 March 2023).
72. China Environmental Statistical Yearbook. Available online: https://www.chinayearbooks.com/china-statistical-yearbook-on-environment-2021.html (accessed on 10 March 2023).
73. Statistical Yearbooks of Each Province. Available online: https://www.chinayearbooks.com (accessed on 10 March 2023).
74. Chen, X.; Zhang, M.; Wang, Y.; Xu, X.; Liu, S.; Ma, L. Coupling and coordination characteristics and influencing factors of the livable environment system for the elderly in China. *Chin. Geogr. Sci.* **2022**, *32*, 1052–1068. [CrossRef]
75. Zlotnik, A.; Roald, L.; Backhaus, S.; Chertkov, M.; Andersson, G. Coordinated scheduling for interdependent electric power and natural gas infrastructures. *IEEE Trans. Power Syst.* **2016**, *32*, 600–610. [CrossRef]
76. Adamo, N.; Al-Ansari, N.; Sissakian, V.; Fahmi, K.J.; Abed, S.A. Climate Change: Droughts and Increasing Desertification in the Middle East, with Special Reference to Iraq. *Engineering* **2022**, *14*, 235–273. [CrossRef]
77. Stverkova, H.; Pohludka, M.; Kurowska-Pysz, J.; Szczepańska-Woszczyna, K. Cross-border enterprepreneurship in euroregion beskydy. [Transgraniczna przedsiębiorczość w euroregionie beskidów]. *Pol. J. Manag. Stud.* **2018**, *18*, 324–337. [CrossRef]
78. Szczepańska-Woszczyna, K.; Gedvilaitė, D.; Nazarko, J.; Stasiukynas, A.; Rubina, A. Assessment of economic convergence among countries in the European Union. *Technol. Econ. Dev. Econ.* **2022**, *28*, 1572–1588. [CrossRef]
79. Rozmiarek, M.; Nowacki, K.; Malchrowicz-Mośko, E.; Dacko-Pikiewicz, Z. Eco-initiatives in municipal cultural institutions as examples of activities for sustainable development: A case study of Poznan. *Sustainability* **2022**, *14*, 682. [CrossRef]
80. Kuzior, A.; Samusevych, Y.; Lyeonov, S.; Krawczyk, D.; Grytsyshen, D. Applying energy taxes to promote a clean, sustainable and secure energy system: Finding the preferable approaches. *Energies* **2023**, *16*, 4203. [CrossRef]
81. Shen, L.; Ochoa, J.J.; Bao, H. Strategies for Sustainable Urban Development—Addressing the Challenges of the 21st Century. *Buildings* **2023**, *13*, 847. [CrossRef]
82. Kusakci, S.; Yilmaz, M.K.; Kusakci, A.O.; Sowe, S.; Nantembelele, F.A. Toward sustainable cities: A sustainability assessment study for metropolitan cities in Turkey via a hybridized IT2F-AHP and COPRAS approach. *Sustain. Cities Soc.* **2022**, *78*, 103655. [CrossRef]
83. Kuzior, A.; Krawczyk, D.; Brożek, P.; Pakhnenko, O.; Vasylieva, T.; Lyeonov, S. Resilience of smart cities to the consequences of the COVID-19 pandemic in the context of sustainable development. *Sustainability* **2022**, *14*, 12645. [CrossRef]
84. Li, L.; Xian, S.; Qi, Z. Planning for Eco-City in China: Policy Mobility in Path Creation of Eco-Zhuhai. *Sustain. Dev. Res.* **2022**, *4*, 27–38. [CrossRef]
85. Rego, R.L. Curitiba 1960s transformations and postmodern ideas. *Int. Plan. Hist. Soc. Proc.* **2022**, *19*, 429–441.
86. Shandas, V.; Hellman, D. Toward an Equitable Distribution of Urban Green Spaces for People and Landscapes: An Opportunity for Portland's Green Grid. In *Green Infrastructure and Climate Change Adaptation: Function, Implementation and Governance*; Springer Nature: Singapore, 2022; pp. 289–301.

87. Gonsalves, S.; Starry, O.; Szallies, A.; Brenneisen, S. The effect of urban green roof design on beetle biodiversity. *Urban Ecosyst.* **2022**, *25*, 205–219. [CrossRef]
88. Yang, P. Urban expansion of Energiewende in Germany: A systematic bibliometric analysis and literature study. *Energy Sustain. Soc.* **2022**, *12*, 52. [CrossRef]
89. Žičkienė, A.; Morkunas, M.; Volkov, A.; Balezentis, T.; Streimikiene, D.; Siksnelyte-Butkiene, I. Sustainable Energy Development and Climate Change Miuon at the Local Level through the Lens of Renewable Energy: Evidence from Lithuanian Case Study. *Energies* **2022**, *15*, 980. [CrossRef]
90. Wong, A.M. Valued waste/wasted value: Waste, value and the labor process in electronic waste recycling in Singapore and Malaysia. *Geogr. Compass* **2022**, *16*, e12616. [CrossRef]
91. Kuzior, A.; Kwilinski, A. Cognitive technologies and artificial intelligence in social perception. *Manag. Syst. Prod. Eng.* **2022**, *30*, 109–115. [CrossRef]

land

MDPI

Article

Assessment of Ecosystem Service Values of Urban Wetland: Taking East Lake Scenic Area in Wuhan as an Example

Zhihao Sun [1,2], Wei Xue [3], Dezhi Kang [2,4] and Zhenghong Peng [2,*]

1 Wuhan Natural Resources Conservation and Utilization Center, Wuhan 430014, China; sun99xt@whu.edu.cn
2 School of Urban Design, Wuhan University, Wuhan 430072, China
3 School of Digital Construction and Explosives Engineering, Jianghan University, Wuhan 430056, China
4 Wuhan Land Arranging Storage Center, Wuhan 430014, China
* Correspondence: pengzhenghong@whu.edu.cn; Tel.: +86-27-6877-306

Abstract: Urban wetlands represent a significant ecosystem type within urban landscapes. The quantitative assessment of their ecological service value holds great significance in guiding and improving the urban habitat. However, due to the insufficient spatial resolution of traditional low-to-medium resolution remote sensing imagery for surface monitoring, previous studies have conducted relatively limited research on the ecosystem services of urban wetlands. In this paper, based on multi-source data including multi-scale remote sensing data, a spatial-temporal fusion model and multiple ecological parameter inversion models were employed to invert three key ecological parameters at high spatial resolution, thereby assessing the ecosystem service values (ESVs) of urban wetlands. Taking the East Lake Scenic Area (ELSA) in Wuhan as an example, the dynamics of its ecosystem services' value components were comparatively analyzed. The results indicate that, while the total value of ecosystem services declined slightly in 2015 compared to 2011, there was a notable increase in their value to CNY 3.219 billion by 2019, which represents a doubling of the total value relative to 2011. This trend could be primarily attributed to a significant rise in cultural services within the region. Specifically, the value of tourism services reached CNY 2.090 billion in 2019, representing a threefold increase compared to 2011. This demonstrates that ecosystem services in the ELSA have been significantly optimized and enhanced through associated ecological projects. Further research should investigate the mechanisms by which urbanization affects these crucial ecosystem services, particularly the characterization of cultural services in urban wetlands, and develop more effective strategies to enhance urban resilience and sustainable development.

Keywords: ecosystem service value; remote sensing; GIS; urban wetland; dynamic change; Wuhan

Citation: Sun, Z.; Xue, W.; Kang, D.; Peng, Z. Assessment of Ecosystem Service Values of Urban Wetland: Taking East Lake Scenic Area in Wuhan as an Example. *Land* **2024**, *13*, 1013. https://doi.org/10.3390/land13071013

Academic Editors: Luca Congedo, Francesca Assennato and Michele Munafò

Received: 25 June 2024
Accepted: 6 July 2024
Published: 8 July 2024

1. Introduction

Wetlands play a crucial role as natural elements in cities, providing a range of ecosystem services that enhance the built environment and human well-being [1–3]. However, the continuous erosion and shrinkage of the wetlands' ecological space has led to a gradual weakening of the ecosystem function and services of the wetlands [4]. It has been estimated that more than 60% of wetlands have been lost since the 20th century [5]. The main cause of urban wetland loss is urbanization [6,7]. The process of urbanization has resulted in both socio-economic development and a gradual reduction in the area of wetlands within the urban environment [8]. In China, the rate of urbanization-induced wetland loss was 2.8 times higher between 2000 and 2010 than between 1990 and 2000 [9]. Given the significant decline in wetland habitats, it is important to conduct a comprehensive and accurate assessment of the benefits that wetlands provide.

As a crucial bridge between ecosystems and socio-economic systems, ecosystem services have been a popular topic in the fields of ecological economics and urban management [10–13]. Urban wetlands are the natural ecosystems that are frequently encountered by urban residents, making them an important component of urban environments [14,15]. The ecosystem services classification most commonly recognized in the context of urban wetlands is that proposed by the United Nations in the 2005 Millennium Ecosystem Assessment (MEA). Based on natural ecological processes and functions, the MEA divided ecosystem services into four major types: support services, production services, regulating services, and cultural services [2]. In comparison to natural wetland, urban wetlands exhibit a number of distinctive characteristics [16,17]. They are often characterized by uneven distribution, smaller areas, and low connectivity of wetland patches. Firstly, these are distinguished by a greater degree of habitat fragmentation and pronounced alterations in microclimatic conditions [18]. Secondly, their social service functions have been significantly enhanced [19]. Thirdly, they are frequently subjected to artificial disturbances and governance, which may be either fully artificial or semi-artificial in nature.

The continuous development of remote sensing technology has led to the widespread use of monitoring methods, such as ecological element identification [20–22] and ecological physicochemical parameter inversion [23–26] based on remote sensing data. Various ecosystem services, such as forests, lakes and grasslands, have been extensively and deeply studied at different scales, including global [27,28], regional [29–31] and watershed [7,32,33]. Although the selected ecosystem services assessed are largely comparable, the valuation is a multifaceted endeavor with diverse approaches and outcomes due to the inherent complexity of ecosystem processes. This is because their functioning depends on various ecosystem processes at different scales, which means that ecosystem services of different types and at different scales will not be identical in nature [34]. Currently, there are two main methods for valuing ecosystem services: the equivalent factorization method and the functional values method [35]. The former is widely used because it only requires land cover data and a small amount of socio-economic data in the region and is relatively simple to calculate. However, the equivalent factorization method fails to account for spatial heterogeneity within ecosystems, which makes it difficult to identify finer spatial value distributions. The latter method is challenging to generalize due to the large amount of empirical data, socio-economic data, and ecological parameters required for the calculation of various ecosystem services [36].

Urban wetlands differ from natural ecosystems in that they have been significantly impacted by human activities. The declining habitat quality of urban wetlands leads to the degradation of their ecological functions, as well as a reduction in biodiversity and species richness [37,38]. Moreover, the diverse topography of urban areas gives rise to a multitude of intricate and diverse ecosystem features, as well as a plethora of distinct ecosystem services offered by urban wetlands. Scholars have conducted research on the ecological service value of certain typical wetlands, such as Sanyang wetland [14], Kilombero wetland [39], New Zealand wetland [40], South Sudan wetland [41], Poyang Lake [42], and Hangzhou Bay Wetland [43]. However, the monitoring accuracy requirements for urban wetlands using traditional low- and medium-resolution remote sensing imagery are difficult to meet due to their relatively small scale [44]. In addition, existing studies on urban wetlands tend to focus on a particular point in time [45,46], which makes it difficult to investigate how changes in their ecosystem services respond to the urbanization process.

Based on the above context, the following three research questions are addressed in the study: (1) How can the value of ecosystem services provided by urban wetlands be accurately assessed? (2) What is the value composition of the ecosystem services provided by urban wetlands? (3) What is the change trend in the value of ecosystem services within urban wetlands in the context of human activities? To answer the above questions, this paper combines remote sensing (RS) and geographic information system (GIS) methods to calculate three key ecological parameters pertaining to urban wetlands at a fine-grained level, and to assess the ecosystem service values (ESVs) of urban wetlands. Taking Wuhan East Lake Scenic Area (ELSA) as an example, this paper explored the value of ecosystem services and their change trends in ELSA for the three years, 2011, 2015, and 2019, during which several infrastructural projects were implemented in this area and had significant impacts on its ecological environment. This analysis offers a detailed understanding of the changing status of ecosystem services under major human activities and to provide some foundations for their protection and development.

2. Materials and Methods

2.1. Study Area

The Wuhan East Lake wetland is located in the eastern section of Wuhan City, occupying the second-largest urban lake area within the city limits. The area is located in the eastern part of the Jianghan Plain and belongs to the subtropical monsoon climate zone. The total area of the region is approximately 62 km^2, with the lake's water area covering about 33 km^2. The lake comprises several sub-lakes, including Guozheng Lake, Tangling Lake, and Lingjiao Lake, which are connected by culverts [47]. In 1982, it was approved by the State Council as one of the first national scenic spots. In 2013, it was rated as a national 5A-level tourist attraction. Thanks to the unique natural and humanistic landscape of East Lake and its superior geographic location, the East Lake Scenic Area provides urban residents with rich ecological wetland corridors along the lake, and expands the public outdoor activity spaces for residents. Currently, it has formed a comprehensive ecological space with a large natural lake as the core, integrating 5A-level tourist attractions, a national wetland park, and national ecological tourism demonstration areas.

As a typical urban wetland, the Wuhan East Lake wetland has experienced a significant change due to urban expansion and development. In this study, the ELSA was selected based on the actual construction land status and the division of administrative units in the region. As shown in Figure 1, the scope covers an area of 61.85 km^2 and comprises eight scenic areas around East Lake wetland: Tingtao, Yuguang, Chuidi, Houhu, Luoyan, Moshan, Yujiashan and Baima scenic area.

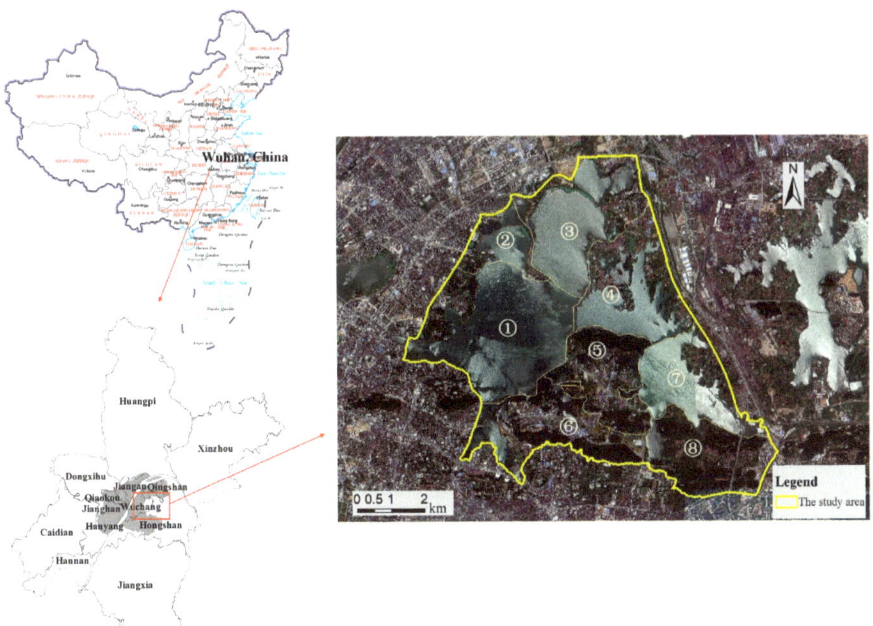

Figure 1. The study area located in the eastern part of Wuhan city center. (Scenic areas: ①Tingtao, ②Yuguang, ③Baima, ④Luoyan, ⑤Moshan, ⑥Yujiashan, ⑦Houhu, ⑧Chuidi.).

2.2. Data Collection

Four main categories of data were utilized in this study: basic geographic data, multi-source remote sensing data and products, hydrometeorological data, and socio-economic data. Basic geographic data consisted of study area boundaries, digital elevation model (DEM) data, and other relevant geo-information. Remote sensing data included GF-1 and Landsat-8 image data, land use data, and month-by-month normalized difference vegetation index (NDVI) data. The GF-1 and Landsat-8 image data were obtained from the geospatial data cloud (https://www.gscloud.cn/, accessed on 10 November 2023) with a spatial resolution of 2 m and 30 m. The land use data were obtained from the Landcover land cover dataset of Tsinghua University in 2017 with a spatial resolution of 10 m [21]. The month-by-month NDVI data were selected from MODIS 13Q1 data from Terra satellite data with a spatial resolution of 250 m and a temporal resolution of 16 days. The hydrometeorological data, including annual precipitation, air temperature, total solar radiation, and water quality and water level data, were provided by the National Ecosystem Observation and Research Network [48]. The socio-economic data such as price-related data, number of tourists, average wage of employees, and consumer price index (CPI) were obtained from the Wuhan Statistical Yearbooks [49–51], National Standards Documents [52], and related research literature.

2.3. Methods

The study employs a structured approach that consists of three key phases. Figure 2 illustrates the methodology framework of this study. First, a comprehensive review of extensive literature was conducted, along with a rigorous data collection effort. Based on the characteristics of urban wetlands and the availability of data, the types of ecosystem services associated with urban wetlands were identified in the study area. Subsequently, the calculation of three key ecological parameters was performed, upon which the quantity of physical quality and value of each ecosystem service was determined in conjunction with relevant data. Finally, a thorough analysis was performed to examine the dynamic changes that occurred over the three year periods, providing valuable insights into the temporal evolution of the urban wetland ecosystem services.

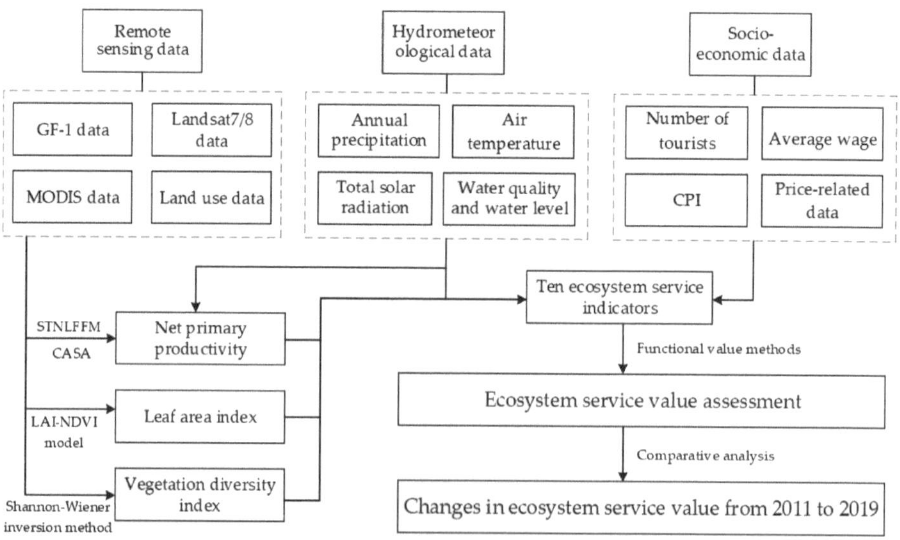

Figure 2. The methodological framework of this study.

2.3.1. Key Ecological Parameter Calculations

1. Net primary productivity

Due to the different electromagnetic radiation characteristics between terrestrial vegetation and water surface, commonly used remote sensing inversion algorithms have difficulty in accurately estimating the true performance of net primary productivity (NPP) in water bodies, leading to severely underestimated water body vegetation growth conditions. Therefore, in the study, the NPP of terrestrial and aquatic ecosystems was calculated separately.

For the terrestrial vegetation part, the spatial–temporal non-local filtering fusion model (STNLFFM) [53] was used to effectively fuse the spatial and temporal complementary information of multi-source remote sensing data, thereby generating monthly high-resolution NDVI spatial distribution data. Considering that the high-resolution image data used in this study has a resolution of 2 m and the MODIS time-series NPP data has a resolution of 250 m, the resolution gap between the two types of data is too large. Therefore, a spatial fusion resolution of 8 m was ultimately determined. Then combined with regional land cover and meteorological data, an improved CASA model was used to calculate the NPP of terrestrial vegetation [54].

For the aquatic plants part, the research on the coverage area of aquatic plants in East Lake by Landsat image inversion conducted by Jiang et al. [47] was used, combined with the average density parameters of aquatic plants in East Lake measured in the relevant literature for estimation.

2. Leaf area index

Leaf area index (LAI) is one of the important indicators for evaluating leaf density of the vegetation canopy. Based on the correlation between LAI and remote sensing surface reflectance, the LAI-NDVI regression model established for green areas in Wuhan by Li et al. [55] was used to retrieve LAI in the region. The regression model used is as follows:

$$LAI = 8.937 - 46.371NDVI + 78.812NDVI^2 - 35.358NDVI^3 \left(R^2 = 0.750\right) \quad (1)$$

where LAI is the average leaf area index within the pixel, and $NDVI$ is the normalized vegetation index within the pixel.

3. Vegetation diversity index

Considering the availability of data in ELSA, the Shannon–Wiener diversity index was employed to quantitatively analyze the ecological diversity characteristics of the vegetation in ELSA. Reference to previous research on the inversion of vegetation species diversity [56], the study adopted the Shannon–Wiener inversion method based on Landsat 8 images. The formula for calculating the average species diversity of vegetation in the East Lake Scenic Area is as follows:

$$H_{avg} = \frac{1}{n}\sum_{i=n}^{n} H_i \quad (2)$$

$$H_i = -0.096NDVI + 0.131TC_2 + 0.143TC_3 + 1.673 \left(R^2 = 0.637\right) \quad (3)$$

where H_i is the Shannon–Wiener diversity index of the i-th pixel, TC_2 and TC_3 are the greenness and humidity components based on the Larson–Jorgensen Cap transform of Landsat multispectral remote sensing images.

The value of biodiversity is referenced to the opportunity cost of species loss under different diversity classes in the Specification for the Assessment of Forest Ecosystem Service Functions, as shown in Table 1.

Table 1. Ranking of different Shannon–Wiener indices and their corresponding values.

Level	Shannon–Wiener Index	Value (CNY/hm^2·a)
1	Index $\geq$ 6	50,000
2	5 $\leq$ Index < 6	40,000
3	4 $\leq$ Index < 5	30,000
4	3 $\leq$ Index < 4	20,000
5	2 $\leq$ Index < 3	10,000
6	1 $\leq$ Index < 2	5000
7	Index < 1	3000

2.3.2. Ecosystem Service Value Assessment

Based on the classification of ecosystem services in the Millennium Ecosystem Assessment, the assessment indicators of ESVs in ELSA were determined by combining with the actual ecological service characteristics of the study area. Considering that the East Lake, as the main urban wetland, has mainly assumed the functions of ecological environment regulation and social and cultural services in recent years, the original production of aquatic products has lost its corresponding ecological value. At the same time, in the forest ecosystem in ELSA, as an important component of urban green spaces, its timber output is not the main ecological function. Therefore, only the value of water storage in the lakes for the surrounding industrial and agricultural water supply is considered in terms of ecological products of wetland and forest, and the economic value contained in the corresponding aquatic products and forests is not calculated.

Therefore, considering the ecological benefits of the lakes and forest ecosystems in the region, 10 indicators, including water supply, carbon sequestration, oxygen release, climate regulation, flood regulation, dust reduction, water quality purification, biodiversity conservation, education and scientific research and tourism services were selected in the study. Based on the connotation of each indicator, appropriate methods for calculating material and value quantity were selected. The selected indicators and relevant methods are shown in Table 2. Furthermore, in order to facilitate the comparison of value quantities across different years, the monetary value calculations for the three years were aligned to 2019 based on the CPI for each year.

Table 2. Evaluation indicators and methods of ecosystem service values of ELSA.

Ecosystem Service	Indicator	Calculation Formula	Description	Evaluation Method
Provision	Water supply	$V_{WS} = \sum_{i=1}^{n} A_i \times H_i \times K_{WS}$	A_i: thearea of the i-th sub-lake (m^2); H_i: the depth of the i-th sub-lake (m); K_{WS}: price of industrial water (CNY/m^3).	Market value method
Regulation	Climate regulation	$V_{CR} = \left(\frac{C_{TP}+C_{Lake}}{81.8} \times 189 \times 24 \times 0.7\right) \times 0.573$ $C_{TP} = L_{TR} \times LAI \times A_{veg} \times D$ $C_{Lake} = L_{Eva} \times A_{Lake} \times C \times \delta T$	L_{TR}: Transpiration heat uptake rate per unit area of leaf blade (J/m^2·d); LAI: leaf area index; A_{veg}: vegetation area (m^2); D: days (d); L_{Eva}: evapotranspiration (mm); A_{Lake}: lake area (m^2); C: specific heat capacity of water. (J/kg·°C); δT: Difference between annual mean temperature and 100 °C (°C)	Alternative engineering method
	Flood regulation	$V_{FR} = \left[A_{Lake} \times (H_{maz} - H_{avg}) + \sum_i A_{TP_i} \times C_{TP\_i} \times P\right] \times K_{AR}$	H_{maz}: annual maximum water level of the lake (m); H_{avg}: average annual water level of lakes (m); A_{TP_i}: area covered by vegetation type i (m^2); C_{TP_i}: rainwater retention rate of vegetation type i; P : total rainfall for the year (mm); K_{AR}: construction cost of artificial reservoirs per unit capacity (CNY/m^3)	Shadow engineering method

Table 2. *Cont.*

Ecosystem Service	Indicator	Calculation Formula	Description	Evaluation Method
	Water purification	$V_{WP} = V_{TOP} + V_{TON}$ $V_{TOP} = (C_{Lake} \times M_{TOP} + C_{AP\_TOP}) \times K_{TOP}$ $V_{TON} = (C_{Lake} \times M_{TON} + C_{AP\_TON}) \times K_{TON}$	C_{Lake}: size of lake storage (m^3); M_{TOP} and M_{TON}: total phosphorus and nitrogen concentration in lake (g/m^3); C_{AP_TOP} and C_{AP_TON}: total phosphorus and nitrogen content of aquatic plants (g); K_{TOP} and K_{TON}: treatment costs for total phosphorus and nitrogen pollutants (CNY/g).	Outcome reference method
	Air purification	$V_{AP} = (A_{Lake} \times L_{DR} + \sum_i A_{TP\_i} LAI_i L_i) \times K_{DR}$	A_{Lake}: lake area (m^2); L_{DR}: dust retention per unit of lake water surface (g/m^2); A_{TP_i}: area covered by vegetation type i (m^2); LAI_i: leaf area index of vegetation type i; L_i: dust retention per unit area of vegetation type i (g/m^2); K_{DR}: charges for dust emissions (CNY/g).	Outcome reference method
	Carbon sequestration and oxygen release	$V_{CS} = (A_{TP} NPP_{TP}/45\% + M_{AP}) \times 1.63 \times K_{CS}$ $V_{OR} = (A_{TP} \times NPP_{TP}/45\% + M_{AP}) \times 1.19 \times K_{OR}$	NPP_{TP}: Annual net primary productivity of terrestrial plants $(g/m^2 \cdot a)$; M_{AP}: Amount of aquatic plant material (g); K_{CS}: carbon tax rate (CNY/g); K_{OR}: industrial oxygen price (CNY/g).	Carbon tax method; shadow engineering method
Support	Biodiversity	$V_{Bio} = A_{Forest} \times K_{DV}$	A_{Forest}: area of forest cover (m^2), K_{DV}: opportunity cost of species loss per unit area of vegetation (CNY/m^2)	Outcome reference method
Cultural	Tourism	$V_{Tour} = V_{TE} + V_{cs} + V_{TOC}$ $V_{TE} = C_{visitors} \times K_{avg\_cost}$ $V_{cs} = V_{TE} \times 40\%$ $V_{TOC} = C_{visitors} \times T_{visitors} \times K_{income} \times 30\%$	V_{TE}: travel expenses; V_{cs}: consumer surplus; V_{TOC}: time opportunity cost; $C_{visitors}$: total number of visitors; K_{avg_cost}: per capita travel costs (CNY); $T_{visitors}$: travel time per capita (h); K_{income}: hourly payroll costs (CNY/h).	Travel cost method
	Scientific research	$V_{SR} = C_{RP} \times K_{ARC}$	C_{RP}: number of academic research papers; K_{ARC}: average research cost per academic paper (CNY/paper)	Outcome reference method

3. Results

3.1. Changes in Land Cover in ELSA from 2010 to 2019

Changes in the ELSA over the years can be observed through the remote sensing image presented in Figure 3. To enhance the visibility of land cover changes, false-color images within the near-infrared (NIR)-red-green spectral band were employed for analysis. In these figures, vegetation was rendered in red, water bodies appeared in darker hues, and the built environment was depicted in lighter colors.

The imagery captured in 2011 highlights a notable phenomenon: the presence of a road traversing the center of the lake, effectively partitioning it into distinct sub-lakes. In 2015, the implementation of the East Lake Tunnel Project is evident. This was executed via the utilization of cofferdam excavation techniques, leading to a distinct demarcation and transformation of East Lake. The observed changes align precisely with the intended direction of the project, spanning from the northwest towards the southeast. By 2019, upon the completion of construction activities, the cofferdam was subsequently removed, and the lakebed was covered with soil. This restoration measure effectively eradicated any visual indications of the previous division created by the project.

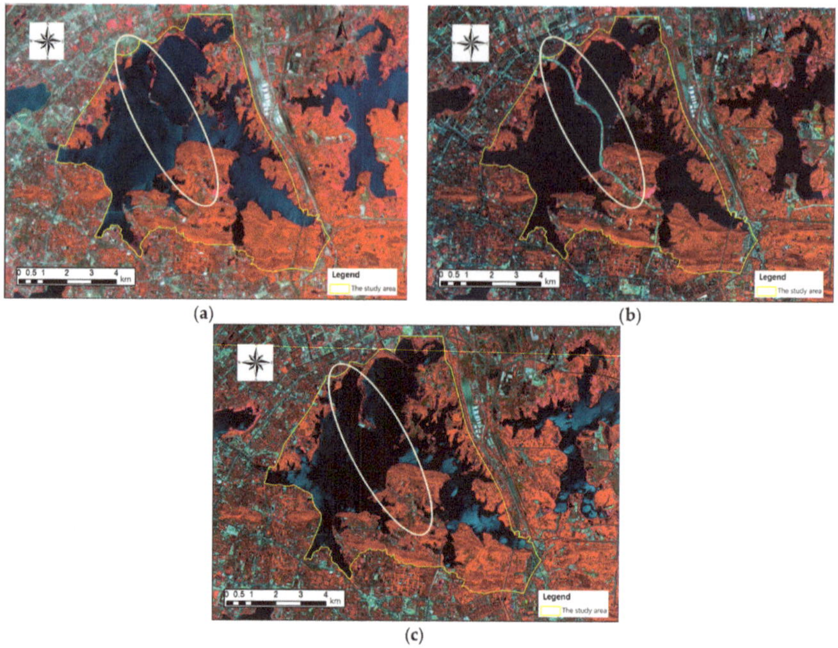

Figure 3. Remote sensing false-color composite images of ELSA from 2011 to 2019. (a) 2011, (b) 2015. (c) 2019.

3.2. Key Ecological Parameters

3.2.1. Net Primary Productivity

Taking 2019 as an example, combining the October 2019 GF-6 remote sensing imagery, and the NDVI temporal products derived from MODIS 13Q1 data with a spatial resolution of 16 days and a temporal resolution of 250 m, the STNLFFM model was used to temporally and spatially fuse the spatial and temporal complementary information from the multi-source remote sensing data, thereby obtaining the month-by-month NDVI distribution information in 2019. The June 2019 NDVI fusion result is shown in Figure 4; it can be seen that the fused results permit a more detail characterization of the distribution of vegetation in the region.

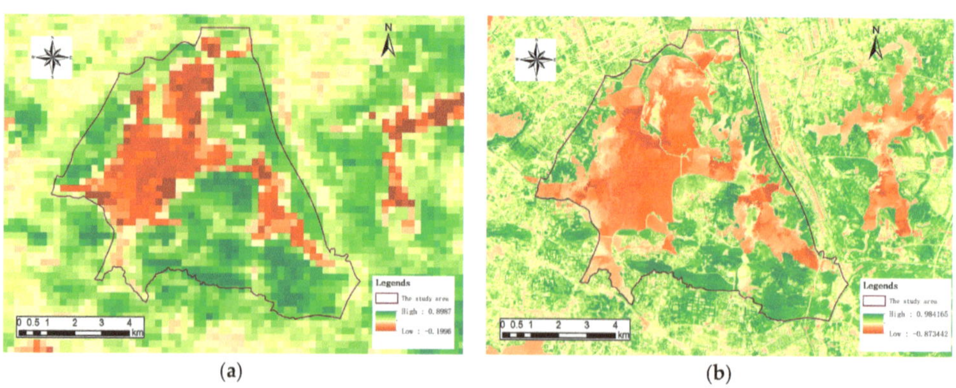

Figure 4. The result of spatial-temporal fusion of high-resolution NDVI in June 2019. (a) Pre-fusion. (b) Post-fusion.

The improved CASA model was employed to calculate the distribution characteristics of net primary productivity (NPP) in the region. As illustrated in Figure 5, in 2019, the maximum value of NPP is 767.45 $gC \cdot m^{-2} \cdot a^{-1}$, the minimum value is 0 $gC \cdot m^{-2} \cdot a^{-1}$, and the average value is 129.52 $gC \cdot m^{-2} \cdot a^{-1}$.

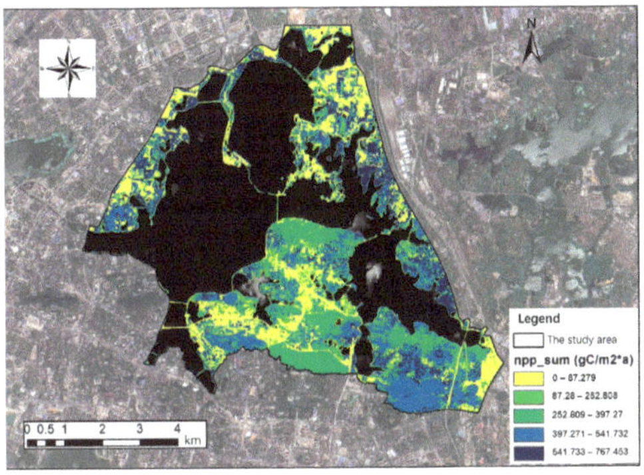

Figure 5. Distribution of net primary productivity within the study area in 2019.

3.2.2. Leaf Area Index

The high-resolution NDVI distribution calculated in Section 3.2.1 was utilized to perform an inversion of the leaf area index (LAI) in the region. This was accomplished through the application of the LAI-NDVI regression model developed by Li Lu et al. The results of this process are presented in Figure 6, which depicts the range of LAI values observed in the study area, spanning from 0 to 4.84. In conjunction with the land cover characteristics, the statistical characteristics of the leaf area index distribution for different vegetation types, including forests, grasslands, and shrubs, were derived (Table 3).

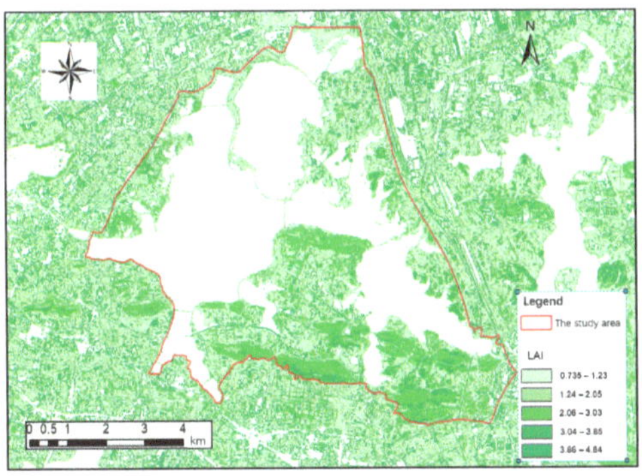

Figure 6. Distribution of leaf area index with in the study area in 2019.

Table 3. Statistic results of leaf area index for different land cover types.

Land Cover	Mean	Std
Forest	2.20	0.83
Grassland	1.96	1.65
Shrub	1.48	0.87

3.2.3. Vegetation Diversity Index

The diversity index calculation necessitates the utilization of tasseled cap transformation based on multispectral bands. However, the current high-resolution remote sensing imagery is devoid of the requisite band information. Consequently, the Landsat-8 images were employed in this study for the inversion calculation of diversity index. A calculation based on the 2019 Landsat imagery data indicates that the region's vegetation diversity is between 1.94 and 2.35.

3.3. *Changes in ESVs in ELSA from 2010 to 2019*

3.3.1. Dynamic Changes in the Total Value of Ecosystem Services

Using the multi-source remote sensing data and socio-economic data for the ELSA for the three periods of 2011, 2015 and 2019, the value of ecosystem services of the ELSA was evaluated to be derived between different years, as shown in Figure 7. It can be seen that the total value of ecosystem services of ELSA was CNY 1.606 billion, CNY 1.588 billion and CNY 3.219 billion for the years of 2011, 2015 and 2019, respectively. The rate of change in its total amount between 2011 and 2015 is −1.13%, with a slight decrease in the total value; the rate of change between 2015 and 2019 is 102.76%, with a significant increase in the total value; and the total rate of change between 2011 and 2019 is 100.47%, with a significant increase in the total value of ecosystem services.

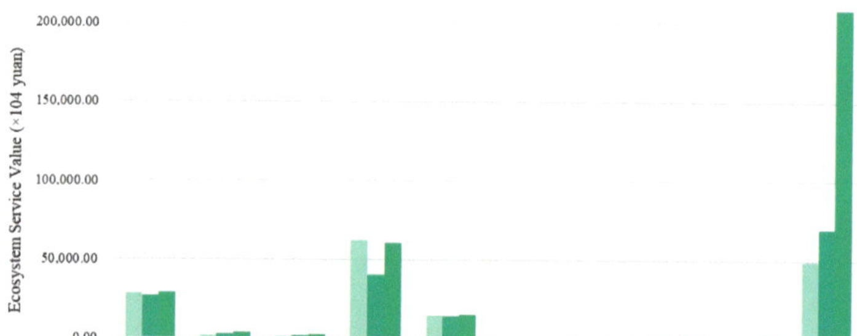

	Water supply	Carbon sequestration	Oxygen release	Climate regulation	Flood regulation	Air purification	Water purification	Biodiversity	Scientific research	Tourism
2011	28,322.22	2,306.78	1,627.49	62,338.69	15,099.07	101.21	37.08	829.71	786.72	49,150.67
2015	26,919.98	3,197.94	2,256.22	40,580.13	14,608.68	74.48	29.65	787.89	321.84	70,005.33
2019	28,726.00	3,900.99	2,752.24	60,190.00	15,351.66	118.50	32.56	971.99	929.76	208,974.14

Figure 7. Ecosystem service values in ELSA from 2011 to 2019.

3.3.2. Dynamic Changes in the Value of Individual Ecosystem Services

Table 4 shows that the ESVs of the water provisioning services remained stable over the three study periods. However, there was a slight decrease of −4.95% in 2015. This change may be related to the impacts on the lake surface caused by the construction of the East Lake Greenway and the East Lake Tunnel during this period.

The value of ecosystem services related to carbon sequestration and oxygen release in regulating services increased continuously at a rate of 69.11% from 2011 to 2019. This suggests a significant improvement in the capacity of terrestrial ecosystems, such as forests and grasslands in the ELSA, to sequester carbon and release oxygen. However, in 2015, the levels of climate regulation, flood storage, dust reduction, and water purification decreased in varying degrees. In 2019, these aspects returned to a higher level.

In terms of support services, biodiversity decreased slightly in 2015 but improved significantly in 2019. The value amount for biodiversity increased by 17.15% in 2019 compared to 2011. This may be related to the disturbance of forests, grasslands, and other ecosystems in the region by human activities. Measures need to be taken to maintain the health and stability of the ecosystems.

Regarding cultural services, there were different trends in scientific research and tourism services over the three study periods. In 2015, scientific research services experienced a significant decline with a rate of change of 59.09%. This decline may be attributed to the constraints imposed by the construction of the East Lake Tunnel and the Greenway Project on related scientific research activities. Tourism services increased significantly during the period, indicating that the demand for such services in ELSA by urban residents had not been affected by the aforementioned projects.

Table 4. Rates of change in ecosystem service values of ELSA from 2011 to 2019.

Ecosystem Service	Service Indicator	2011–2015	2015–2019	2011–2019
Supply service	Water supply	−4.95	6.71	1.43
Regulation service	Carbon sequestration	38.63	21.98	69.11
	Oxygen release	38.63	21.98	69.11
	Climate regulation	−34.90	48.32	−3.45
	Flood regulation	−3.25	5.09	1.67
	Air purification	−26.41	59.10	17.08
	Water purification	−20.04	9.81	−12.19
Support service	Biodiversity	−5.04	23.37	17.15
Culture service	Scientific research	−59.09	188.89	18.18
	Tourism	42.43	198.51	325.17
Total		−1.13	102.76	100.47

3.3.3. Dynamic Changes in the Composition of the Value of Ecosystem Services

Figure 8 shows the proportion of individual ESVs in the total value volume. It is evident from the figure that the composition of the ESVs of ELSA underwent significant changes during the period of 2011–2019. In 2011, climate regulation services had the highest value share in ELSA ecosystem services, followed by recreation, water supply, and flood regulation. This emphasizes the significant importance of climate regulation services. However, in 2015, the value of tourism services increased rapidly, surpassing all other ecosystem services in value share. In contrast, the value share of climate regulation, water supply, and flood regulation decreased accordingly. By 2019, the value volume of tourism services had increased even further, with its value share reaching 65.21%. The results indicate that tourism services have become the most important type of ecosystem services in ELSA. Furthermore, the proportion of services such as climate regulation, water supply, and flood regulation continue to decrease. This reinforces the importance of tourism services in the ELSA ecosystem.

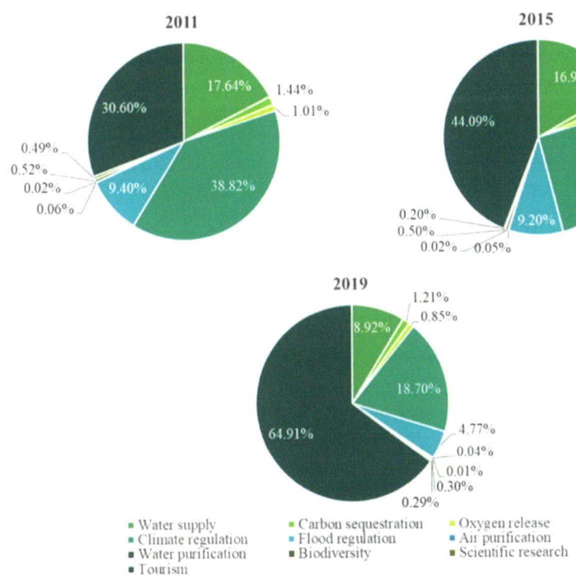

Figure 8. Changes in the share of ecosystem service values of the ELSA in different years.

4. Discussion

4.1. The Differences in ESVs between Urban and Natural Wetlands

In this study, multi-source data including remote sensing data, ground observation data, and socio-economic data were used. In order to capture the ecological characteristics of urban wetlands at finer scales, multiple remote sensing data at different scales were integrated for the inversion of high-resolution ecological parameters. In terms of the composition of the ecosystem services, ELSA's cultural services accounted for a relatively high proportion of ecosystem services in all three time periods, with a peak of 65.21% observed in 2019. In other studies, regulatory and support services are typically identified as the main ecosystem service components on natural wetlands [4,40,43]. This observation further confirms that urban residents have a greater need for cultural services related to urban wetlands compared to natural wetlands.

4.2. Changes in ESVs in the ELSA Region

Based on the above analysis, it can be clearly observed that the value of the ecosystem services of the ELSA has changed significantly between 2011 and 2019. In 2011, its total value was CNY 16.06×10^8, of which the value of climate regulating services accounted for the largest proportion. In 2015, due to the construction of the East Lake Tunnel Project by means of cofferdam open excavation, the catchment area of the East Lake was slightly reduced, resulting in a decrease in the value of some ecosystem services. However, with the completion of the construction project and the implementation of the related ecological restoration project, the ecosystem function of the lake water was effectively restored, leading to an increase in the amount of corresponding ESVs. By 2019, the total ESVs of the ELSA reached CNY 32.19×10^8. Therefore, although the construction of the East Lake Tunnel Project had a certain impact on the lake's watershed, resulting in a reduction in some ecosystem services, the lake's ecosystem function has been restored through effective ecological restoration measures, which in turn has increased the total value of ecosystem services.

There was a gradual increase in carbon sequestration and oxygen release over the three periods. Additionally, tourism also showed a similar trend. Although the reduction in water surface area appears to have resulted in a decline in aquatic vegetation, the overall carbon sequestration capacity and oxygen release exhibited a rising tendency, suggesting an enhanced provision of these two ecosystem services by terrestrial vegetation. The growth in the tourism service's volume is primarily attributed to the annual increase in the number of tourists. This implies that while the construction of the East Lake Tunnel Project may have affected the ecological environment and playing conditions in the lake area of East Lake, the land area's ecological space may have been the primary attraction for residents during this period. Moreover, the East Lake Greenway project has enhanced the quality of the ecological space in the area, resulting in a doubling in the number of visitors to the region in 2019 compared to 2015. This has led to a significant increase in the volume of cultural services.

4.3. The Culture Service Value in the ELSA Region

Urban wetlands, as the main open space accessible to residents, have garnered extensive scholarly attention for their cultural services, particularly in regard to their impact on resident health and social interactions [57,58]. For instance, studies in several cities have shown that regular visits to urban wetlands by residents positively correlate with improved mental health and reduced stress levels [59]. The green spaces and natural habitats within these wetlands provide residents with a respite from the urban hustle and bustle, fostering a sense of relaxation and well-being. Additionally, these spaces often host community events, festivals, and other gatherings, promoting interactions between residents of diverse backgrounds.

In this paper, the values of tourism services and scientific research were assessed based on the number of tourist visits and the number of published academic papers, respectively. Comparing various ecosystem services between 2011 and 2019, it was shown that tourism services were the fastest-growing service type within the region as the wetlands were constructed and environmentally enhanced. Moreover, the utilization of perceived behavior data derived from the participation GIS and social media platforms has emerged as a research focus for quantifying the value of cultural services [60,61].

Consequently, a deeper examination of cultural services within the ELSA area would be both significant and necessary. Unlike the trade-off relationships in previous studies [62–64], ELSA's ecosystem service was optimized and enhanced through the significant enhancement of cultural services while maintaining the original amount of natural ecosystem services. This process reveals the synergistic relationships of the East Lake region in terms of ecosystem services, further highlighting the interaction between human activities and natural ecosystems. Therefore, it is important to continue focusing on maintaining the stability and health of the natural ecosystem while also fully utilizing the potential of cultural services and promoting their further enhancement in the region.

4.4. Limitations and Future Works

Although this study conducted a refined inversion of ecosystem services in urban wetlands, the type and refinement of ecosystem services still need to be further improved due to the lack of data and the incompleteness of the model. For example, in the inversion of NPP, the terrestrial NPP was accurately inverted in the study, and the NPP of aquatic plants was measured by acquiring the inversion results from related literature, which may lead to the wrong estimation of aquatic plant productivity. Furthermore, due to the lack of data on plant and animal species in the study area, particularly aquatic organisms, the diversity value was calculated based on terrestrial vegetation data, which resulted in an underestimation of the region's diversity. Additionally, the cultural services indicator can be refined by incorporating specific activities of residents in the region. In future studies, further data collection can be conducted to construct more refined inverse models of ecological parameters. This foundation allows for further investigation into the impact of urbanization on the evolution of various ecosystem services.

5. Conclusions

The assessment of ESVs in urban wetlands is crucial for understanding their ecological and social importance and for informing decision-making in urban planning and management. The present study has provided a comprehensive assessment of the changes in the ESVs of the ELSA in Wuhan, China over a three study periods from 2011 to 2019. The utilization of remote sensing and geographic information system techniques has enabled a fine-grained quantitative analysis of key ecological parameters, thereby revealing the dynamic nature of urban lake wetlands and their associated ecosystem services. The results indicate that, despite a slight decline in the total value of ecosystem services in 2015 compared to 2011, the value experienced a notable increase by 2019, reaching CNY 3.219 billion. This significant growth can be primarily attributed to the remarkable enhancement of tourism services in the ELSA. This outcome highlights the importance of urban wetlands in providing valuable recreational opportunities for city residents.

The optimization and enhancement of the ESVs in the ELSA demonstrate the potential for urban wetlands to serve as valuable natural resources, not only for biodiversity conservation but also for human well-being. The study further highlights the necessity for continued monitoring and management of urban wetlands in order to ensure their sustainability and the optimal utilization of their ecosystem services. In light of the rapid urbanization and growing demand for natural resources, it is important to achieve a balance between development and conservation, ensuring the protection of urban wetlands and the preservation of their ecological functions.

Author Contributions: Conceptualization, Z.S. and D.K.; data curation, Z.S.; formal analysis, Z.S.; funding acquisition, Z.S., W.X. and Z.P.; investigation, Z.S. and D.K.; methodology, Z.S.; project administration, Z.P.; supervision, D.K. and Z.P.; validation, Z.S.; visualization, Z.S. and D.K.;

writing—original draft, Z.S.; writing—review and editing, Z.S., W.X. and Z.P. All authors have read and agreed to the published version of the manuscript.

Funding: This research was funded by the National Natural Science Foundation of China (grant number 51978535), Hubei Post doc Innovation and Application Funding (grant number 202320) and Hubei Provincial Natural Science Foundation Program Youth Project (grant number 2023AFB510).

Data Availability Statement: The original contributions presented in the study are included in the article, further inquiries can be directed to the corresponding author.

Conflicts of Interest: The authors declare no conflicts of interest.

References

1. King, D.M.; Wainger, L.A.; Bartoldus, C.C.; Wakeley, J.S. *Expanding Wetland Assessment Procedures: Linking Indices of Wetland Function with Services and Values*; Defense Technical Information Center: Fort Belvoir, VA, USA, 2000.
2. Millennium Ecosystem Assessment. *Ecosystems and Human Well-Being*; World Resources Institute: Washington, DC, USA, 2005; ISBN 1-56973-597-2.
3. Maltby, E.; Acreman, M.C. Ecosystem Services of Wetlands: Pathfinder for a New Paradigm. *Hydrol. Sci. J.* **2011**, *56*, 1341–1359. [CrossRef]
4. Adekola, O.; Mitchell, G. The Niger Delta Wetlands: Threats to Ecosystem Services, Their Importance to Dependent Communities and Possible Management Measures. *Int. J. Biodivers. Sci. Ecosyst. Serv. Manag.* **2011**, *7*, 50–68. [CrossRef]
5. Davidson, N.C. How Much Wetland Has the World Lost? Long-Term and Recent Trends in Global Wetland Area. *Mar. Freshw. Res.* **2014**, *65*, 934. [CrossRef]
6. Hu, S.; Niu, Z.; Chen, Y.; Li, L.; Zhang, H. Global Wetlands: Potential Distribution, Wetland Loss, and Status. *Sci. Total Environ.* **2017**, *586*, 319–327. [CrossRef]
7. Huang, Q.; Zhao, X.; He, C.; Yin, D.; Meng, S. Impacts of Urban Expansion on Wetland Ecosystem Services in the Context of Hosting the Winter Olympics: A Scenario Simulation in the Guanting Reservoir Basin, China. *Reg. Environ. Chang.* **2019**, *19*, 2365–2379. [CrossRef]
8. Xu, Z.; Peng, J.; Qiu, S.; Liu, Y.; Dong, J.; Zhang, H. Responses of Spatial Relationships between Ecosystem Services and the Sustainable Development Goals to Urbanization. *Sci. Total Environ.* **2022**, *850*, 157868. [CrossRef] [PubMed]
9. Mao, D.; Wang, Z.; Wu, J.; Wu, B.; Zeng, Y.; Song, K.; Yi, K.; Luo, L. China's Wetlands Loss to Urban Expansion. *Land Degrad. Dev.* **2018**, *29*, 2644–2657. [CrossRef]
10. Fisher, B.; Turner, R.K.; Morling, P. Defining and Classifying Ecosystem Services for Decision Making. *Ecol. Econ.* **2009**, *68*, 643–653. [CrossRef]
11. Braat, L.C. The Value of the Ecosystem Services Concept in Economic and Biodiversity Policy. In *Ecosystem Services*; Elsevier: Amsterdam, The Netherlands, 2013; pp. 97–103. ISBN 978-0-12-419964-4.
12. Ouyang, Z.; Zheng, H.; Xiao, Y.; Polasky, S.; Liu, J.; Xu, W.; Wang, Q.; Zhang, L.; Xiao, Y.; Rao, E.; et al. Improvements in Ecosystem Services from Investments in Natural Capital. *Science* **2016**, *352*, 1455–1459. [CrossRef]
13. Costanza, R.; de Groot, R.; Braat, L.; Kubiszewski, I.; Fioramonti, L.; Sutton, P.; Farber, S.; Grasso, M. Twenty Years of Ecosystem Services: How Far Have We Come and How Far Do We Still Need to Go? *Ecosyst. Serv.* **2017**, *28*, 1–16. [CrossRef]
14. Tong, C.; Feagin, R.A.; Lu, J.; Zhang, X.; Zhu, X.; Wang, W.; He, W. Ecosystem Service Values and Restoration in the Urban Sanyang Wetland of Wenzhou, China. *Ecol. Eng.* **2007**, *29*, 249–258. [CrossRef]
15. Zhou, L.; Guan, D.; Huang, X.; Yuan, X.; Zhang, M. Evaluation of the Cultural Ecosystem Services of Wetland Park. *Ecol. Indic.* **2020**, *114*, 106286. [CrossRef]
16. Radford, K.G.; James, P. Changes in the Value of Ecosystem Services along a Rural–Urban Gradient: A Case Study of Greater Manchester, UK. *Landsc. Urban Plan.* **2013**, *109*, 117–127. [CrossRef]
17. Datri, L.A.; Lopez, M.; Buchter, S.; Pazcel, E.M.; Gandini, M. Functional Urban Wetlands in Dysfunctional Cities. *Curr. Landsc. Ecol. Rep.* **2024**, *9*, 21–30. [CrossRef]
18. Manzo, L.M.; Epele, L.B.; Grech, M.G.; Kandus, P.; Miserendino, M.L. Wetland Genesis Rules Invertebrate Spatial Patterns at Patagonian Ponds (Santa Cruz, Argentina): A Multiscale Perspective. *Ecol. Eng.* **2019**, *126*, 43–54. [CrossRef]
19. Das, A.; Das, M.; Gupta, R. Comparison of Ecosystem Services Provided by an Urban and a Riverine Wetland: A Multi-Scale Evaluation from Lower Gangetic Plain, Eastern India. *Environ. Sci. Pollut. Res.* **2022**, *29*, 79529–79544. [CrossRef] [PubMed]
20. Cao, J.; Leng, W.; Liu, K.; Liu, L.; He, Z.; Zhu, Y. Object-Based Mangrove Species Classification Using Unmanned Aerial Vehicle Hyperspectral Images and Digital Surface Models. *Remote Sens.* **2018**, *10*, 89. [CrossRef]
21. Gong, P.; Liu, H.; Zhang, M.; Li, C.; Wang, J.; Huang, H.; Clinton, N.; Ji, L.; Li, W.; Bai, Y.; et al. Stable Classification with Limited Sample: Transferring a 30-m Resolution Sample Set Collected in 2015 to Mapping 10-m Resolution Global Land Cover in 2017. *Sci. Bull.* **2019**, *64*, 370–373. [CrossRef]
22. Chen, C.; Chen, H.; Liang, J.; Huang, W.; Xu, W.; Li, B.; Wang, J. Extraction of Water Body Information from Remote Sensing Imagery While Considering Greenness and Wetness Based on Tasseled Cap Transformation. *Remote Sens.* **2022**, *14*, 3001. [CrossRef]

23. Mohammadi, J.; Shataee, S. Possibility Investigation of Tree Diversity Mapping Using Landsat ETM+ Data in the Hyrcanian Forests of Iran. *Remote Sens. Environ.* **2010**, *114*, 1504–1512. [CrossRef]
24. Tian, F.; Qiu, G.; Lü, Y.; Yang, Y.; Xiong, Y. Use of High-Resolution Thermal Infrared Remote Sensing and "Three-Temperature Model" for Transpiration Monitoring in Arid Inland River Catchment. *J. Hydrol.* **2014**, *515*, 307–315. [CrossRef]
25. Zhu, Y.; Liu, K.; Liu, L.; Myint, S.; Wang, S.; Liu, H.; He, Z. Exploring the Potential of WorldView-2 Red-Edge Band-Based Vegetation Indices for Estimation of Mangrove Leaf Area Index with Machine Learning Algorithms. *Remote Sens.* **2017**, *9*, 1060. [CrossRef]
26. Torresani, M.; Rocchini, D.; Sonnenschein, R.; Zebisch, M.; Marcantonio, M.; Ricotta, C.; Tonon, G. Estimating Tree Species Diversity from Space in an Alpine Conifer Forest: The Rao's Q Diversity Index Meets the Spectral Variation Hypothesis. *Ecol. Inform.* **2019**, *52*, 26–34. [CrossRef]
27. De Groot, R.; Brander, L.; Van Der Ploeg, S.; Costanza, R.; Bernard, F.; Braat, L.; Christie, M.; Crossman, N.; Ghermandi, A.; Hein, L.; et al. Global Estimates of the Value of Ecosystems and Their Services in Monetary Units. *Ecosyst. Serv.* **2012**, *1*, 50–61. [CrossRef]
28. Keil, P.; Chase, J.M. Global Patterns and Drivers of Tree Diversity Integrated across a Continuum of Spatial Grains. *Nat. Ecol. Evol.* **2019**, *3*, 390–399. [CrossRef] [PubMed]
29. Huang, B.; Li, R.; Ding, Z.; O'Connor, P.; Kong, L.; Xiao, Y.; Xu, W.; Guo, Y.; Yang, Y.; Li, R.; et al. A New Remote-Sensing-Based Indicator for Integrating Quantity and Quality Attributes to Assess the Dynamics of Ecosystem Assets. *Glob. Ecol. Conserv.* **2020**, *22*, e00999. [CrossRef]
30. Xu, C.; Jiang, W.; Huang, Q.; Wang, Y. Ecosystem Services Response to Rural-Urban Transitions in Coastal and Island Cities: A Comparison between Shenzhen and Hong Kong, China. *J. Clean. Prod.* **2020**, *260*, 121033. [CrossRef]
31. Wang, Y.; Chang, Q.; Fan, P.; Shi, X. From Urban Greenspace to Health Behaviors: An Ecosystem Services-Mediated Perspective. *Environ. Res.* **2022**, *213*, 113664. [CrossRef] [PubMed]
32. Hamel, P.; Hamann, M.; Kuiper, J.J.; Andersson, E.; Arkema, K.K.; Silver, J.M.; Daily, G.C.; Guerry, A.D. Blending Ecosystem Service and Resilience Perspectives in Planning of Natural Infrastructure: Lessons from the San Francisco Bay Area. *Front. Environ. Sci.* **2021**, *9*, 601136. [CrossRef]
33. Ashrafi, S.; Kerachian, R.; Pourmoghim, P.; Behboudian, M.; Motlaghzadeh, K. Evaluating and Improving the Sustainability of Ecosystem Services in River Basins under Climate Change. *Sci. Total Environ.* **2022**, *806*, 150702. [CrossRef]
34. Ehrenfeld, J.G. Evaluating Wetlands within an Urban Context. *Ecol. Eng.* **2000**, *15*, 253–265. [CrossRef]
35. Li, F.; Wang, F.; Liu, H.; Huang, K.; Yu, Y.; Huang, B. A Comparative Analysis of Ecosystem Service Valuation Methods: Taking Beijing, China as a Case. *Ecol. Indic.* **2023**, *154*, 110872. [CrossRef]
36. Zhao, Q.; Wang, Q. Water Ecosystem Service Quality Evaluation and Value Assessment of Taihu Lake in China. *Water* **2021**, *13*, 618. [CrossRef]
37. Mondal, B.; Dolui, G.; Pramanik, M.; Maity, S.; Biswas, S.S.; Pal, R. Urban Expansion and Wetland Shrinkage Estimation Using a GIS-Based Model in the East Kolkata Wetland, India. *Ecol. Indic.* **2017**, *83*, 62–73. [CrossRef]
38. Das, A.; Basu, T. Assessment of Peri-Urban Wetland Ecological Degradation through Importance-Performance Analysis (IPA): A Study on Chatra Wetland, India. *Ecol. Indic.* **2020**, *114*, 106274. [CrossRef]
39. Koko, I.A.; Misana, S.B.; Kessler, A.; Fleskens, L. Valuing Ecosystem Services: Stakeholders' Perceptions and Monetary Values of Ecosystem Services in the Kilombero Wetland of Tanzania. *Ecosyst. People* **2020**, *16*, 411–426. [CrossRef]
40. Schallenberg, M.; de Winton, M.D.; Verburg, P.; Kelly, D.J.; Hamill, K.D.; Hamilton, D.P. Ecosystem Services of Lakes. In *Ecosystem Services in New Zealand: Conditions and Trends*; Manaaki Whenua Press: Lincoln, New Zealand, 2013; pp. 203–225.
41. Mulatu, D.W.; Ahmed, J.; Semereab, E.; Arega, T.; Yohannes, T.; Akwany, L.O. Stakeholders, Institutional Challenges and the Valuation of Wetland Ecosystem Services in South Sudan: The Case of Machar Marshes and Sudd Wetlands. *Environ. Manag.* **2022**, *69*, 666–683. [CrossRef] [PubMed]
42. Liu, C.; Liu, Y.; Giannetti, B.F.; Almeida, C.M.V.B.; Sevegnani, F.; Li, R. Spatiotemporal Differentiation and Mechanism of Anthropogenic Factors Affecting Ecosystem Service Value in the Urban Agglomeration around Poyang Lake, China. *Ecol. Indic.* **2023**, *154*, 110733. [CrossRef]
43. Lin, W.; Xu, D.; Guo, P.; Wang, D.; Li, L.; Gao, J. Exploring Variations of Ecosystem Service Value in Hangzhou Bay Wetland, Eastern China. *Ecosyst. Serv.* **2019**, *37*, 100944. [CrossRef]
44. De Valck, J.; Beames, A.; Liekens, I.; Bettens, M.; Seuntjens, P.; Broekx, S. Valuing Urban Ecosystem Services in Sustainable Brownfield Redevelopment. *Ecosyst. Serv.* **2019**, *35*, 139–149. [CrossRef]
45. Zhang, B.; Shi, Y.; Liu, J.; Xu, J.; Xie, G. Economic Values and Dominant Providers of Key Ecosystem Services of Wetlands in Beijing, China. *Ecol. Indic.* **2017**, *77*, 48–58. [CrossRef]
46. Díaz-Pinzón, L.; Sierra, L.; Trillas, F. The Economic Value of Wetlands in Urban Areas: The Benefits in a Developing Country. *Sustainability* **2022**, *14*, 8302. [CrossRef]
47. Jiang, Y.; Chen, X.; Yang, X. Changes of Aquatic Plants in Donghu Lake of Wuhan Based 1990-2020 Landsat Images. *Chin. J. Plant Ecol.* **2022**, *46*, 1551–1561. [CrossRef]
48. National Ecosystem Research Network of China Chinese National Ecosystem Observation Research Network Science and Technology Resource Service System. Available online: http://www.cnern.org.cn. (accessed on 31 October 2023).
49. Wuhan Municipal Statistics Bureau. *Wuhan Statistical Yearbook—2011*; Wuhan Municipal Statistics Bureau: Wuhan, China, 2011.

50. Wuhan Municipal Statistics Bureau. *Wuhan Statistical Yearbook—2015*; Wuhan Municipal Statistics Bureau: Wuhan, China, 2015.
51. Wuhan Municipal Statistics Bureau. *Wuhan Statistical Yearbook—2019*; Wuhan Municipal Statistics Bureau: Wuhan, China, 2019.
52. *GB/T38582—2020*; Forest Ecosystem Service Function Assessment Specification. State Administration for Market Regulation: Beijing, China, 2020.
53. Cheng, Q.; Liu, H.; Shen, H.; Wu, P.; Zhang, L. A Spatial and Temporal Nonlocal Filter-Based Data Fusion Method. *IEEE Trans. Geosci. Remote Sens.* **2017**, *55*, 4476–4488. [CrossRef]
54. Zhu, W.; Pan, Y.; Zhang, J. Estimation of Net Primary Productivity of Chinese Terrestrial Vegetation Based on Remote Sensing. *Chin Jour Plan Ecolo* **2007**, *31*, 413–424. [CrossRef]
55. Li, L.; Zhou, G.; Yao, C. Study on the Green Volume of Different Types of Urban Green Spaces. *Chin. Landsc. Archit.* **2015**, *31*, 17–21.
56. Zhang, C.; Li, W.; Hu, Z.; Zhu, J.; Yan, Q. Distribution Pattern of Tree Layer Species Diversity Based on RS and GIS: A Case Study of Water Conservation Forests in Montane Regions of Eastern Liaoning Province of China. *Chin. J. Ecol.* **2009**, *28*, 1749.
57. Pedersen, E.; Weisner, S.E.B.; Johansson, M. Wetland Areas' Direct Contributions to Residents' Well-Being Entitle Them to High Cultural Ecosystem Values. *Sci. Total Environ.* **2019**, *646*, 1315–1326. [CrossRef]
58. Wang, K.; Sun, Z.; Cai, M.; Liu, L.; Wu, H.; Peng, Z. Impacts of Urban Blue-Green Space on Residents' Health: A Bibliometric Review. *Int. J. Environ. Res. Public Health* **2022**, *19*, 16192. [CrossRef]
59. Stigsdotter, U.K.; Ekholm, O.; Schipperijn, J.; Toftager, M.; Kamper-Jørgensen, F.; Randrup, T.B. Health Promoting Outdoor Environments—Associations between Green Space, and Health, Health-Related Quality of Life and Stress Based on a Danish National Representative Survey. *Scand. J. Public Health* **2010**, *38*, 411–417. [CrossRef]
60. Ko, H.; Son, Y. Perceptions of Cultural Ecosystem Services in Urban Green Spaces: A Case Study in Gwacheon, Republic of Korea. *Ecol. Indic.* **2018**, *91*, 299–306. [CrossRef]
61. Lingua, F.; Coops, N.C.; Griess, V.C. Valuing Cultural Ecosystem Services Combining Deep Learning and Benefit Transfer Approach. *Ecosyst. Serv.* **2022**, *58*, 101487. [CrossRef]
62. Jopke, C.; Kreyling, J.; Maes, J.; Koellner, T. Interactions among Ecosystem Services across Europe: Bagplots and Cumulative Correlation Coefficients Reveal Synergies, Trade-Offs, and Regional Patterns. *Ecol. Indic.* **2015**, *49*, 46–52. [CrossRef]
63. Luo, Q.; Zhou, J.; Zhang, Y.; Yu, B.; Zhu, Z. What Is the Spatiotemporal Relationship between Urbanization and Ecosystem Services? A Case from 110 Cities in the Yangtze River Economic Belt, China. *J. Environ. Manag.* **2022**, *321*, 115709. [CrossRef]
64. Zhou, Z.X.; Li, J.; Guo, Z.Z.; Li, T. Trade-Offs between Carbon, Water, Soil and Food in Guanzhong-Tianshui Economic Region from Remotely Sensed Data. *Int. J. Appl. Earth Obs. Geoinf.* **2017**, *58*, 145–156. [CrossRef]

land

MDPI

Article

Coupling Coordination Relationship and Spatiotemporal Heterogeneity between Urbanization and Ecosystem Services in the Songhua River Basin

He Bai [1,2], Yuanyuan Chen [1,2], Shaohan Wang [1,2], Rui Chu [1,2], Jiyuan Fang [1,2], Huina Zhang [1,2], Shuhan Xing [1,2], Lei Wang [1,2] and Dawei Xu [1,2,*]

1 College of Landscape Architecture, Northeast Forestry University, Harbin 150040, China; bh@nefu.edu.cn (H.B.)
2 Key Lab for Garden Plant Germplasm Development & Landscape Eco-Restoration in Cold Regions of Heilongjiang Province, Harbin 150040, China
* Correspondence: xudw@nefu.edu.cn

Abstract: Rapid urbanization in the Songhua River Basin (SRB), a crucial ecological barrier in China and Northeast Asia, has led to the degradation of ecosystem service functions and a decline in their value, thereby posing a significant threat to regional ecological security. Clarifying the complex coupling coordination relationship between urbanization and ecosystem services (ESs) and identifying the spatiotemporal heterogeneity of their interactions will facilitate the high-quality and coordinated development of urbanization and ESs in the SRB. This study employed a systems approach, treating urbanization and ESs as overarching systems and delineating different aspects of urbanization and ecosystem service functions as subsystems within these systems. The spatiotemporal characteristics of urbanization and the ecosystem service value (ESV) in the SRB from 1985 to 2021 were revealed. The coupling coordination relationship and the spatiotemporal heterogeneity of the interactions between urbanization and ESs in the SRB at both the system and subsystem levels were analyzed using the coupling coordination degree (CCD) model and the spatiotemporal geographically weighted regression (GTWR) model. The findings indicated that during the study period: (1) The urbanization index of SRB rose from 0.09 to 0.34, while the ESV experienced a decrease from 2091.42×10^7 CNY to 2002.44×10^7 CNY. (2) The coupling coordination degree (CCD) between urbanization and ESs in the SRB at both the system and subsystem levels increased significantly, generally transitioning from the moderately unbalanced to the basically balanced stage. Areas with high CCD values were mainly distributed in ecological function areas and low-level urbanized areas, while areas with low CCD values were mainly distributed in grassland ecological degradation areas, ecologically fragile areas, resource-dependent old industrial cities, and highly urbanized areas. (3) The subsystems of urbanization had an overall negative impact on Ess, with varying trends, but the spatial distribution pattern of the interactions remained relatively stable. Conversely, the subsystems of ESs all exhibited a trend of initially strengthening and then weakening their negative impacts on urbanization, and the spatial distribution pattern was highly correlated with the spatial distribution pattern of ESV in the SRB.

Keywords: urbanization; ecosystem services; coupling coordination relationship; spatiotemporal heterogeneity; Songhua River Basin

Citation: Bai, H.; Chen, Y.; Wang, S.; Chu, R.; Fang, J.; Zhang, H.; Xing, S.; Wang, L.; Xu, D. Coupling Coordination Relationship and Spatiotemporal Heterogeneity between Urbanization and Ecosystem Services in the Songhua River Basin. *Land* **2024**, *13*, 938. https://doi.org/10.3390/land13070938

Academic Editors: Luca Congedo, Francesca Assennato and Michele Munafò

Received: 12 June 2024
Revised: 23 June 2024
Accepted: 26 June 2024
Published: 27 June 2024

1. Introduction

Owing to escalating global urbanization, the proportion of the global urban population is projected to rise to 68% by 2050, up from 56% in 2021 [1]. Urbanization exerts multidimensional pressures on ecosystems, resulting in the degradation of ecosystem functions, a loss of ecosystem services (ESs), and a decline in their value [2,3], thereby jeopardizing regional sustainable development [4]. The Millennium Ecosystem Assessment

(MA) report indicates that over 60% of global ESs are declining, and this trend is unlikely to be effectively reversed in the next 50 years [5]. Therefore, it is imperative to enhance our understanding of the evolution of ESs during the urbanization process, as well as the coupling coordination relationship and interactions between urbanization and ESs. This understanding is pivotal in alleviating the disruption of ESs induced by urbanization and fostering the high-quality and coordinated development of urbanization and ESs.

ESs encompass the tangible or intangible benefits derived from the structures, functions, and processes within ecosystems [6,7]. The ecosystem service value (ESV) serves as a key indicator for measuring ESs [8]. Various monetary and non-monetary methodologies have been suggested for assessing ESs, typically categorized into two groups: raw data-based approaches and unit value-based approaches [9]. Raw data-based methodologies utilize ecological process models and data on ecological components to compute ESV [10]. However, they are mainly applicable to small-scale areas or individual ESs [11,12], and their relevance and specificity render a comprehensive assessment of ESV challenging. The unit value-based approach allows for a thorough evaluation of ESV by considering the comparative potential of diverse ecosystems. It is more suitable for ecological asset assessment at large spatial scales [13–15]. Costanza et al. [7] initially proposed the equivalent factor method for evaluating ESV. However, this method does not entirely align with the actual ecosystem situation in China [16]. Therefore, Xie et al. [17–19] adapted the equivalent factor method by conducting a questionnaire survey involving 200 ecologists and established an ecosystem service valuation system suitable for China. However, this assessment method neglects the amount of biomass in the ecosystem. Correcting the ESV by utilizing the plant net primary productivity (NPP) index can yield more accurate assessment results [20]. Urbanization exerts a pivotal force on ecosystem service trade-offs and synergies [21], supply and demand [22], and value assessment [23]. However, most previous studies ignored other socio-economic indicators and focused only on one specific dimension of urbanization (e.g., land use change, urbanization rate, or nighttime lighting data) and were unable to evaluate urbanization comprehensively, holistically, and scientifically from a multidimensional perspective.

Urbanization and ESs evolve separately and interact with each other. Based on systemic thinking, they can be regarded as two independent but interrelated systems, interacting and influencing each other through their own natural and human elements, with complex coupling coordination relationships and interactions. Scholars have carried out abundant studies on the complex relationships between urbanization and ESs [24–28]. Domestic and international scholars primarily adopt the grey correlation degree model [29], the pressure-state-response (PSR) model [30], and the coupling coordination degree (CCD) model [31] to study the coupling coordination relationship between urbanization and ecosystems. Among these, the CCD model can comprehensively analyze the nonlinear relationship between the two systems and has been widely used to analyze the coordination effect between urbanization and ESs. However, most studies have only explored the linkages between urbanization and ESs from a system-level perspective and lack in-depth exploration of the coupling coordination relationship between urbanization and ESs at the subsystem level [31,32]. Additionally, the CCD model cannot effectively analyze the interaction and influence between urbanization and ESs from the perspective of spatiotemporal heterogeneity. Therefore, spatiotemporal models must be introduced to provide a more refined interpretation of the characteristics and intensity of the interactions between urbanization and ESs. Existing studies on the interaction between urbanization and ESs have mainly used the system dynamics (SD) model [33], the STIRPAT model [34], the geographical detector technique [35], the OLS least squares method [36], the geographically weighted regression (GWR) model [37,38], the spatiotemporal geographically weighted regression (GTWR) model [39], and other methods. Among these, the GTWR model introduces the temporal dimension into the geographically weighted regression model [40], effectively addressing the traditional GWR model's insufficient consideration of "time–space" non-stationarity and multiscale effects, as well as the problem of poor fit, thereby better explaining the spatiotemporal heterogeneity of the variables and the dependent variable [41]. It provides an important analytical tool for solving the problem of limited samples of cross-sectional data and addressing "time–space" non-stationarity. Many scholars have demonstrated the effectiveness of the GTWR model [42,43]. However, most of these studies focus on the overall impact of urbanization on ESs while neglecting the impact of urbanization on ESs at the subsystem level and the impact of ESs on urbanization at the subsystem level [44–47]. In short, there is a paucity of research on the coupling coordination relationship between urbanization and ESs at both system and subsystem levels, as well as on the spatiotemporal heterogeneity of their interactions. This oversight hampers our capacity to devise tailored policies that effectively synchronize ecological conservation endeavors with urbanization.

Given the aforementioned research deficiencies, the study developed a comprehensive three-dimensional urbanization index evaluation system based on demographic, spatial, and economic subsystems. The study employed the NPP-modified equivalent factor method to calculate ESV, examined the spatiotemporal evolution characteristics of urbanization and ESV by combining with spatial autocorrelation analysis, and utilized the CCD model and the GTWR model to analyze the complex coupling coordination relationship between urbanization and ESs and the spatiotemporal heterogeneity of their interactions at both the system and subsystem levels from 1985 to 2021 in the Songhua River Basin (SRB). The objective of this study is to furnish a scientific foundation and novel perspectives for coordinating ecological protection and the high-quality development of urbanization in the SRB and other similar basins, with regional representativeness and long-term strategic significance.

2. Materials and Methods

2.1. The Study Area

The SRB, located between 41°42′ N and 51°38′ N latitude and 119°52′ E and 132°31′ E longitude, is situated in the north-central part of Northeast China (Figure 1a). Covering a basin area of 55.46×10^4 km^2, it is the third largest basin in China after the Yangtze River and the Yellow River [48]. This basin encompasses the northeastern forest belt within China's "Two ecological barriers and three shelters" (TEBTS), a strategic ecological security pattern, and incorporates five nationally important ecological function zones, principally contributing to water conservation and biodiversity protection [49]. It holds strategic importance in regulating the water cycle and local climate of Northeast Asia, comprehensively revitalizing Northeast China, and safeguarding the country's ecological security [50]. Due to its abundant natural resources, energy, and mineral resources, the SRB serves as a significant industrial, forestry, animal husbandry, and food production base in China. The study area encompasses 27 prefecture-level cities spanning four provinces: Heilongjiang, Jilin, Liaoning, and the Inner Mongolia Autonomous Region. It includes most of the resource-based, old, industrial cities in northeast China based on administrative units (Figure 1b). The SRB lies in the cold-temperate zone, illustrated by a gradual rise in elevation from the Songnen Plain in the central area to the mountain system located at the border (Figure 1c). The region experiences a distinct continental monsoon climate, characterized by low annual average temperatures and an uneven spatiotemporal distribution of precipitation. The spatial distribution of the annual average net primary productivity (NPP) of vegetation exhibited an uneven distribution pattern of "high in the southeast, low in the southwest, high in the surrounding area, and low in the middle of the country" (Figure 1d). Land use in the SRB exhibited clear spatial distribution patterns, with farmland predominantly situated in the Songnen Plain and the Sanjiang Plain, forestland mainly concentrated in the western and eastern regions, and grassland sporadically scattered across the central and southwestern regions (Figure 1e). The SRB, with its high forest cover and large per capita arable land, bears significant responsibility for the pattern of national food security and ecological security [51], ensuring that "food production increases" and "ecological barriers are not destroyed" will be the main focuses of its development. Between 1985 and 2021, the per capita GDP of the SRB increased from 0.12×10^4 to 4.74×10^4 CNY, and the urban population surged from 6.12 × 10^7 to 7.09 × 10^7, marked by a rise in the urban population proportion from 51.18% to 55.37%. The rapid socio-economic development has markedly heightened the level of urbanization. However, as the demographic structure has aged, the industrial structure has heavily relied on resource-based industries, and some of these industries continue to decline [52]. The rapid expansion of urban space in the SRB over the past 30 years has diminished the ESs and ecological carrying capacity of the basin [53]. The black soil in the SRB has deteriorated [54], and the grassland has experienced degradation and salinization [55]. Furthermore, there have been issues such as lake wetland degradation and shrinkage [56], frequent floods and droughts [57], and other emerging ecological problems that have posed a challenge to the SRB's high-quality development. Previous studies on the SRB have mainly focused on climate change, habitat quality, pollution assessment, and land cover change, with fewer studies on the dynamic monitoring and quantification of ESV over long time series, the complex coupling coordination relationship between urbanization and ESs, and the spatiotemporal heterogeneity of their interactions.

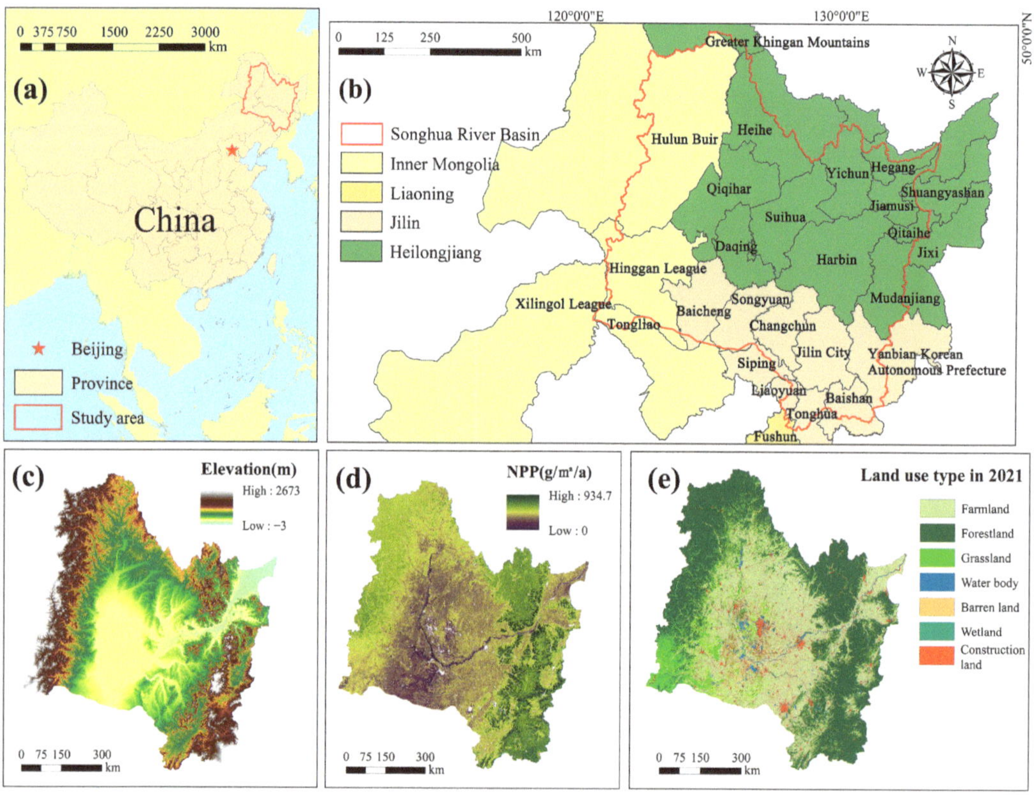

Figure 1. The study area. (**a**) Location of SRB in China; (**b**) administrative divisions; (**c**) elevation; (**d**) spatial distribution of annual average NPP in the SRB; (**e**) land use type.

2.2. Data Sources

The years 1985, 1990, 1995, 2000, 2005, 2010, 2015, and 2021 were chosen as the cross-sectional years. The data used in this study include administrative boundary data, digital elevation model (DEM) data, land use data, grain production data and annual vegetation net primary productivity (NPP) data used to modify the equivalence factor of ESV, and various types of socio-economic data (e.g., percentage of the nonagricultural population, percentage of employees in the secondary and tertiary sectors, paved road area per capita, percentage of urban built-up area in the total area of a city, GDP per capita, total retail sales of consumer goods per capita, etc.) used to measure urbanization. All raw spatial raster data were converted to uniform projected coordinates (Krasovsky_1940_Albers) and a spatial resolution of 100 m × 100 m after resampling and projection transformation in ArcGIS 10.8. The details and sources of the data are shown in Table 1.

Table 1. Details and sources for the datasets used in this study.

Data Name	Detail	Data Source
Administrative boundaries	Shapefile, polygon	Data Center for Resources and Environmental Sciences, Chinese Academy of Sciences (http://www.resdc.cn, accessed on 12 November 2023)
Digital elevation model (DEM)	Rater, 30 m	
LULC data	Rater, 30 m	The Annual China Land Cover Dataset (CLCD), produced by Jie Yang and Xin Huang's team from Wuhan University, based on Landsat images from Google Earth Engine (GEE)
Grain production data	Text, county level	China Yearbook of Agricultural Price Survey
Net primary productivity (NPP)	Rater, 500 m	The "MODIS/006/MOD17A3HGF" dataset on the Google Earth Engine (GEE) cloud computing platform (https://developers.google.cn/earth-engine/, accessed on 15 November 2023)
Socioeconomic data	Text, county level	China City Statistical Yearbook; Statistical Yearbook of four provinces: Heilongjiang, Jilin, Liaoning, and the Inner Mongolia Autonomous Region

2.3. *Methods*

2.3.1. Evaluation of Urbanization

Urbanization is multidimensional and heterogeneous. It encompasses a range of complex spatial evolution processes, including urban demographic agglomeration, economic development, and urban land expansion [58]. The core element of urbanization is the increase in the urban demographic ratio, with capital agglomeration serving as its driving force. The alteration in the land use structure and land use mode represents the spatial manifestation of urbanization. Establishing an evaluation index system is crucial to measuring urbanization. Since demographic, economic, and spatial urbanization are coordinated and mutually supportive, the comprehensive index method better reflects the overall characteristics and level of urbanization compared to the single-index approach based on a single dimension of land use or population.

This study referenced existing research findings [59,60]. The assessment encompassed the comprehensive urbanization index across three dimensions: demographic, spatial distribution, and economic indicators. Three primary indicators (demographic urbanization index (DUI), spatial urbanization index (SUI), and economic urbanization index (EUI)) were chosen to assess the urbanization index (UI). Meanwhile, based on the influencing factors of comprehensive urbanization, the characteristics of regional development, and the scientific and acquisition feasibility of data acquisition, nine secondary indicators, which are widely used to reflect China's urbanization process, were selected to evaluate each urbanization index. To mitigate the impact of diverse quantitative data on the comprehensive assessment, this study initially standardized the raw data values of the aforementioned indicators (see Equation (1)). Subsequently, it applied the entropy weight method to distribute weights to each indicator, drawing upon the information entropy obtained from the dispersion of these indicators [61]. This approach aimed to uphold objectivity and precision. Finally, the comprehensive urbanization index was calculated using the linear weighted sum method and integrated into a comprehensive index (see Equation (2)). This process formed the SRB urbanization comprehensive evaluation index (Table 2).

$$U'_{ij} = \frac{U_{ij} - min(U_j)}{max(U_j) - min(U_j)} \tag{1}$$

Table 2. Comprehensive evaluation index system of the urbanization level.

Primary Index	Weight	Secondary Index	Weight
Demographic urbanization index (DUI)	0.16	Population density (persons km^{-2})	0.10
		Percentage of nonagricultural population (%)	0.04
		Percentage of employees in the secondary and tertiary sectors (%)	0.02
Spatial urbanization index (SUI)	0.35	Paved road area per capita (m^2)	0.13
		Percentage of urban built-up area in the total area of a city (%)	0.12
		Building density (%)	0.10
Economic urbanization index (EUI)	0.50	GDP per capita (CNY)	0.19
		Total retail sales of consumer goods per capita (CNY)	0.18
		Night light values (cd m^{-2})	0.13

Here, U'_{ij} represents the standardized value of urbanization indicator j (i.e., DUI, SUI, and EUI); U_{ij} denotes the raw value of urbanization indicator j in year I; max (U_j) and min (U_j) are the maximum and minimum values of urbanization indicator j across all years, respectively.

$$UI_i = \sum_{j=1}^{3} U'_{ij} \times w_j \tag{2}$$

where UI_i is the composite urbanization index in year i and w_j is the weight of urbanization indicator j.

2.3.2. Evaluation of Ecosystem Services Value

Various scholars have classified ecosystems in diverse ways. In this study, ESs were categorized into four main categories and nine subcategories based on the principles of comprehensiveness (including multiple types of ESs), dominant function (expressing the natural environmental traits of the study region), feasibility, and existing classifications proposed by the Millennium Ecosystem Assessment [5], Costanza et al. [7], and Xie et al. [18]. This classification was integrated with the geographic and environmental characteristics of the SRB, along with studies conducted in northeast China [62]. The application of the equivalent factor method to compute ESV in the SRB involved three important aspects: first, the determination of the value of a standardized equivalent factor in the SRB; second, the construction of a table of ESV per unit area in the SRB; and third, the classification of land use and the determination of the area of each type of land use. The standard equivalent factor of ESV was determined based on the annual yield per hectare of the farmland ecosystem, which represented the contribution to ESs. For different land types, the corresponding ecological value equivalent could be calculated, and subsequently, the land area could be used to evaluate the ESV. The equation for determining the value of a standardized equivalent factor is as follows:

$$D = 1/7 \times \sum_{f=1}^{s} P_f \times Q_f \tag{3}$$

where D represents the value of an equivalent factor—specifically, ESV per unit area of farmland; $1/7$ signifies that the economic worth of the equivalent factor of ESV in the SRB equals $1/7$ of the average grain yield's market value. This implies that the economic value offered by natural ecosystems, which do not require artificial inputs, amounts to $1/7$ of the economic worth of the grain production services per unit area of existing farmland. Additionally, f denotes the main grain species in the SRB, encompassing rice, wheat, and corn; s stands for the number of crop species; P_f signifies the grain yield per unit area of crop

f, calculated by dividing the yield of crop f by the arable area of crop f; and Q_f represents the grain price of crop f. To ameliorate the ramifications of fluctuations in grain prices and inflation regarding the estimation outcomes, the grain price was set as the average price of the three major grain species of the SRB in 2021. By substituting into Equation (3), the standard equivalent factor in the SRB was determined to be 2204.41 CNY/hm^2.

Equations (4) and (5) were utilized to determine the ESV per unit area in the SRB. The "Ecosystem service equivalent value per unit area" proposed by Xie et al. presents the unit value of ecosystem service in China based on the national average state. However, the actual ESV in the SRB was contingent upon the biomass quantity within its ecosystem. To address the heterogeneity, complexity, and dynamics of ESV, this study chose plant net primary productivity (NPP) as the SRB biomass correction factor [63]. NPP reflected the organic matter production capacity of vegetation communities in the natural environment. In conjunction with the grain yield correction method, the ESV coefficient per unit area underwent adjustment to better synchronize the outcomes with the ecological characteristics of the SRB. The calculation equation is as follows:

$$VC_{ik} = D \times M_{ik} \times N \tag{4}$$

$$N = \frac{NPP_s}{NPP_c} \tag{5}$$

where VC_{ik} represents the modified ESV per unit area (CNY/hm^2) of the ith ecosystem service provided by land use type k in the ecosystem; M_{ik} denotes the equivalence coefficient of the ith ecosystem service provided by the different land use types k; N is the adjustment coefficient; NPP_s represents the net primary productivity of vegetation in the SRB; and NPP_c represents the national net primary productivity of vegetation. NPP_s is the NPP in the SRB, while NPP_c is the NPP in the entire country. Taking into account the attributes of land resources in the SRB and the prevailing circumstances, coupled with China's land use classification system, Current Land Use Classification [64], the land utilization data were reclassified into seven distinct categories: farmland, forestland, grassland, water body, barren land, wetland, and construction land using ArcGIS 10.8. The average regulation coefficient N value of the SRB from 1985 to 2021 was calculated to be 0.97. From the VC results, the table of ESV per unit area of the SRB was obtained (Table 3), and the individual ESV and total ESV were calculated in the following equations:

$$ESV_i = \sum_{k=1}^{n} A_k \times VC_{ik} \tag{6}$$

$$ESV = \sum_{i=1}^{m} ESV_i \tag{7}$$

where ESV_i and ESV are the ESV subtype and total ESV, respectively. A_k denotes the area of land use type k, and m signifies the number of cities.

Table 3. ESV per unit area of different land use types in the SRB (CNY/hm^2).

ESs		Farmland	Forestland	Grassland	Wetland	Water Body	Barren Land
Provision service (PSV)	Food production (FP)	2132.66	703.78	917.04	767.76	1130.31	42.65
	Raw materials (RM)	831.74	6355.33	767.76	511.84	746.43	85.31
	Subtotal	2964.40	7059.11	1684.80	1279.60	1876.74	127.96
Regulating service (RSV)	Gas regulation (GR)	1535.52	9213.09	3198.99	5139.71	1087.66	127.96
	Climate regulation (CR)	2068.68	8679.93	3326.95	28,897.55	4393.28	277.25
	Water regulation (WR)	1642.15	8722.58	3241.64	28,662.96	40,030.04	149.29
	Waste disposal (WD)	2964.40	3668.18	2815.11	30,710.32	31,670.01	554.49
	Subtotal	8210.74	30,283.78	12,582.70	93,410.54	77,181.00	1108.98

Table 3. *Cont.*

ESs		Farmland	Forestland	Grassland	Wetland	Water Body	Barren Land
Support service (SSV)	Soil conservation (SC)	3135.01	8573.30	4777.16	4244.00	874.39	362.55
	Biodiversity maintenance (BM)	2175.31	9618.30	3988.08	7869.52	7315.03	853.06
	Subtotal	5310.33	18,191.60	8765.24	12,113.51	8189.42	1215.62
Cultural service (CSV)	Aesthetic landscape provision (ALP)	362.55	4435.93	1855.41	10,002.18	9469.01	511.84
Total ESV		16,848.02	59,970.42	24,888.15	116,805.83	96,716.17	2964.40

2.3.3. Spatial Autocorrelation Analysis

Spatial autocorrelation analysis enables the assessment of aggregation and dispersion effects within regional units [65]. To further explore whether there are certain patterns or correlations within the spatiotemporal relationships between urbanization and ESV that interact at global and local scales, this study utilized the GeoDa 1.22 [66]. We used spatial autocorrelation analysis, which included single-variable spatial autocorrelation as well as bivariate spatial autocorrelation analyses (both global and local). The calculation equations are as follows.

$$I = \frac{\sum_{i=1}^{n} \sum_{j=1}^{n} W_{ij}(x_i - \overline{x})(y_j - \overline{y})}{S^2 \sum_{i=1}^{n} \sum_{j=1}^{n} W_{ij}} \quad (8)$$

In the equation, I represents the global bivariate Moran's I, which signifies the comprehensive correlation between urbanization and ESV at the global scale, with a value range of $[-1, 1]$. When $I > 0$, it signifies a positive spatial correlation, with a stronger correlation represented by larger values. When $I < 0$, it signifies a negative spatial correlation, with greater spatial variability reflected in larger values; $I = 0$ suggests that the spatial distribution pattern tends to be random, with no spatial correlation. Here, $n = 27$; W_{ij} denotes the spatial weighting matrix; x_i and y_j represent the urbanization and ESV of different cities, respectively; and S^2 represents the variance of all samples.

$$I_i = Z_i \sum_{j=1}^{n} W_{ij} Z_j \quad (9)$$

where I_i represents the local bivariate Moran's I. Z_i and Z_j denote the normalized values of urbanization and ESV observed in spatial units i and j, respectively. The LISA cluster maps can yield four distinct clustering patterns: high–high, high–low, low–high, and low–low. High–high indicates high values of urbanization and a high ESV, while low–low indicates a low urbanization index and low ESV; both patterns are spatially positively correlated. Conversely, low–high and high–low are spatially negatively correlated.

2.3.4. The Coupling Coordination Degree Model

To enhance clarity regarding the coupling coordination relationship of urbanization and ESs at the both system and subsystem levels in the SRB, this study utilized the coupling coordination degree (CCD) model to quantitatively assess the coordination degree of the interactive coupling between ESs and urbanization. Initially, the coupling degree (C) between urbanization and ESs was calculated using Equation (10).

$$C = \sqrt{(U_1 \times U_2)/((U_1 + U_2)/2)^2} \quad (10)$$

Here, C represents the level of coupling between urbanization and ESV, which reflects the degree of interaction between the systems [67]. The value of C varies between 0 and 1, with a higher value of C indicating a stronger interaction between urbanization and ESV. U_1 and U_2 represent the integrated assessment values of the urbanization index and ESV in the study area, respectively. Both U_1 and U_2 are normalized. It should be noted that high coupling values differ from high levels of coupling. Coupling only reflects the level of correlation between systems rather than indicating the level of coordinated development; the actual development of these two systems may be poor, but the level of coupling is high. Therefore, the degree of coordination between urbanization and ESV as well as the degree of coupling coordination were measured using Equations (11) and (12), based on the coupling degree analysis.

$$T = \alpha U_1 + \beta U_2 \tag{11}$$

$$D = \sqrt{C \times T} \tag{12}$$

In the equation, T encapsulates the overarching influence and extent of coordination among the subsystems, assessing the extent of beneficial coupling among the coupling relationships within the systems. It represents the healthy development of the system and mirrors the caliber of coordination. [68]. α and β denote the weighting coefficients of the two systems. Given that current research indicates equal importance between urbanization and ESs, the values of the α group and β group are both set to 0.5. D indicates the coupling coordination degree (CCD) of the urbanization and ESV [69], ranging from 0 to 1, reflecting the dynamic trend of the subsystem from disorder and incoherence towards orderly coordination.

Drawing from previous studies [70,71], we categorized the CCD between urbanization and ESV into five levels: seriously unbalanced (0.0–0.2), moderately unbalanced (0.2–0.4), basically balanced (0.4–0.6), moderately balanced (0.6–0.8), and highly balanced (0.8–1.0). To further elucidate the intrinsic connections between urbanization and ESV, each type of CCD was further classified into three subtypes: urbanization lagging, systematically balanced, and ESV lagging (Table 4).

Table 4. The standard of the coupling coordination degree.

Range of D Value	Type of CCD
$0.8 < D \leq 1$	Highly balanced
$0.6 < D \leq 0.8$	Moderately balanced
$0.4 < D \leq 0.6$	Basically balanced
$0.2 < D \leq 0.4$	Moderately unbalanced
$0 < D \leq 0.2$	Seriously unbalanced

2.3.5. The Spatiotemporal Geographically Weighted Regression Model

This study used the GTWR model to explore the spatiotemporal heterogeneity of urbanization and ESs at both the system and subsystem levels in the SRB. The GTWR model was calculated using Equation (13).

$$Y_i = \beta_0(u_i, v_i, t_i) + \sum_{k=1}^{p} \beta_k(u_i, v_i, t_i) X_{ik} + \varepsilon_i \tag{13}$$

In the equation, Y_i denotes the explanatory variables in the ith sample city; (u_i, v_i, t_i) denotes the spatiotemporal coordinates of the ith sample city; μ_i, v_i, and t_i are the longitude, latitude, and time of the ith sample city, respectively; β_0 (u_i, v_i, t_i) denotes the regression intercept, i.e., the constant term in the model, of the ith sample city; X_{ik} denotes the regression coefficients of the kth explanatory variable in the ith sample city; p denotes the number of explanatory variables; β_k (u_i, v_i, t_i) denotes the regression coefficient of the kth explanatory variable in the ith sample city; and ε_i is the residual term of the model.

The GTWR model employs regression coefficients to indicate the effect of each factor on the research system. A regression coefficient β greater than 0 indicates a positive correlation with the explanatory variables, i.e., a facilitating effect, while a coefficient less than 0 indicates an inhibiting effect. The greater the absolute value of the regression coefficient β, the greater the intensity of the effect. This study's GTWR model analysis was primarily based on ArcGIS 10.8, using AICc optimization settings, and the GTWR plug-in developed by Huang et al. [40].

3. Results

3.1. Spatiotemporal Characteristics of Urbanization

Figure 2 displays the changes in various urbanization indicators of the SRB. From 1985 to 2021, the level of urbanization in the SRB changed dramatically, with the UI increasing from 0.09 in 1985 to 0.34 in 2021. From 1985 to 2005, the UI in the SRB experienced slow growth. Starting around 2005, aided by the implementation of the Northeast Revitalization Strategy, urbanization within the SRB embarked upon a phase of accelerated advancement, with the rate of urbanization increasing significantly. The growth rate of UI slowed down from 2015 to 2021, indicating that urbanization in the SRB entered a phase of deceleration, focusing on improving quality and stability after a period of prosperity. The DUI, SUI, and EUI increased by 0.08, 0.22, and 0.33, respectively, during the study period. Among them, SUI maintained a stable growth trend; however, EUI showed a significant growth trend until 2021 and declined in 2021 due to China's strict blockade and quarantine measures during the COVID-19 epidemic, severely disrupting business activities in the SRB and across China. The DUI experienced slow growth throughout the study period, significantly lagging behind SUI and EUI, which aligns with the observed phenomenon of "industry–city separation" in the basin. Since the initiation of the Northeast Revitalization Strategy, urban economies within the SRB have shown signs of development, accompanied by a continuous expansion in urban spatial dimensions. However, challenges persist in resource-dependent, old, industrial cities, characterized by outdated property rights systems and industrial structures. These factors have hampered industrial growth vitality and have failed to adequately attract population concentration. Consequently, significant population outflows have ensued, contributing to sluggish overall population growth within the region. Based on the weights of the three primary urbanization indicators, SUI and EUI significantly contributed to the UI (0.84), signifying a substantial shift in urbanization within the SRB from rapid population agglomeration to a high-quality growth model driven by spatial expansion and economic development.

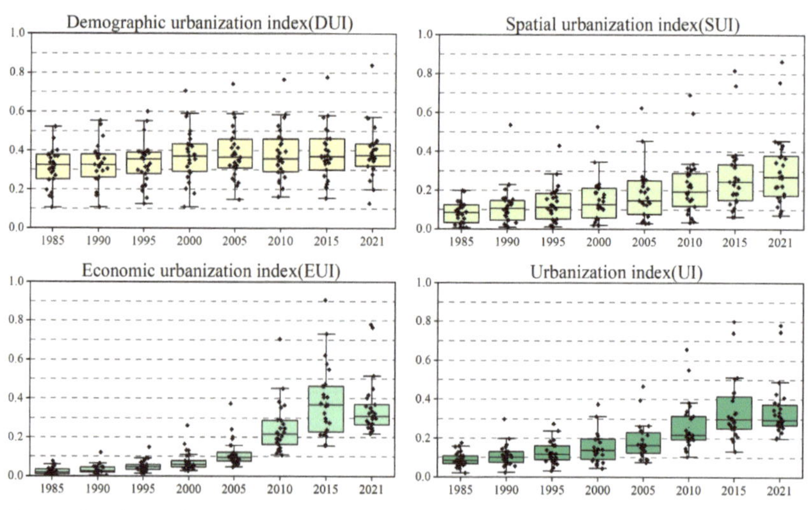

Figure 2. Levels of different urbanization indicators.

The spatial arrangement of urbanization within the SRB demonstrated significant heterogeneity, characterized by an uneven spatial pattern of "high in the middle–low in the surroundings", with the core city of the Harbin-Changchun City Cluster (HCCC) serving as the center of gravity. The temporal evolution demonstrated the Matthew effect, wherein "the strong are always stronger and the weak are always weaker" (Figure 3). The UI of all cities in the SRB increased significantly from 1985 to 2021. However, notable regional disparities existed in the rate of urbanization expansion, with urbanization progressing more rapidly in the city clusters located in the central part of the region, characterized by a stronger developmental foundation. Conversely, urbanization advanced at a slower pace in the northwestern eco-functional area, southeastern mountainous area, and ecologically fragile area in the western part of the region, which had a weaker developmental foundation. Based on the LISA cluster and significance analysis, the high–high and low–low agglomerations of the urbanization in the SRB exhibited relatively stable spatial patterns. The high–high agglomerations were primarily concentrated in the HCCC, including cities such as Daqing, Harbin, and Changchun. Local variations in spatial patterns demonstrated the time-varying characteristics of high–high agglomerations, which expanded and contracted over time. The low–low agglomerations were predominantly situated in the municipalities of the Inner Mongolia Autonomous Region. Overall, the rate of urbanization decreased from central cities and economically developed areas towards the periphery, consequently augmenting the spatial heterogeneity of urbanization.

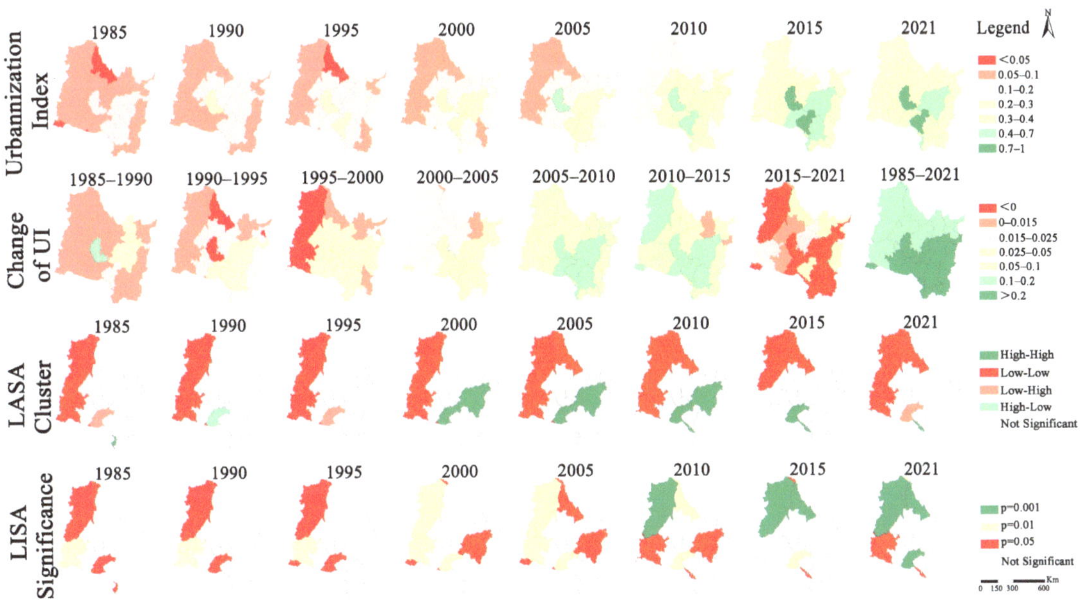

Figure 3. Spatiotemporal change and LISA cluster of urbanization in the SRB.

3.2. Spatiotemporal Characteristics of Land Use and ESV

The main land use types in the SRB were farmland and forestland, which accounted for 45.90% (25.46 × 10^4 km^2) and 42.49% (23.57 × 10^4 km^2) of the basin area, respectively. Following them were grassland and construction land, comprising 6.07% (3.36 × 10^4 km^2) and 3.74% (2.07 × 10^4 km^2) of the basin area, respectively. Conversely, the area of water bodies and barren land was relatively small, constituting 1.39% (0.77 × 10^4 km^2) and 0.40% (0.22 × 10^4 km^2), respectively. Wetland accounted for the smallest proportion, at 0.02% (0.01 × 10^4 km^2). With the advancement of urbanization and the implementation of land improvement projects in the SRB, there was an increasing trend in the area of farmland and construction land, while the area of grassland, forestland, wetland, and barren land showed a decreasing trend from 1985 to 2021. The farmland area increased the most, with a cumulative increase of 1.56 × 10^4 km^2, representing an average rate of

6.54%, followed by construction land, with an increase of 1.20 × 10^4 km^2. The grassland, forestland, wetland, and barren land areas decreased, with decreases of 1.47 × 10^4 km^2, 1.02 × 10^4 km^2, 0.15 × 10^4 km^2, and 0.12 × 10^4 km^2, respectively (Figure 4). This suggests that due to China's agricultural development policy aimed at enhancing food security, local governments and agricultural producers in the SRB continued to convert ecological land into farmland, leading to serious deforestation and land clearing. Meanwhile, economic development and urbanization have precipitated a rise in the extent of construction land, resulting in the encroachment of other land areas. However, no change in the water body was evident during the same period.

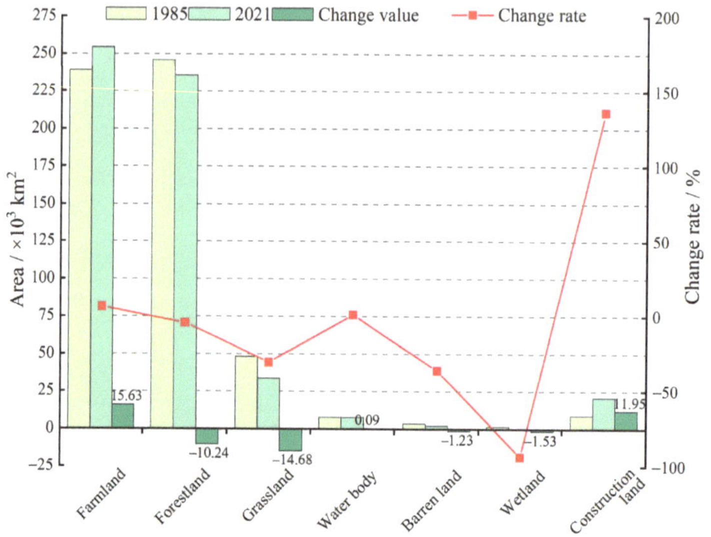

Figure 4. Land use dynamics in the SRB from 1985 to 2021.

Throughout the period from 1985 to 2021, the total ESV in the SRB experienced considerable change, decreasing from 2091.42 × 10^7 CNY in 1985 to 2002.44 × 10^7 CNY in 2021, representing a decrease of 81.61 × 10^7 CNY. During this time frame, the total ESV exhibited a sharp decrease from 1985 to 2000, followed by a slow increase year by year from 2000 to 2010, but remained on a decreasing trend from 2010 to 2021 (Table 5). The years 2000 and 2010 are significant time points for changes in ESV in the SRB, which are closely related to changes in land use transition patterns. Since 1985, large-scale deforestation activities in the SRB have significantly expanded the amount of potential farmland, resulting in significant decreases in forestland and grassland, thereby having major impacts on the ecosystem. Given this background, food security has once again emerged as a core concern of the Chinese government. In 2010, despite the execution of the grain capacity building project, the increase in ESV attributable to the expanding area of farmland was not sufficient to compensate for the decrease attributed to the reduction in the area of other ecological land. Consequently, the ESV persisted in showing a decreasing trend after 2010. Following 2015, with the implementation of the national food security strategy and the establishment of high-standard farmland, the rate of decline in ESV markedly decelerated.

Regarding individual ESs, each of the four types exhibited different degrees of decline during the study period. The order of each ES value from highest to lowest was regulating service > support service > provision service > cultural service. This indicated that the greater forest cover in the SRB rendered regulating service the core function of ecosystem service and determined the overall trend of ESV. From the perspective of subtypes for ESs, aside from food production (FP), which increased during the study period, the value of other services exhibited varying degrees of decreasing trends, with the water regulation

(WR) function experiencing the largest value loss (up to 15.17 × 10^7 CNY). This suggested that the increase in the cultivated land area in the SRB contributed to the increase in food production (FP). Nevertheless, the proliferation of farmland and the expansion of construction land encroached upon additional land spaces, resulting in the degradation of other functions of the ecosystem in the SRB, such as soil conservation (SC) and climate regulation (CR). Thus, the decrease in ESV in the SRB was primarily attributed to the expansion of farmland and construction land, resulting in the loss of forestland and grassland.

Table 5. ESV changes in the SRB from 1985 to 2021 (↑ indicates an increase in ESV; ↓ indicates a decrease in ESV).

ESs		ESV (10^7 CNY)							
		1985	1990	1995	2000	2005	2010	2015	2021
PSV	FP	73.70	74.05	75.43	75.18	74.28	73.84	74.68	74.85
	RM	180.55	180.77	178.48	176.09	175.98	176.20	174.05	174.13
	Subtotal	254.24	254.82	253.91	251.27	250.26	250.04	248.73	248.98
RSV	GR	280.42	279.73	274.82	271.99	272.47	272.54	268.51	267.90
	CR	287.14	285.41	279.26	275.97	276.10	276.04	272.71	272.18
	WR	304.65	303.04	295.45	288.47	289.16	290.60	289.54	289.48
	WD	204.00	202.66	199.06	194.68	194.24	194.75	196.14	196.25
	Subtotal	1076.20	1070.85	1048.60	1031.10	1031.97	1033.92	1026.90	1025.81
SSV	SC	310.30	309.51	305.55	303.43	303.59	303.14	299.62	298.73
	BM	314.93	314.04	308.66	305.24	305.72	305.83	302.09	301.38
	Subtotal	625.23	623.56	614.21	608.67	609.31	608.97	601.70	600.11
CSV	ALP	135.74	134.83	130.91	128.67	129.31	129.71	127.93	127.53
Total ESV		2091.42	2084.05	2047.62	2019.71	2020.86	2022.64	2005.26	2002.44
ESs		Change rate (%)							
		1985–1990	1990–1995	1995–2000	2000–2005	2005–2010	2010–2015	2015–2021	1985–2021
PSV	FP	0.48 ↑	1.86 ↑	−0.33 ↓	−1.20 ↓	−0.59 ↓	1.14 ↑	0.23 ↑	1.57 ↑
	RM	0.12 ↑	−1.27 ↓	−1.34 ↓	−0.06 ↓	0.12 ↑	−1.22 ↓	0.05 ↑	−3.55 ↓
	Subtotal	0.23 ↑	−0.36 ↓	−1.04 ↓	−0.40 ↓	−0.09 ↓	−0.52 ↓	0.10 ↑	−2.07 ↓
RSV	GR	−0.24 ↓	−1.76 ↓	−1.03 ↓	0.18 ↑	0.02 ↑	−1.48 ↓	−0.23 ↓	−4.46 ↓
	CR	−0.60 ↓	−2.15 ↓	−1.18 ↓	0.05 ↑	−0.02 ↓	−1.21 ↓	−0.19 ↓	−5.21 ↓
	WR	−0.53 ↓	−2.50 ↓	−2.36 ↓	0.24 ↑	0.50 ↑	−0.36 ↓	−0.02 ↓	−4.98 ↓
	WD	−0.65 ↓	−1.78 ↓	−2.20 ↓	−0.22 ↓	0.26 ↑	0.72 ↑	0.06 ↑	−3.80 ↓
	Subtotal	−0.50 ↓	−2.08 ↓	−1.67 ↓	0.08 ↑	0.19 ↑	−0.68 ↓	−0.11 ↓	−4.68 ↓
SSV	SC	−0.26 ↓	−1.28 ↓	−0.69 ↓	0.05 ↑	−0.15 ↓	−1.16 ↓	−0.30 ↓	−3.73 ↓
	BM	−0.28 ↓	−1.71 ↓	−1.11 ↓	0.16 ↑	0.03 ↑	−1.22 ↓	−0.23 ↓	−4.30 ↓
	Subtotal	−0.27 ↓	−1.50 ↓	−0.90 ↓	0.11 ↑	−0.06 ↓	−1.19 ↓	−0.26 ↓	−4.02 ↓
CSV	ALP	−0.68 ↓	−2.91 ↓	−1.71 ↓	0.50 ↑	0.31 ↑	−1.37 ↓	−0.31 ↓	−6.05 ↓
Total ESV		−0.35 ↓	−1.75 ↓	−1.36 ↓	0.06 ↑	0.09 ↑	−0.86 ↓	−0.14 ↓	−4.25 ↓

Spatially, the dispersion of ESV within the SRB showcased pronounced spatial heterogeneity (Figure 5). High-value ESV predominantly occurred in the northwestern part of the SRB, which was notably abundant in ecological resources as an important forest resource reserve area in China. Being located at the border of China, where development is restricted, resulted in higher ESVs in this region. Conversely, a low-value ESV was prevalent in the Songnen Plain, located in the central part of the SRB. Being the main grain-producing area in China, this region experienced a high population concentration and agricultural production activities. The conversion from dryland to rice to boost grain production resulted in high water demand, leading to the loss and degradation of wetlands and grasslands. As a result, the ESV remained low. Regarding spatial evolution, the ESV displayed a general degradation trend at the municipal level over the study period. During this time, 85.19% of cities witnessed a decline in ESV, with variations in the degree of decline from city to city.

Among them, the ESV in Heihe City decreased the most, by 19.42×10^7 CNY, which was closely related to environmental problems such as regional soil and water loss caused by agricultural over-reclamation, forestland over-logging, grassland over-grazing, and mine over-mining in the process of rapid urbanization. The results from the LISA cluster and significance indicated that there was minimal alteration observed in the spatial dispersion of ESV within the SRB over time, but the overall spatial distribution pattern remained stable. The high–high agglomeration areas were mainly distributed in northeastern Inner Mongolia and northwestern Heilongjiang Province, characterized by high vegetation cover and rich biodiversity, while the areas exhibiting low–low agglomeration were predominantly situated in the southern region of Jilin Province, where industrial production activities were more prevalent. The disturbance of the landscape by human activities further intensified the spatial heterogeneity of the distribution of ESV in the SRB.

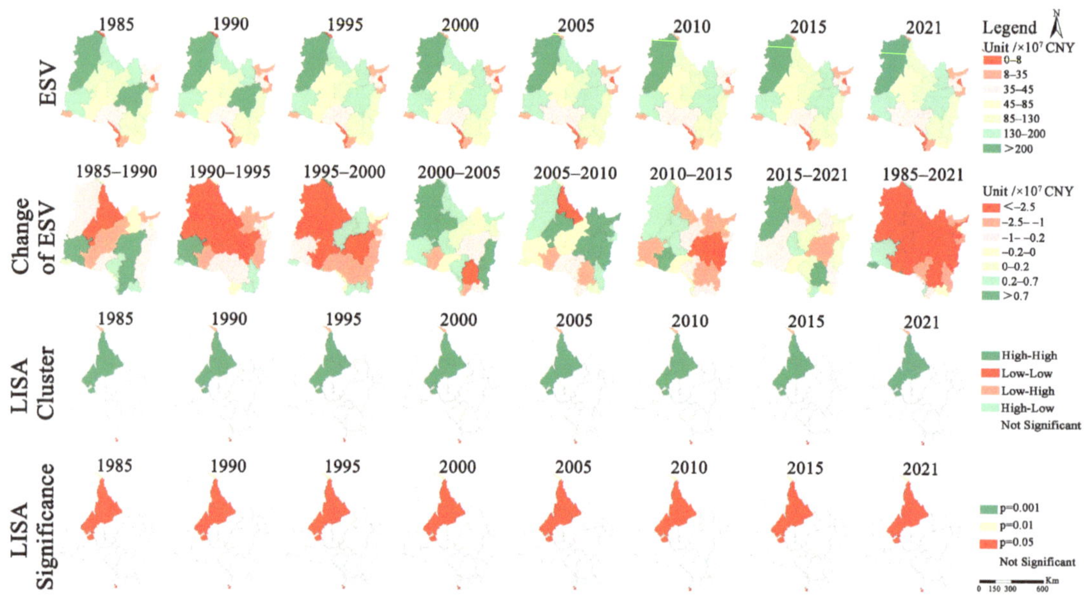

Figure 5. Spatiotemporal change and LISA cluster of ESV in the SRB.

3.3. The Coupling Coordination Relationship between Urbanization and ESs

3.3.1. The Temporal Differentiation Characteristics of the CCD in the SRB

After revealing the spatiotemporal evolution characteristics of urbanization and ESs, the CCD model was employed to analyze the coupling coordination relationship between urbanization and ESs at both the system and subsystem levels in the SRB (Figure 6). The results indicated that the coupling coordination degree (CCD) between urbanization and ESs at both the system and subsystem levels showed an increasing trend in the SRB from 1985 to 2021, transitioning from the moderately unbalanced to the basically balanced stage. It showed that SRB made significant achievements in improving the coupling coordination relationship between urbanization and ESs. Specifically, the CCD of urbanization and ESs increased from 0.30 to 0.44, representing an increase of 43.16%. At different stages of urban development, variations existed in the intensity and coordination of the interaction between urbanization and ESs, as well as in the development trends. The upward trend of the CCD was relatively flat from 1985 to 2005, and the CCD level increased significantly from 2005 to 2021. In 1985, merely 12% of urban areas in the SRB achieved the CCD indicative of a basically balanced stage. The main reason was the backwardness of economic development in the SRB during this period. Moreover, the industrial composition of the SRB during this period was dominated by heavy industry, resulting in the region being energy-consuming, resource-dependent, and environmentally burdensome. This resulted in the substantial degradation of the ecological environment. By 2005, due to the implementation of policy support from the National Northeast Revitalization Strategy and proactive measures by local governments, the urbanization development level and environmental management capacity of the SRB had rapidly improved. Consequently, 41% of the cities reached the basically balanced stage, and the number of cities with a seriously unbalanced stage dropped to zero. By 2021, with the rise in the ecological civilization as a national strategy, the SRB implemented measures such as returning farmland to forests, restoring grasslands, and remediating polluting enterprises. These ecological construction projects and environmental management efforts yielded remarkable achievements. As a result, 56% of the cities were at the basically balanced stage, and 7% reached the moderately balanced stage.

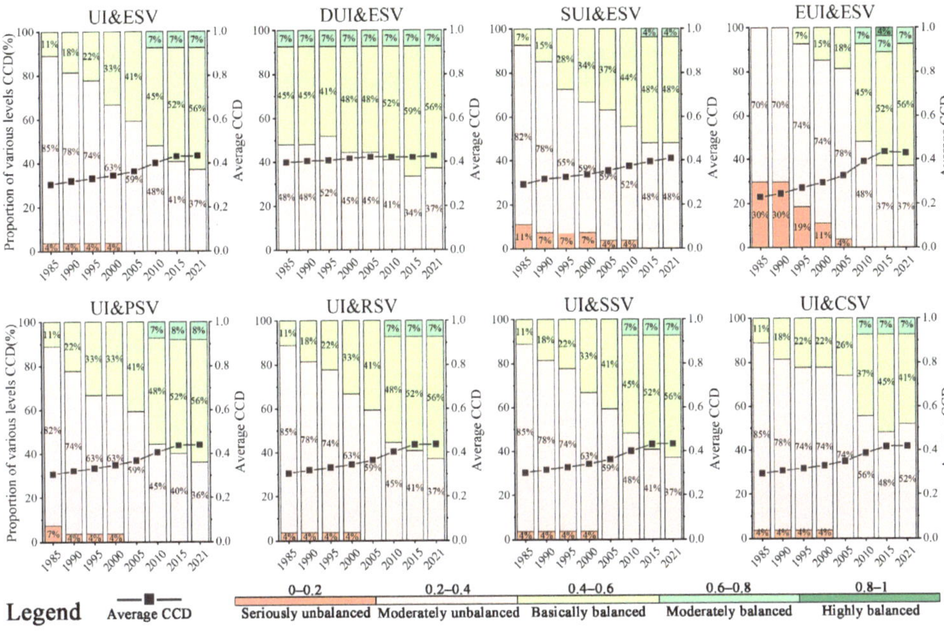

Figure 6. The temporal changes in the CCD between urbanization and ESs in the SRB from 1985 to 2021.

3.3.2. The Spatial Differentiation Characteristics of the CCD in the SRB

Figure 7 shows the spatial differentiation characteristics of the CCD in the SRB from 1985 to 2021. The results indicated that the CCD between urbanization and ESs improved significantly at both the system and subsystem levels in the SRB, with seriously unbalanced and moderately unbalanced municipalities contracting and basically balanced and moderately balanced municipalities expanding. There was an overall shift from seriously unbalanced and moderately unbalanced to basically balanced, moderately balanced, and even highly balanced. Overall, most cities in the SRB were still in the transitional phase of coupling coordination development. Except for Hulunbuir, where the CCD between economic urbanization and ESs reached the highly balanced stage in 2015, the various types of CCD between urbanization and ESs in most cities were still at the basically balanced and moderately balanced stages by 2021. In addition, there was considerable variation in the CCD between different cities, with marked variations in the spatial distribution. Spatially, the areas with low CCD values in the SRB were mainly concentrated in grassland ecological degradation areas and ecologically fragile areas like Tongliao City, resource-dependent, old, industrial cities such as Hegang, Jiamusi, Jixi, and Tonghua City, as well as highly urbanized areas like Liaoyuan City. Tongliao City, known for agriculture and animal husbandry, is a typical area of grassland ecological degradation and is ecologically fragile. Transitional grazing practices there had resulted in severe degradation, with grassland resources becoming degraded, sandy, and highly salinized. The delayed benefits of grassland ecological management contributed to a relatively low CCD. Concerning resource-dependent, old, industrial cities like Tonghua City and Jixi City, their single industrial structure, high energy consumption, and high emissions have significantly and negatively impacted ecosystem structures, processes, and functions. Consequently, these areas exhibited a relatively low CCD. The areas with high CCD values were mainly concentrated in the Daxing'an and Xiaoxing'an mountains forest ecological function area, the Hulunbuir grassland meadow ecological function area, and the Changbai Mountain forest ecological function area, which are regions with low levels of urbanization, as well as in Harbin, strategically positioned as the "Northeast Asia Regional Center City". The differing levels of urbanization and ESV development across various cities presented a challenge in promoting the high-quality coordinated development of urbanization and ESs in the SRB.

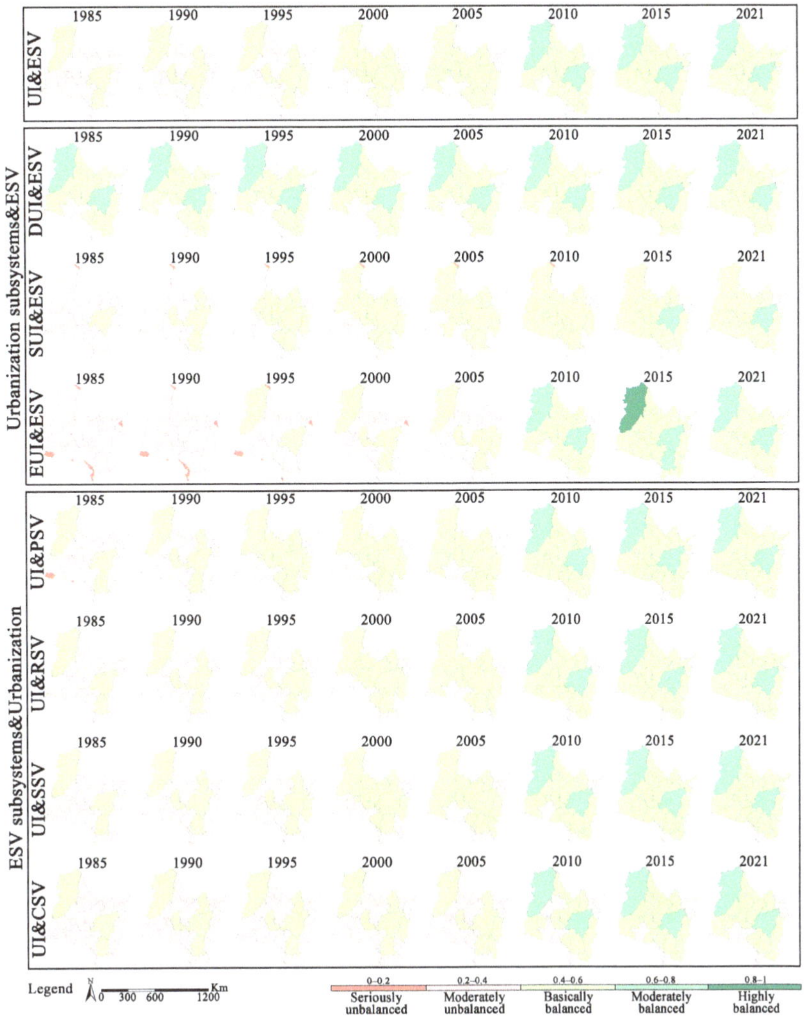

Figure 7. The spatial differentiation characteristics of CCD in the SRB from 1985 to 2021.

3.4. Spatiotemporal Heterogeneity in the Interaction between Urbanization and ESs

3.4.1. Data Verification

This study employed the GTWR model. To further explore the impacts of various urbanization subsystems on ESs, we analyzed the nine secondary indicators (X1–X9) of the three primary indicators (demographic, spatial, and economic urbanization) as explanatory variables, with ESs as the dependent variable. The spatiotemporal heterogeneity of the impacts of each subsystem on ESs was analyzed. The spatiotemporal heterogeneity of the impacts of each ES subsystem on urbanization in the SRB from 1985 to 2021 was also analyzed, with urbanization as the dependent variable and the four subsystems of ecosystem services (Y1–Y4) as the explanatory variables. Before performing the GTWR model calculations, all variables were standardized to prevent pseudo-regression. Subsequently, SPSS 29 was employed to test the multicollinearity of all standardized variables using regression analysis. The selected subsystems of urbanization and ESs met the variance inflation factor (VIF) criterion of <10 and passed the significance test. Detailed model calculation results and parameters are presented in Table 6. The parameters indicated that the model effectively measured the spatiotemporal heterogeneity of the influence of explanatory variables on dependent variables.

Table 6. Explanation of the GTWR model parameters.

Dependent Variable	R^2	Bandwidth	Residual Squares	Sigma	AIC_C	Spatiotemporal Distance Ratio
Urbanization	0.9234	0.1150	0.5331	0.0488	−528.4480	0.5418
Ecosystem Services	0.9782	0.1221	0.1842	0.0287	−658.2960	0.2731

3.4.2. Temporal Heterogeneity in the Interaction between Urbanization and ESs

Figure 8a–i illustrate the temporal evolution of the regression coefficients for the impact of each subsystem of urbanization on ESs. Overall, SRB urbanization exhibited a generally negative impact on ESs. Specifically, the temporal heterogeneity characteristics of each urbanization subsystem's impact on ESs in the SRB from 1985 to 2021 varied significantly. The positive impact of population density (X1) on ESs gradually decreased and then stabilized; the negative impact of the percentage of the nonagricultural population (X2) on ESs increased; the impact of the percentage of employees in the secondary and tertiary sectors (X3) on ESs gradually changed from positive to negative; the positive impact of the per capita paved road area (X4) on ESs decreased and then increased; the positive impact of the percentage of urban built-up area (X5) on ESs increased; the negative impact of building density (X6) on ESs first decreased and then increased; the impact of GDP per capita (X7) on ESs shifted from positive to negative over time; the impact of per capita total retail sales of consumer goods (X8) on ESs gradually changed from negative to positive; and the negative impact of night light values (X9) on ESs gradually weakened. There were significant differences in the direction and intensity of each urbanization subsystem's influence on ESs, with varying trends, reflecting the complexity of urbanization's impact on ESs.

Figure 8j–m illustrate the temporal evolution of the regression coefficients for the impact of each subsystem of ESs on urbanization. The results indicated that the regression coefficients of the impacts of each subsystem of ESs on urbanization in the SRB from 1985 to 2021 were all negative, suggesting that all subsystems of ESs negatively impacted urbanization. After 2005, the absolute value of the regression coefficients decreased over time, indicating a declining intensity of the negative impacts of the subsystems of ESs on urbanization, suggesting significant progress in improving ecosystem quality in the SRB.

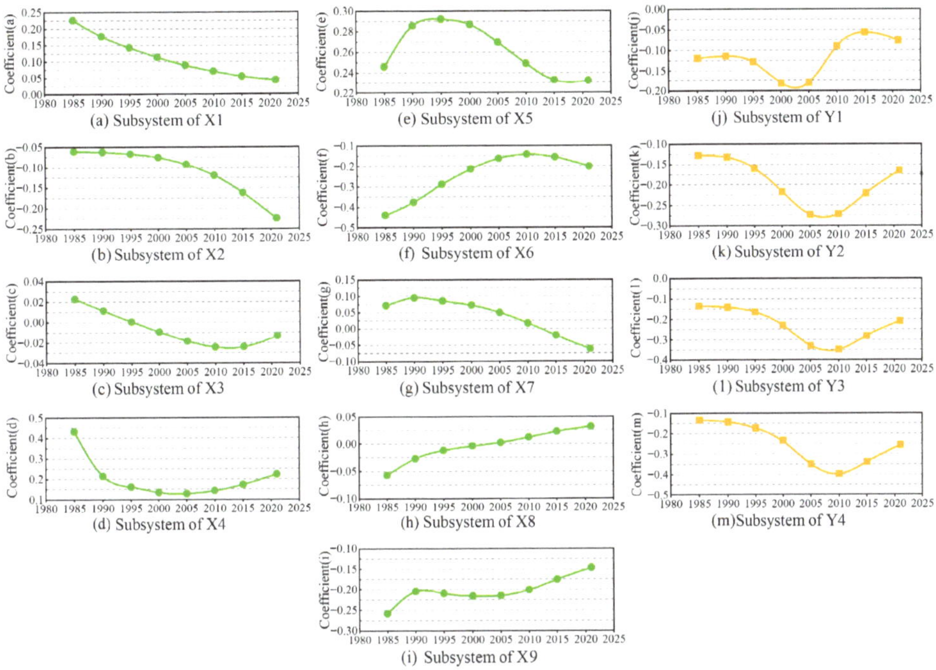

Figure 8. Temporal heterogeneity in the interaction between urbanization and ESs.

3.4.3. Spatial Heterogeneity in the Interaction between Urbanization and ESs

The GTWR model quantified the spatial heterogeneity characteristics of the impacts of various urbanization subsystems on ESs. Using the natural breaks method, the GTWR regression coefficients were classified and visualized. Positive impacts were divided into high positive impact, moderate positive impact, and low positive impact, while negative impacts were categorized into high negative impact, moderate negative impact, and low negative impact, as shown in Figure 9. Overall, urbanization had a generally negative impact on ESs, but the influence of urbanization subsystems on ESs was multifaceted, with varying directions and intensities across different regions. Over the past three decades, the spatial distribution pattern of the impacts of urbanization subsystems on ESs remained relatively stable, except for occasional fluctuations between 2000 and 2005. This pattern exhibited significant north–south and east–west spatial differentiation.

Specifically, the population density (X1) had both positive and negative impacts on ESs, with a balanced proportion of both. Negative impact areas were mainly concentrated in the ecologically fragile regions in the western and eastern parts of the SRB, while positive impact areas were primarily distributed in the high-ESV areas in the north and resource-dependent, old, industrial cities in the south. The percentage of the nonagricultural population (X2) had a negative impact, mainly in the northern SRB, with positive impacts primarily in the southern SRB. The impact of the percentage of employees in the secondary and tertiary sectors (X3) on ESs was mainly characterized by low positive and low negative impacts, with negative-impact areas concentrated in the ecologically fragile regions of the western and eastern SRB. The per capita paved road area (X4) generally had a positive impact on ESs, with the impact strength gradually weakening from north to south. The percentage of the urban built-up area (X5) showed complex spatial heterogeneity, with high-positive-impact areas mainly in the northern SRB. The building density (X6) had a high negative impact mainly in the high-ESV areas in the northern SRB. The areas with positive impacts of GDP per capita (X7) on ESs were mainly concentrated in resource-dependent, old, industrial cities in the southern SRB. The per capita total retail sales of consumer goods (X8) mainly had a low positive impact on ESs. The areas with positive impacts of night light values (X9) on ESs gradually shifted towards the northern SRB.

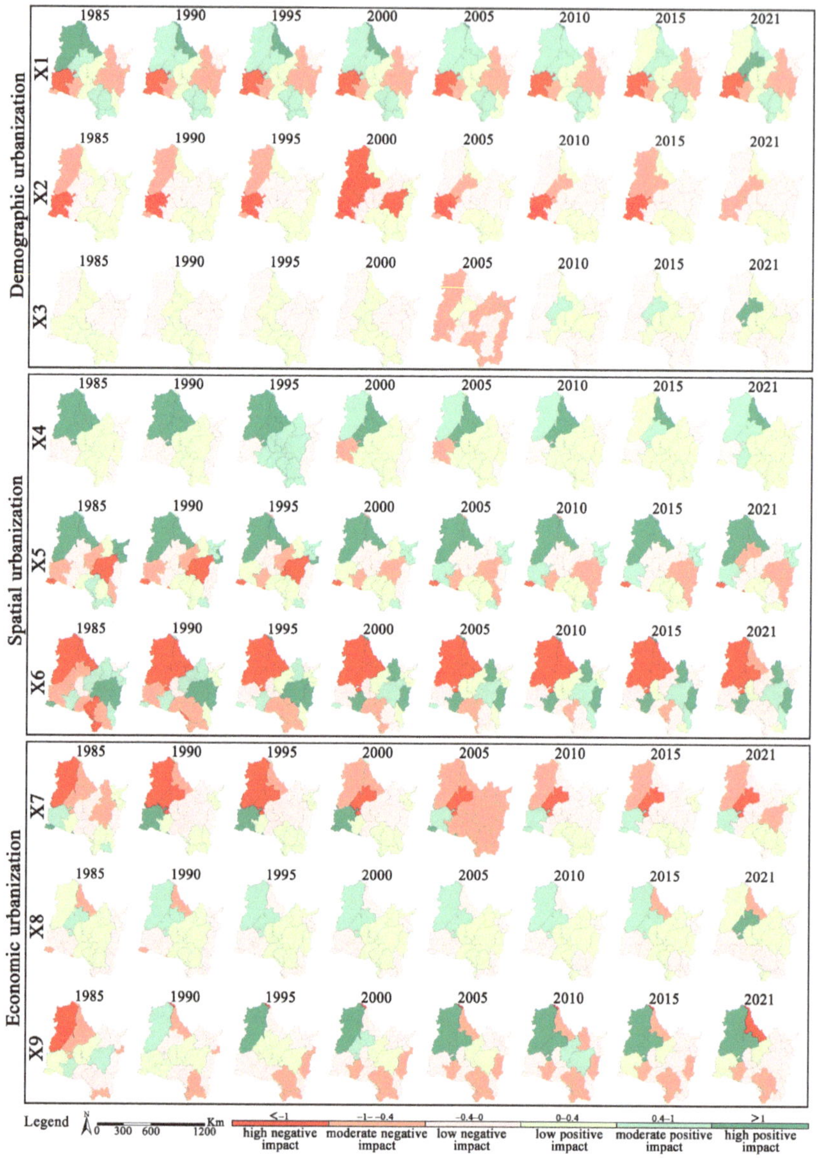

Figure 9. Spatial heterogeneity characteristics of the impact of each subsystem of urbanization on ESs in the SRB from 1985 to 2021.

Additionally, a comparative analysis of the GTWR regression coefficients revealed that the ecosystems in the northwestern SRB were more sensitive and had a higher response to urbanization changes, indicating a necessity for the enhanced protection and improvement of the ecological spaces in this region.

The spatial differentiation characteristics of the impacts of various ESs subsystems on urbanization are illustrated in Figure 10. The results indicated that the spatial patterns of and intensity changes in the impacts of various ESs subsystems on comprehensive urbanization had exhibited similar trends. Over time, the spatial heterogeneity of the impacts of various ESs subsystems on urbanization had increased. From 1985 to 2010, the scope of the negative impacts of various ESs subsystems on comprehensive urbanization had decreased, while the intensity of these impacts had increased. Conversely, the scope and intensity of the positive impacts of various ESs subsystems on urbanization had expanded gradually. From 2010 to 2021, the intensity and spatial patterns of both types of impacts had occasionally fluctuated but had remained generally stable. In resource-dependent, old, industrial cities like Daqing, Baicheng, and Songyuan, the urbanization process experienced prolonged and intense negative impacts from various ESs subsystems. The rapid expansion of urbanization in these cities had caused significant disturbances to the structure and function of ecosystems, resulting in encroachment and degradation. In the vast surrounding areas, forest, grassland, and farmland ecosystems had transformed into artificial or semi-artificial ecosystems, leading to a decline in ESV. This decline in ESV had further constrained urbanization development.

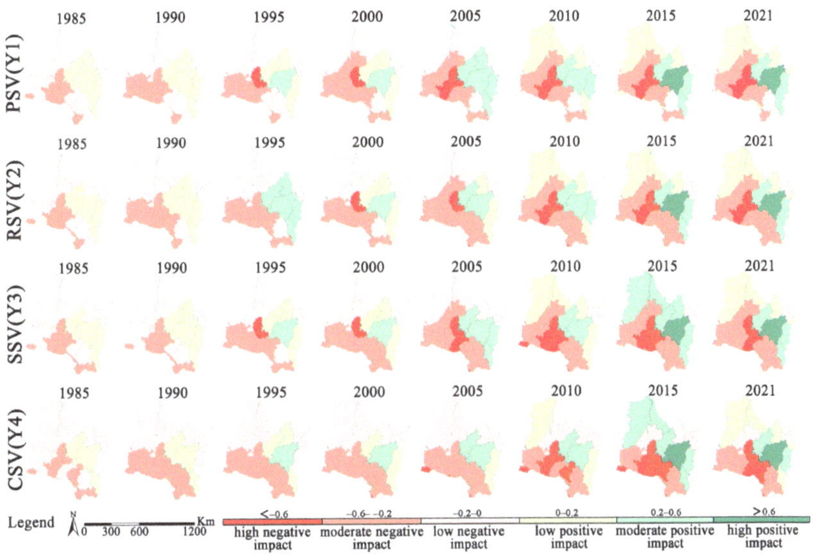

Figure 10. Spatial heterogeneity characteristics of the impact of each subsystem of ESs on urbanization in the SRB from 1985 to 2021.

Notably, the spatial distribution pattern of the impacts of the four ESs subsystems on urbanization was closely correlated with the spatial distribution of ESV. In areas with low ESVs, the impacts of ESs subsystems on urbanization had been primarily negative, whereas in areas with high ESVs, the impacts had been primarily positive. This finding indicated that urbanization development had been significantly constrained by the ecological carrying capacity, as higher ESVs had favored urbanization development.

4. Discussion

4.1. The Coupling Coordination Relationship between Urbanization and ESs

This study utilized the CCD model to analyze the coupling coordination relationship between urbanization and ESs in the SRB at both the system and subsystem levels and explored the spatial differentiation characteristics. From 1985 to 2021, the CCD between urbanization and ESs in various cities of SRB showed an upward trend at both the system and subsystem levels. This indicated that after experiencing a period of rough economic growth, the contradiction between the supply and demand of urban construction land in the SRB forced continuous improvements in resource utilization efficiency. Urbanization was promoted through refined and efficient land use, with increased attention to ecological protection, leading to improvements in the CCD level. However, the overall CCD had not yet reached a satisfactory level, indicating considerable room for improvement.

The areas with high CCD values between urbanization and ESs in the SRB were mainly concentrated in ecological function areas and regions with low levels of urbanization. Among them, HulunBuir boasted the highest ESV, but due to its low level of urbanization, the coordination between urbanization and ESs had not yet reached a highly balanced stage [31]. The city of Harbin, being the economic and social center of Northeast China, had actively improved land use efficiency and focused on industrial structure adjustment and layout optimization in response to sustainable development principles. By limiting industries with high environmental costs and a high consumption of raw materials and energy, supporting green industries, and developing urbanization and ecological protection in parallel, Harbin had effectively enhanced the stability and balanced development capability between urbanization and ESs. Despite its high level of urbanization, Harbin had achieved the CCD of a moderately balanced stage.

The areas with low CCD values between urbanization and ESs in the SRB were mainly concentrated in grassland ecological degradation areas, ecologically fragile areas, resource-dependent, old, industrial cities, and highly urbanized areas. Resource-dependent, old, industrial cities face acute problems such as depleted natural resources, severe environmental pollution, and economic growth stagnation, inevitably leading to increased conflicts between urbanization and ESs [32,72]. As the ESV decreases, the pace of urbanization development gradually slows, accompanied by population loss and economic decline, making it increasingly difficult to improve the level of CCD. Resource-dependent, old, industrial cities such as Jiamusi and Hegang are located in the Sanjiang Plain wetland ecological functional area with a low population density and large wetland areas. If the local government effectively enhances environmental management, introduces new production technologies, and optimizes the industrial structure from the source, process, and end-of-pipe emissions, the conflicts between urbanization and ESs could be reduced, potentially improving the CCD level. The city of Tongliao, a grassland ecological degradation area with a large population but low per capita economic indicators, faces dual pressures of ecological environmental protection and socio-economic growth due to underdeveloped secondary and tertiary industries and slower overall urbanization compared to urban agglomerations.

Urbanization and ESs are interdependent and mutually restrictive, with the dynamic changes in their CCD reflecting the amalgamated impacts of various processes, encompassing economic advancement, the urbanization process, alterations in ecosystems, and policy orientations [8,73]. Against the backdrop of "Revitalizing the old industrial base in Northeast China", the relevant governments in the SRB have adopted a series of policy measures to seek coordinated development paths between urbanization and ecological protection. In addition to considering assessments of environmental carrying capacity and land suitability, it is also worth considering incorporating the CCD development into future national spatial planning.

4.2. Spatiotemporal Heterogeneity in the Interaction between Urbanization and ESs

The spatiotemporal heterogeneity characteristics of the interaction between urbanization and ESs in the SRB from 1985 to 2021 revealed the complexity of the direction and intensity of their influences. The overall impact of various urbanization subsystems on ESs in the SRB was generally negative. Specifically, demographic urbanization and spatial urbanization trends showed a weakening positive impact and a strengthening negative impact on ESs. In the early years, the level of demographic urbanization in the SRB was low. The continuous concentration of the urban population could effectively alleviate issues such as resource consumption and environmental pollution. However, as the diversified demands for food, water, and land resources from demographic urbanization increased, the rapid expansion of cropland and construction land at the expense of ecological land, such as forestland and grassland, placed immense pressure on the carrying capacity of resources and the environment, exacerbating the loss of ESs. Spatial urbanization led to the expansion of construction land and road networks, which increased surface runoff, fragmented habitat patches, disrupted the ecosystem balance, and reduced ESs capacity. With the growth of the regional economy and social productivity in the SRB, residents' environmental awareness increased, and the demand for a high-quality environment grew. High-pollution, high-emission, and high-energy consumption industries, previously reliant on resource development and infrastructure construction, transitioned toward green development. Concurrently, the government invested more in ecological restoration and protection, adjusting policies to enhance the ecosystem's resilience against the negative impacts of urbanization, so the negative impact of economic urbanization on ESs gradually weakened. Notably, the shift in trends of urbanization subsystems' impacts on ESs indicated that the interaction between urbanization subsystems and ESs had critical thresholds. The intensity of urbanization expansion beyond these thresholds suppressed ESs capacity. Furthermore, the interaction between urbanization and ESs was delayed. The decline in ESV did not immediately constrain urbanization development, and optimizing urbanization patterns did not immediately improve ESV [32].

The negative impacts of various ESs subsystems on urbanization in the SRB initially increased and then decreased during the study period, indicating that the disordered expansion of construction land due to rapid urbanization in the past disrupted material cycles and energy flows, weakened the carrying capacity of resources and the environment, fragmented habitat patches, and undermined regional ecological security, thus constraining urbanization development. However, with the implementation of land use policies such as permanent basic farmland and ecological redlines, the optimization of landscape patterns and the enhancement of ecological connectivity in the SRB improved the ecosystem's resilience to external disturbances, maintained the stability of ESs, and ensured their sustainable supply, thus gradually reducing the negative impacts of ESs on urbanization. The spatial distribution patterns of ESs subsystems' impacts on urbanization were highly correlated with the spatial distribution patterns of ESV in the SRB. In the western part of the SRB, particularly in the Horqin Grassland ecological functional area of Hinggan League, strategic positioning had restricted large-scale industrial agglomeration and the expansion of built-up areas, leading to continued negative impacts of ESs subsystems on urbanization. With the rise of China's ecological civilization strategy after 2010, the impacts of ESs subsystems on urbanization in the northwestern SRB had shifted from negative to positive, indicating that the interaction between urbanization and ESs was influenced not only by ecological carrying capacity but also by policy factors [74].

Urbanization exacerbated the demand for land resources and infrastructure through population growth, economic development, and spatial expansion. This led to drastic changes in regional land-use types, structures, and patterns, disrupting ecosystem structure, function, and quality, thereby significantly inhibiting ESs [75–79]. Conversely, ecosystems had a limit to the services they could provide. When the external disturbances from urbanization expansion exceed the carrying capacity of the ecosystem to maintain its ecological balance, the imbalance between the supply and demand of ESs triggers environmental problems. This, in turn, constrains and limits urbanization development through factors such as resource endowment, population displacement, capital competition, and policy intervention, thereby affecting regional coordinated development [80–82]. Different types of urbanization influenced each other, indirectly affecting ESs. In summary, there existed complex interactions between urbanization and ESs, with ESs possibly being constrained by other factors such as rainfall [83], topography [84], and soil type [85].

4.3. Policy Implications

This study demonstrated that the coupling coordination relationship and interaction between urbanization and ESs in the SRB exhibited spatiotemporal heterogeneity, with national policies significantly influencing their coordination. Therefore, the government and policymakers in the SRB should consider high-quality coordination as a goal, adjusting urbanization planning and formulating differentiated ecological restoration and environmental development policies based on the characteristics of coupling coordination relationships in different regions. Given this, we propose region-specific suggestions for coordinating urbanization and ESs in the SRB.

For ecological function zones and regions with low levels of urbanization, efforts should focus on mitigating urban expansion impacts on ecosystems. This involves scientifically allocating land resources between ecological and non-ecological uses, reasonably limiting urbanization scale and intensity, and strictly controlling disorderly expansion. Additionally, black soil protection should be strengthened, high-standard farmland construction promoted, and the transition to sustainable agriculture and food diversity enhanced to increase agricultural systems' resilience and alleviate pressure on other ecosystems [86,87]. Shifts in agricultural water use from extensive to conservation-intensive methods should be encouraged, and the upper limits of groundwater resource utilization in the Songnen Plain and Sanjiang Plain should be strictly observed. Further support for ecological economic shelter forests should be provided through funding and technology to improve reforested lands. Residents should adopt ecological lifestyles, accelerate the development of productive service industries such as eco-leisure tourism, and expand the market for green products and services.

For grassland ecological degradation areas and ecologically fragile zones, efforts should focus on the conservation of grassland and wetland resources to maintain their ecosystem functions. Utilizing the advantage of extensive wetlands in the SRB could improve the wetland protection system, promote the restoration of degraded wetlands, maintain wetland biodiversity, and improve the wetland protection and management mechanism. Simultaneously, under the premise of grassland ecological security, the protection and rational use of grassland resources should be ensured. Basic grassland protection, grass-livestock balance, and rotational grazing systems must be strictly enforced, alongside efforts to restore degraded grasslands, develop ecological restoration technologies, and implement targeted protection and restoration for various degradation types, such as desertification, salinization, and severe degradation. These measures aim to improve overall grassland vegetation cover and restore the lost ecosystem functions.

For resource-dependent, old, industrial cities and highly urbanized areas, efforts should focus on improving land use efficiency and balanced development capacity. Gradual adjustments should be made to the unreasonable land use patterns formed during the early stages of urbanization to mitigate the adverse impacts on ecosystems, optimizing urbanization development patterns to enhance the ecological carrying capacity. Coordinated planning for urbanization development and ecological restoration should be strengthened, emphasizing urban forest construction and biodiversity protection [88]. Maximizing the retention and maintenance of existing forest and lake natural ecosystems' self-regulation, self-purification, and self-restoration capabilities is crucial to avoiding further habitat loss [89]. Both natural and artificial restoration methods should be used to restore river and lake ecosystems, and the comprehensive control of soil erosion in mining wastelands, sloping farmlands in black soil areas, and erosion gullies should be promoted. Emphasis should be placed on transforming resource-dependent industries, fostering alternative industries, accelerating the completion of resource industry chains, diversifying the single industrial structure, and forming a green industrial system aligned with sustainable development principles. Alongside economic transformation, equal emphasis should be placed on improving urban functions and environmental protection, promoting breakthrough development in both the economy and ecology, and enhancing the sustainable development capacity of resource-dependent, old, industrial cities.

The SRB's ecosystems involve multiple stakeholders and administrative units, requiring full consideration of the regional coupling coordination characteristics to effectively coordinate the interests of various stakeholders and guide specific ecological activities. Possible interventions in the SRB include establishing effective communication and cooperation mechanisms, developing green ecological corridors to improve ecological connectivity in areas with declining ESs, strengthening the flow of ESs, and continually enhancing ecosystem functions and the ecological carrying capacity of the SRB. Future efforts should focus on coordinating economic and social development, land use, and the ecological environment, enhancing the targeting and feasibility of planning, and promoting urbanization towards a verdant, low-emission, innovation-centric, and synergistic development paradigm.

4.4. Limitations and Future Work

This study analyzed the spatiotemporal evolution characteristics of urbanization and ESs, exploring the complex coupling coordination relationship and spatiotemporal heterogeneity of their interactions. However, it does have some shortcomings. Initially, this study only selected three urbanization indicators—economic, demographic, and spatial—which have some limitations. Future research should explore more scientific methods for measuring urbanization and develop a more comprehensive urbanization assessment indicator system. Second, this study employed a static equivalent factor method to evaluate ESV, which failed to account for the spatiotemporal heterogeneity of ecosystems [90]. Although the standard equivalent factor of ESV was modified in this study according to the SRB's NPP, the land use type data and the standard equivalent factor of ESV are still affected by data precision and errors in earlier statistical data. Thus, we mitigated the negative impact on the precision of ecosystem service valuation by preserving the proportion of changes between the data in each period through the examination of longitudinal data series. Lastly, the relatively coarse spatial resolution of the urban scale as the spatial study unit in this paper could potentially have overlooked, excluded, or distorted the interaction between urbanization and ESs at different scales. Moving forward, we should embrace a more detailed multi-scale perspective to comprehensively investigate the impact of urbanization on ESs across various scales, their coupling coordination relationships, and scale effects. This approach will facilitate the acquisition of more scientifically rigorous and comprehensive results underpinning future research on urbanization and ESs.

5. Conclusions

This study revealed the spatiotemporal evolution characteristics of urbanization and ESs in the SRB from 1985 to 2021, based on multi-source data analysis. A multidimensional urbanization evaluation index system, combined with the modified equivalence factor method and spatial autocorrelation analysis, facilitated a comprehensive analysis of the spatiotemporal evolution characteristics of urbanization and ESs in the SRB. The innovation of this study lies in its comprehensive perspective, assessing the coupling coordination relationship between urbanization and ESs in the SRB over an extended period at both the system and subsystem levels. Additionally, by employing the GTWR model, the study treated different subsystems of urbanization and ESs as explanatory variables to explore the spatiotemporal heterogeneity characteristics of their interactions. The study yielded the following conclusions:

From 1985 to 2021, the urbanization index in the SRB increased from 0.09 to 0.34. The development of urbanization underwent a significant transition from rapid growth driven by population aggregation to a high-quality growth model dominated by spatial expansion and economic development. During this period, the ESV in the SRB decreased by 88.98×10^7 CNY, with all subtypes of ecosystem services, except for food production services, experiencing a decrease. Both urbanization and ESV exhibited significant spatial heterogeneity in their distributions. The decrease in ESV in the SRB was mainly attributed to the expansion of farmland and construction land, leading to the loss of forestland and grassland.

From 1985 to 2021, the CCD between urbanization and ESs in the SRB at both the system and subsystem levels showed significant improvement. The overall trend transitioned from moderately unbalanced to basically balanced, indicating an improving trajectory. Spatially, the results of this study showed that areas with high CCD values were mainly distributed in ecological functional zones and low-urbanization areas, while areas with low CCD values were primarily located in grassland ecological damage zones, ecologically fragile areas, resource-dependent, old, industrial cities, and highly urbanized areas. Most cities in the SRB remain in the transitional stage of CCD development, with significant differences among cities, posing a challenge to promoting the high-quality, coordinated development of urbanization and ESs in the SRB.

From 1985 to 2021, the impacts and trends of the various urbanization subsystems on ESs in the SRB varied, exhibiting significant spatial heterogeneity in the coefficients but maintaining a relatively stable spatial distribution pattern. The northwestern part of the SRB exhibited a higher response to urbanization changes. Conversely, all ESs subsystems showed a trend of initially strengthening and then weakening negative impacts on urbanization. The spatial distribution patterns of the positive and negative impacts of ESs subsystems on urbanization were closely correlated with the spatial distribution pattern of ESV in the SRB. The interaction between urbanization and ESs is influenced not only by resource endowment and ecological carrying capacity but also by policy and additional factors.

This study, grounded in the basin's actual conditions, provided a practical guide for promoting high-quality urbanization development and high-level ecological protection in the SRB by elucidating the evolutionary characteristics of urbanization and ecosystem services and their coupling coordination relationship and interaction. The models employed in this study can also be applied to urbanization and ecosystem high-quality coordinated development in other similar basin regions, offering a scientific basis and decision-making reference for future national spatial planning.

Author Contributions: Conceptualization, H.B. and D.X.; methodology, H.B. and Y.C.; software, H.B. and Y.C.; validation, H.B. and S.W.; formal analysis, H.B. and R.C.; investigation, H.B. and J.F.; resources, D.X.; data curation, H.B. and H.Z.; writing—original draft preparation, H.B; writing—review and editing, H.B. and Y.C.; visualization, H.B. and S.X.; supervision, D.X.; project administration, D.X.; funding acquisition, L.W. All authors have read and agreed to the published version of the manuscript.

Funding: This research was funded by the National Natural Science Foundation of China (grant number 42171246).

Data Availability Statement: Data available on request due to restrictions (e.g., privacy, legal or ethical reasons): The data presented in this study are available on request from the corresponding author (accurately indicate status).

Acknowledgments: The authors are particularly grateful to all researchers and institutes for providing data for this study. The authors are also very grateful to the editors and reviewers for their comments and suggestions for improving this study.

Conflicts of Interest: The authors declare no conflicts of interest.

References

1. Martins, J.O.; Sharifi, A. World Cities Report 2022: Envisaging the Future of Cities. United Nations Human Settlements Programme: New York, NY, USA, 2022; p. XV. Available online: https://www.un-ilibrary.org/content/books/9789210028592 (accessed on 25 November 2023).
2. Ouyang, X.; Tang, L.; Wei, X.; Li, Y. Spatial Interaction between Urbanization and Ecosystem Services in Chinese Urban Agglomerations. *Land Use Pol.* **2021**, *109*, 105587. [CrossRef]
3. Pham, K.T.; Lin, T.-H. Effects of Urbanisation on Ecosystem Service Values: A Case Study of Nha Trang, Vietnam. *Land Use Pol.* **2023**, *128*, 106599. [CrossRef]
4. Xu, Z.; Peng, J.; Liu, Y.; Qiu, S.; Zhang, H.; Dong, J. Exploring the Combined Impact of Ecosystem Services and Urbanization on SDGs Realization. *Appl. Geogr.* **2023**, *153*, 102907. [CrossRef]
5. Millennium Ecosystem Assessment (MEA). *Ecosystems and Human Well-Being: Synthesis*; Island Press: Washington, DC, USA, 2005.
6. Daily, G. *Nature's Services: Societal Dependence on Natural Ecosystems*; Island Press: Washington, DC, USA, 1997.
7. Costanza, R.; d'Arge, R.; de Groot, R.; Farber, S.; Grasso, M.; Hannon, B.; Limburg, K.; Naeem, S.; O'Neill, R.V.; Paruelo, J.; et al. The Value of the World's Ecosystem Services and Natural Capital. *Nature* **1997**, *387*, 253–260. [CrossRef]
8. Mao, D.; He, X.; Wang, Z.; Tian, Y.; Xiang, H.; Yu, H.; Man, W.; Jia, M.; Ren, C.; Zheng, H. Diverse Policies Leading to Contrasting Impacts on Land Cover and Ecosystem Services in Northeast China. *J. Clean. Prod.* **2019**, *240*, 117961. [CrossRef]

9. Jiang, W. Ecosystem Services Research in China: A Critical Review. *Ecosyst. Serv.* **2017**, *26*, 10–16. [CrossRef]
10. Zhang, Y.; Liu, Y.; Zhang, Y.; Liu, Y.; Zhang, G.; Chen, Y. On the Spatial Relationship between Ecosystem Services and Urbanization: A Case Study in Wuhan, China. *Sci. Total Environ.* **2018**, *637–638*, 780–790. [CrossRef]
11. Lee, D.-K. Analysis of the Potential Value of Cultural Ecosystem Services: A Case Study of Busan City, Republic of Korea. *Ecosyst. Serv.* **2024**, *65*, 101596. [CrossRef]
12. Lopez-Rivas, J.D.; Cardenas, J.-C. What Is the Economic Value of Coastal and Marine Ecosystem Services? A Systematic Literature Review. *Mar. Pol.* **2024**, *161*, 106033. [CrossRef]
13. Xing, L.; Xue, M.; Wang, X. Spatial Correction of Ecosystem Service Value and the Evaluation of Eco-Efficiency: A Case for China's Provincial Level. *Ecol. Indic.* **2018**, *95*, 841–850. [CrossRef]
14. Wu, J.; Wang, G.; Chen, W.; Pan, S.; Zeng, J. Terrain Gradient Variations in the Ecosystem Services Value of the Qinghai-Tibet Plateau, China. Glob. *Ecol. Conserv.* **2022**, *34*, e02008. [CrossRef]
15. Zhou, Z.; Sun, X.; Zhang, X.; Wang, Y. Inter-Regional Ecological Compensation in the Yellow River Basin Based on the Value of Ecosystem Services. *J. Environ. Manag.* **2022**, *322*, 116073. [CrossRef] [PubMed]
16. Xie, G.; Lu, C.; Leng, Y.; Zheng, D.; Li, S. Ecological assets valuation of the Tibetan Plateau. *J. Nat. Resour.* **2003**, *18*, 189–196. (In Chinese) [CrossRef]
17. Xie, G.; Zhen, L.; Lu, C.; Xiao, Y.; Chen, C. Expert Knowledge Based Valuation Method of Ecosystem Services in China. *J. Nat. Resour.* **2008**, *23*, 911–919. (In Chinese) [CrossRef]
18. Xie, G.; Zhang, C.; Zhang, L.; Chen, W.; Li, S. Improvement of the Evaluation Method for Ecosystem Service Value Based on Per Unit Area. *J. Nat. Resour.* **2015**, *30*, 1243–1254. (In Chinese) [CrossRef]
19. Xie, G.; Lin, Z. Research on evaluation and accounting of ecosystem assets and services. *China Soft Sci.* **2024**, *39*, 388–393. (In Chinese) [CrossRef]
20. Costanza, R.; de Groot, R.; Braat, L.; Kubiszewski, I.; Fioramonti, L.; Sutton, P.; Farber, S.; Grasso, M. Twenty Years of Ecosystem Services: How Far Have We Come and How Far Do We Still Need to Go? *Ecosyst. Serv.* **2017**, *28*, 1–16. [CrossRef]
21. Luo, Q.; Zhou, J.; Zhang, Y.; Yu, B.; Zhu, Z. What Is the Spatiotemporal Relationship between Urbanization and Ecosystem Services? A Case from 110 Cities in the Yangtze River Economic Belt, China. *J. Environ. Manag.* **2022**, *321*, 115709. [CrossRef]
22. Deng, C.; Liu, J.; Liu, Y.; Li, Z.; Nie, X.; Hu, X.; Wang, L.; Zhang, Y.; Zhang, G.; Zhu, D.; et al. Spatiotemporal Dislocation of Urbanization and Ecological Construction Increased the Ecosystem Service Supply and Demand Imbalance. *J. Environ. Manag.* **2021**, *288*, 112478. [CrossRef]
23. Yuan, Y.; Chen, D.; Wu, S.; Mo, L.; Tong, G.; Yan, D. Urban Sprawl Decreases the Value of Ecosystem Services and Intensifies the Supply Scarcity of Ecosystem Services in China. *Sci. Total Environ.* **2019**, *697*, 134170. [CrossRef]
24. Yang, M.; Gao, X.; Siddique, K.H.M.; Wu, P.; Zhao, X. Spatiotemporal Exploration of Ecosystem Service, Urbanization, and Their Interactive Coercing Relationship in the Yellow River Basin over the Past 40 Years. *Sci. Total Environ.* **2023**, *858*, 159757. [CrossRef] [PubMed]
25. Cheng, Y.; Kang, Q.; Liu, K.; Cui, P.; Zhao, K.; Li, J.; Ma, X.; Ni, Q. Impact of Urbanization on Ecosystem Service Value from the Perspective of Spatio-Temporal Heterogeneity: A Case Study from the Yellow River Basin. *Land* **2023**, *12*, 1301. [CrossRef]
26. Cumming, G.; Buerkert, A.; Hoffmann, E.; Schlecht, E.; von Cramon-Taubadel, S.; Tscharntke, T. Implications of agricultural transitions and urbanization for ecosystem services. *Nature* **2014**, *515*, 50–57. [CrossRef] [PubMed]
27. Delphin, S.; Escobedo, F.J.; Abd-Elrahman, A.; Cropper, W.P. Urbanization as a Land Use Change Driver of Forest Ecosystem Services. *Land Use Pol.* **2016**, *54*, 188–199. [CrossRef]
28. Zhou, T.; Chen, W.; Wang, Q.; Li, Y. Urbanisation and Ecosystem Services in the Taiwan Strait West Coast Urban Agglomeration, China, from the Perspective of an Interactive Coercive Relationship. *Ecol. Indic.* **2023**, *146*, 109861. [CrossRef]
29. Chen, M.; Lu, Y.; Ling, L.; Wan, Y.; Luo, Z.; Huang, H. Drivers of Changes in Ecosystem Service Values in Ganjiang Upstream Watershed. *Land Use Pol.* **2015**, *47*, 247–252. [CrossRef]
30. Kang, P.; Chen, W.; Hou, Y.; Li, Y. Spatial-temporal risk assessment of urbanization impacts on ecosystem services based on pressure-status-response framework. *Sci. Rep.* **2019**, *9*, 16806. [CrossRef] [PubMed]
31. Na, L.; Zhao, Y.; Guo, L. Coupling Coordination Analysis of Ecosystem Services and Urbanization in Inner Mongolia, China. *Land* **2022**, *11*, 1870. [CrossRef]
32. Zhu, S.; Huang, J.; Zhao, Y. Coupling Coordination Analysis of Ecosystem Services and Urban Development of Resource-Based Cities: A Case Study of Tangshan City. *Ecol. Indic.* **2022**, *136*, 108706. [CrossRef]
33. Li, Y.; Li, J.; Chu, J. Research on Land-Use Evolution and Ecosystem Services Value Response in Mountainous Counties Based on the SD-PLUS Model. *Ecol. Evol.* **2022**, *12*, e9431. [CrossRef]
34. Long, X.; Ji, X.; Ulgiati, S. Is Urbanization Eco-Friendly? An Energy and Land Use Cross-Country Analysis. *Energy Policy* **2017**, *100*, 387–396. [CrossRef]
35. Qiu, Z.; Guan, Y.; Zhou, K.; Kou, Y.; Zhou, X.; Zhang, Q. Spatiotemporal Analysis of the Interactions between Ecosystem Services in Arid Areas and Their Responses to Urbanization and Various Driving Factors. *Remote Sens.* **2024**, *16*, 520. [CrossRef]
36. Li, J.; Wang, J.; Zhou, W. Different Impacts of Urbanization on Ecosystem Services Supply and Demand across Old, New and Non-Urban Areas in the ChangZhuTan Urban Agglomeration, China. *Landsc. Ecol.* **2024**, *39*, 107. [CrossRef]
37. Yu, Q.; Feng, C.-C.; Shi, Y.; Guo, L. Spatiotemporal Interaction between Ecosystem Services and Urbanization in China: Incorporating the Scarcity Effects. *J. Clean Prod.* **2021**, *317*, 128392. [CrossRef]

38. Wang, C.; Wang, L.; Zhan, J.; Liu, W.; Teng, Y.; Chu, X.; Wang, H. Spatial Heterogeneity of Urbanization Impacts on Ecosystem Services in the Urban Agglomerations along the Yellow River, China. *Ecol. Eng.* **2022**, *182*, 106717. [CrossRef]
39. Wang, S.; Liu, Z.; Chen, Y.; Fang, C. Factors Influencing Ecosystem Services in the Pearl River Delta, China: Spatiotemporal Differentiation and Varying Importance. *Resour. Conserv. Recycl.* **2021**, *168*, 105477. [CrossRef]
40. Huang, B.; Wu, B.; Barry, M. Geographically and temporally weighted regression for modeling spatio-temporal variation in house prices. *Int. J. Geogr. Inf. Sci.* **2010**, *24*, 383–401. [CrossRef]
41. Ariken, M.; Zhang, F.; Chan, N.W.; Kung, H. Coupling Coordination Analysis and Spatio-Temporal Heterogeneity between Urbanization and Eco-Environment along the Silk Road Economic Belt in China. *Ecol. Indic.* **2021**, *121*, 107014. [CrossRef]
42. Liang, L.; Wang, Z.; Li, J. The Effect of Urbanization on Environmental Pollution in Rapidly Developing Urban Agglomerations. *J. Clean Prod.* **2019**, *237*, 117649. [CrossRef]
43. Shi, T.; Yang, S.; Zhang, W.; Zhou, Q. Coupling Coordination Degree Measurement and Spatiotemporal Heterogeneity between Economic Development and Ecological Environment—Empirical Evidence from Tropical and Subtropical Regions of China. *J. Clean Prod.* **2020**, *244*, 118739. [CrossRef]
44. Seto, K.C.; Güneralp, B.; Hutyra, L.R. Global Forecasts of Urban Expansion to 2030 and Direct Impacts on Biodiversity and Carbon Pools. *Proc. Natl. Acad. Sci. USA* **2012**, *109*, 16083–16088. [CrossRef] [PubMed]
45. Wang, J.; Zhou, W.; Pickett, S.T.A.; Yu, W.; Li, W. A Multiscale Analysis of Urbanization Effects on Ecosystem Services Supply in an Urban Megaregion. *Sci. Total Environ.* **2019**, *662*, 824–833. [CrossRef] [PubMed]
46. Fang, C.; Liu, H.; Li, G. International Progress and Evaluation on Interactive Coupling Effects between Urbanization and the Eco-Environment. *J. Geogr. Sci.* **2016**, *26*, 1081–1116. [CrossRef]
47. Chen, W.; Chi, G. Urbanization and Ecosystem Services: The Multi-Scale Spatial Spillover Effects and Spatial Variations. *Land Use Pol.* **2022**, *114*, 105964. [CrossRef]
48. Sun, X.; Zhou, Q.; Wang, Y.; Ren, W. Influence of Hydro-Geomorphology, Land-Use and Riparian Zone Characteristics on Herbicide Occurrence and Distribution in Sediments in Songhua River Basin, Northeastern China. *Geoderma* **2013**, *193*, 156–164. [CrossRef]
49. Niu, T.; Yu, J.; Yue, D.; Yang, L.; Mao, X.; Hu, Y.; Long, Q. The Temporal and Spatial Evolution of Ecosystem Service Synergy/Trade-Offs Based on Ecological Units. *Forests* **2021**, *12*, 992. [CrossRef]
50. Liang, F.; Liang, C.; Li, X.; Bai, J.; Mu, Y.; Xie, X. Simulated Priority Protection Pattern for the Wetlands in Songhua River Basin Based on Systematic Conservation Planning. *Wetl. Sci.* **2022**, *20*, 56–64. (In Chinese) [CrossRef]
51. Feng, Y.; Guo, Y.; Chen, X.; Liu, M.; Shen, Y. Classification of Major Crops Using MODIS Data in the Songhua River Basin. *Chin. J. Eco-Agric.* **2023**, *31*, 1602–1612. (In Chinese) [CrossRef]
52. Zhang, Y.; Zhao, F.; Zhang, J.; Wang, Z. Fluctuation in the Transformation of Economic Development and the Coupling Mechanism with the Environmental Quality of Resource-Based Cities—A Case Study of Northeast China. *Resour. Policy* **2021**, *72*, 102128. [CrossRef]
53. Shen, Y.; Cao, H.; Tang, M.; Deng, H. The Human Threat to River Ecosystems at the Watershed Scale: An Ecological Security Assessment of the Songhua River Basin, Northeast China. *Water* **2017**, *9*, 219. [CrossRef]
54. Liu, J.; Liu, B.; Liu, H.; Zhang, F. Long-Term Cultivation Drives Soil Carbon, Nitrogen, and Bacterial Community Changes in the Black Soil Region of Northeastern China. *Land Degrad. Dev.* **2024**, *35*, 428–441. [CrossRef]
55. Tang, Q.; Hua, L.; Cao, Y.; Jiang, L.; Cai, C. Human Activities Are the Key Driver of Water Erosion Changes in Northeastern China. *Land Degrad. Dev.* **2024**, *35*, 62–75. [CrossRef]
56. Mo, X.; Liu, S.; Meng, D.; Lin, Z. Exploring the Interannual and Spatial Variations of ET and GPP with Climate by a Physical Model and Remote Sensing Data in a Large Basin of Northeast China. *Int. J. Climatol.* **2014**, *34*, 1945–1963. [CrossRef]
57. Wang, H.; Wang, S.; Shu, X.; He, Y.; Huang, J. Increasing Occurrence of Sudden Turns From Drought to Flood Over China. *J. Geophys. Res. Atmos.* **2024**, *129*, e2023JD039974. [CrossRef]
58. Bai, X.; Shi, P.; Liu, Y. Society: Realizing China's Urban Dream. *Nature* **2014**, *509*, 158–160. [CrossRef] [PubMed]
59. Xing, L.; Zhu, Y.; Wang, J. Spatial Spillover Effects of Urbanization on Ecosystem Services Value in Chinese Cities. *Ecol. Indic.* **2021**, *121*, 107028. [CrossRef]
60. Shi, Y.; Feng, C.-C.; Yu, Q.; Han, R.; Guo, L. Contradiction or Coordination? The Spatiotemporal Relationship between Landscape Ecological Risks and Urbanization from Coupling Perspectives in China. *J. Clean Prod.* **2022**, *363*, 132557. [CrossRef]
61. Guo, X.; Fang, C. Integrated Land Use Change Related Carbon Source/Sink Examination in Jiangsu Province. *Land* **2021**, *10*, 1310. [CrossRef]
62. Riao, D.; Zhu, X.; Tong, Z.; Zhang, J.; Wang, A. Study on Land Use/Cover Change and Ecosystem Services in Harbin, China. *Sustainability* **2020**, *12*, 6076. [CrossRef]
63. Wei, H.; Xue, D.; Huang, J.; Liu, M.; Li, L. Identification of Coupling Relationship between Ecosystem Services and Urbanization for Supporting Ecological Management: A Case Study on Areas along the Yellow River of Henan Province. *Remote Sens.* **2022**, *14*, 2277. [CrossRef]
64. Natural Resources and Territory Spatial Planning. *Current Land Use Classification (GB/T 21010-2017)*; China Zhijian Publishing House: Beijing, China, 2017. (In Chinese)
65. Anselin, L.; Getis, A. Spatial Statistical Analysis and Geographic Information Systems. *Ann. Reg. Sci.* **1992**, *26*, 19–33. [CrossRef]
66. Anselin, L. A Test for Spatial Autocorrelation in Seemingly Unrelated Regressions. *Econ. Lett.* **1988**, *28*, 335–341. [CrossRef]

67. Tu, D.; Cai, Y.; Liu, M. Coupling Coordination Analysis and Spatiotemporal Heterogeneity between Ecosystem Services and New-Type Urbanization: A Case Study of the Yangtze River Economic Belt in China. *Ecol. Indic.* **2023**, *154*, 110535. [CrossRef]
68. Hu, Y.; Wu, T.; Guo, L.; Zhang, S. Spatiotemporal Relationships between Ecosystem Health and Urbanization on the Tibetan Plateau from a Coupling Coordination Perspective. *Land* **2023**, *12*, 1635. [CrossRef]
69. Zhao, W.; Shi, P.; Wan, Y.; Yao, Y. Coupling and Coordination Relationship between Urbanization Quality and Ecosystem Services in the Upper Yellow River: A Case Study of the Lanzhou-Xining Urban Agglomeration, China. *Land* **2023**, *12*, 1085. [CrossRef]
70. Ariken, M.; Zhang, F.; Liu, K.; Fang, C.; Kung, H.-T. Coupling Coordination Analysis of Urbanization and Eco-Environment in Yanqi Basin Based on Multi-Source Remote Sensing Data. *Ecol. Indic.* **2020**, *114*, 106331. [CrossRef]
71. Li, W.; Wang, Y.; Xie, S.; Cheng, X. Coupling Coordination Analysis and Spatiotemporal Heterogeneity between Urbanization and Ecosystem Health in Chongqing Municipality, China. *Sci. Total Environ.* **2021**, *791*, 148311. [CrossRef]
72. Wan, L.; Ye, X.; Lee, J.; Lu, X.; Zheng, L.; Wu, K. Effects of Urbanization on Ecosystem Service Values in a Mineral Resource-Based City. *Habitat Int.* **2015**, *46*, 54–63. [CrossRef]
73. Guo, X.; Fang, C.; Mu, X.; Chen, D. Coupling and Coordination Analysis of Urbanization and Ecosystem Service Value in Beijing-Tianjin-Hebei Urban Agglomeration. *Ecol. Indic.* **2022**, *137*, 108782. [CrossRef]
74. Luo, Q.; Luo, Y.; Zhou, Q.; Song, Y. Does China's Yangtze River Economic Belt Policy Impact on Local Ecosystem Services? *Sci. Total Environ.* **2019**, *676*, 231–241. [CrossRef]
75. Peng, J.; Tian, L.; Liu, Y.; Zhao, M.; Hu, Y.; Wu, J. Ecosystem Services Response to Urbanization in Metropolitan Areas: Thresholds Identification. *Sci. Total Environ.* **2017**, *607–608*, 706–714. [CrossRef] [PubMed]
76. Tian, Y.; Jiang, G.; Zhou, D.; Li, G. Systematically Addressing the Heterogeneity in the Response of Ecosystem Services to Agricultural Modernization, Industrialization and Urbanization in the Qinghai-Tibetan Plateau from 2000 to 2018. *Sci. Total Environ.* **2021**, *285*, 125323. [CrossRef]
77. Salvati, L.; Zambon, I.; Chelli, F.M.; Serra, P. Do Spatial Patterns of Urbanization and Land Consumption Reflect Different Socioeconomic Contexts in Europe? *Sci. Total Environ.* **2018**, *625*, 722–730. [CrossRef] [PubMed]
78. Guan, X.; Wei, H.; Lu, S.; Su, H. Mismatch Distribution of Population and Industry in China: Pattern, Problems and Driving Factors. *Appl. Geogr.* **2018**, *97*, 61–74. [CrossRef]
79. Dadashpoor, H.; Azizi, P.; Moghadasi, M. Land Use Change, Urbanization, and Change in Landscape Pattern in a Metropolitan Area. *Sci. Total Environ.* **2019**, *655*, 707–719. [CrossRef] [PubMed]
80. Feng, X.; Fu, B.; Piao, S.; Wang, S.; Ciais, P.; Zeng, Z.; Lü, Y.; Zeng, Y.; Li, Y.; Jiang, X.; et al. Revegetation in China's Loess Plateau Is Approaching Sustainable Water Resource Limits. *Nat. Clim. Chang.* **2016**, *6*, 1019–1022. [CrossRef]
81. Fang, W.; An, H.; Li, H.; Gao, X.; Sun, X. Urban Economy Development and Ecological Carrying Capacity: Taking Beijing City as the Case. *Energy Procedia* **2017**, *105*, 3493–3498. [CrossRef]
82. Xiao, R.; Lin, M.; Fei, X.; Li, Y.; Zhang, Z.; Meng, Q. Exploring the Interactive Coercing Relationship between Urbanization and Ecosystem Service Value in the Shanghai–Hangzhou Bay Metropolitan Region. *J. Clean Prod.* **2020**, *253*, 119803. [CrossRef]
83. Dai, X.; Wang, L.; Li, X.; Gong, J.; Cao, Q. Characteristics of the Extreme Precipitation and Its Impacts on Ecosystem Services in the Wuhan Urban Agglomeration. *Sci. Total Environ.* **2023**, *864*, 161045. [CrossRef]
84. Ye, Y.; Bryan, B.A.; Zhang, J.; Connor, J.D.; Chen, L.; Qin, Z.; He, M. Changes in Land-Use and Ecosystem Services in the Guangzhou-Foshan Metropolitan Area, China from 1990 to 2010: Implications for Sustainability under Rapid Urbanization. *Ecol. Indic.* **2018**, *93*, 930–941. [CrossRef]
85. Ma, S.; Wang, L.-J.; Zhao, Y.-G.; Jiang, J. Coupling Effects of Soil and Vegetation from an Ecosystem Service Perspective. *Catena* **2023**, *231*, 107354. [CrossRef]
86. Kremen, C. Reframing the Land-Sparing/Land-Sharing Debate for Biodiversity Conservation. *Ann. N. Y. Acad. Sci.* **2015**, *1355*, 52–76. [CrossRef] [PubMed]
87. Hou, X.; Liu, J.; Zhang, D.; Zhao, M.; Xia, C. Impact of Urbanization on the Eco-Efficiency of Cultivated Land Utilization: A Case Study on the Yangtze River Economic Belt, China. *J. Clean Prod.* **2019**, *238*, 117916. [CrossRef]
88. Nichiforel, L.; Duduman, G.; Scriban, R.E.; Popa, B.; Barnoaiea, I.; Drăgoi, M. Forest Ecosystem Services in Romania: Orchestrating Regulatory and Voluntary Planning Documents. *Ecosyst. Serv.* **2021**, *49*, 101276. [CrossRef]
89. Chuai, X.; Huang, X.; Wu, C.; Li, J.; Lu, Q.; Qi, X.; Zhang, M.; Zuo, T.; Lu, J. Land Use and Ecosystems Services Value Changes and Ecological Land Management in Coastal Jiangsu, China. *Habitat Int.* **2016**, *57*, 164–174. [CrossRef]
90. Cao, S.; Yu, Z.; Zhang, J.; Feng, F.; Xu, D.; Mu, X. Cost–Benefit Analysis of Ecosystem Services in China. *Ecol. Eng.* **2018**, *125*, 143–148. [CrossRef]

land

Article

Impact of Urbanization on Ecosystem Service Value from the Perspective of Spatio-Temporal Heterogeneity: A Case Study from the Yellow River Basin

Yonghui Cheng [1,2], Qi Kang [2], Kewei Liu [1], Peng Cui [1,*], Kaixu Zhao [1], Jianwei Li [1], Xue Ma [1] and Qingsong Ni [1]

[1] College of Urban and Environmental Sciences, Northwest University, Xi'an 710127, China; chengyonghui1@stumail.nwu.edu.cn (Y.C.); xdlkw@nwu.edu.cn (K.L.); zhaokaixu@stumail.nwu.edu.cn (K.Z.); jwli@nwu.edu.cn (J.L.); 202221226@stumail.nwu.edu.cn (X.M.); cindyni@stumail.nwu.edu.cn (Q.N.)
[2] Northwest Branch, Beijing Tsinghua Tongheng Urban Planning and Design Institute, Xi'an 710076, China; kangqi@thupdi.com
* Correspondence: cuipeng@nwu.edu.cn

Abstract: Ecosystem services are the beneficial goods and services that ecosystems provide to humans. Urbanization is an important feature of human social development. While promoting economic and social development, it also brings about land degradation, resource depletion, environmental pollution and other problems, intensifying the transformation of natural ecosystems into semi-natural and artificial ecosystems, ultimately leading to the loss of ecosystem service functions and declining value. The study of the impact of urbanization on the value of ecosystem services is of critical importance for the conservation of ecosystems and sustainable development. This study examined the spatio-temporal patterns of urbanization's impacts on ecosystem service value in the Yellow River Basin from the perspective of spatio-temporal heterogeneity. Findings: (1) Both the ecosystem service value (ESV) and urbanization level (UL) in the Yellow River Basin were on the rise on the whole, but they were significantly spatially negatively correlated and mainly characterized by the high–low spatial clustering of "low ESV–high UL" and "high ESV–low UL". This negative correlation was gradually weakened with the transformation of the urbanization development mode and ecological restoration projects in the Yellow River Basin. (2) The impacts of the five urbanization subsystems on the value of ecosystem services were diverse. Landscape urbanization had a negative impact on the value of ecosystem services in all regions; economic urbanization and innovation urbanization changed from having a negative to a positive impact; and demographic urbanization and social urbanization had both a positive and a negative impact. (3) To promote the coordinated development of ecological environmental protection and urbanization in the YRB, this paper proposes to change the urbanization development model, implement ecological restoration by zoning, and formulate classified development plans. This study compensates for the shortcomings of current studies that ignore the different impacts of urbanization subsystems on ecosystem service value and lack sufficient consideration of the spatio-temporal heterogeneity characteristics of urbanization and ESVs, enriches the theoretical understanding of the interrelationships between natural and human systems in basin areas, and provides a scientific basis for the rational formulation of urban planning and ecological protection policies in the region, which is of great theoretical and practical significance.

Keywords: ecosystem service value; urbanization; spatio-temporal heterogeneity; geographically weighted regression; Yellow River Basin

Citation: Cheng, Y.; Kang, Q.; Liu, K.; Cui, P.; Zhao, K.; Li, J.; Ma, X.; Ni, Q. Impact of Urbanization on Ecosystem Service Value from the Perspective of Spatio-Temporal Heterogeneity: A Case Study from the Yellow River Basin. *Land* **2023**, *12*, 1301. https://doi.org/10.3390/land12071301

Academic Editors: Luca Congedo, Francesca Assennato and Michele Munafò

Received: 24 May 2023
Revised: 24 June 2023
Accepted: 26 June 2023
Published: 28 June 2023

1. Introduction

Ecosystem services refer to material subsistence and services provided by ecosystems to human society [1,2]. Ecosystem service value (ESV) is the monetary value of tangible or intangible benefits that humans derive directly or indirectly from ecosystems [3] and can be quantified by models such as InVEST, ARIES, or MIMES. ESV is a reflection of the potential of ecosystems to provide material subsistence and services for people [4], and it serves as an important measure to evaluate the quality of ecosystem services and social sustainability [5]. The enhancement of human well-being and the sustainable development of the region will not be possible without the guarantee of ecosystem services [6]. However, with the rapid economic development and great social progress in recent years, the strong interference [7,8] of human activities in the ecosystem has led to the decline of many ecosystems around the world [9], blocking the improvement of living standards [10–12]. Unprecedented urbanization is taken as the main human activity leading to global ecosystem changes [13–15]. Urbanization has caused ecological and environmental problems, such as land degradation, resource depletion, and environmental pollution, due to the expansion of construction land, demographic concentration, and economic growth, thus leading to degraded ecosystem functions. For example, urban land expansion has eroded a large amount of ecological land, such as farmland, forest land, and grassland, leading to structural changes in land use [16], resulting in a dramatic decrease in the net primary productivity [17], carbon sequestration capacity [18], and hydrological regulation capacity [19] of ecosystems. In addition, urban demographic agglomeration and social industrialization have brought about a significant increase in the intensity of human activities, intensifying the consumption of natural resources such as water and minerals [20] and the massive discharge of pollutants such as exhaust gases, wastewater, and waste into the ecosystem [21]. The resulting reduction in the self-purification capacity and anti-disturbance capacity of the ecosystem has triggered a series of ecological and environmental degradation problems, such as the over-exploitation of water resources, shortages of non-renewable energy, soil and water pollution, atmospheric pollution, and climate change. In short, urbanization has intensified the transformation of natural ecosystems into semi-natural and artificial ecosystems [22], changing the structure and processes of ecosystems, leading to the loss of ecosystem services and a decrease in the value of ecosystem services. The question of how to mitigate the disturbance of ecosystem services by urbanization has emerged as a major problem that calls for an immediate solution across the region to maintain sustainable development. Therefore, the impact of urbanization on ESV is one of the major components of current research, and its analysis will help to provide a deeper understanding of the relationship between natural and human systems while laying a theoretical basis for the formulation of ecological protection and sustainable development policies.

Academia has engaged in a detailed discussion of the relationship between urbanization and ESV. Some scholars have examined the coordination of the two in space and time using Pearson correlation methods [23], gray correlation analysis (GCA) models [24], coupled coordination (CCD) models [25], telecoupling coordination degree (LTCCD) models [26], and decoupling models [27]. Moreover, urbanization's impact on ESV and its mechanism of action have gained the most attention in the field [28–30]. The complex interaction between the two is mainly explored in a series of studies using ordinary least squares (OLS) models [31], panel regression models [32], geographically weighted regression (GWR) models [33], curve estimation [34], and segmented linear regression [35]. The findings are not identical and some even contradict each other. According to studies, there are both linear (e.g., positively related [36] and negatively related [37]) and non-linear (e.g., inverted U-shaped [38], N-shaped [39], and double exponential curve [40]) relationships between urbanization and ESV. Specifically, urbanization may improve ecosystem services as a result of industrial structure optimization and urban management enhancement, while resulting in the degradation of land use systems [41], biodiversity reductions [42], landscape fragmentation [43], net primary productivity (NPP) declines [44], and reduced human welfare and benefits [45] due to land expansion, demographic growth, energy consumption, and pollutant emissions [20]. In turn, ecosystem services play a supportive or inhibitory role in urbanization [46,47]. In summary, urbanization and ESV are interdependent and mutually restricted [48].

Although these studies have revealed the complex relationship between the UL and ESV [49], there are some limitations. The existing literature focuses on studies of ESV, including assessment and prediction [50], synergies and trade-offs [51], and supply–demand balance [52]. However, the evaluation of the level of urbanization is one-sided and lacks diversity; specifically, most studies examine the scale growth of urbanization in terms of demographic agglomeration [53], land expansion [54], or economic development [55] in a single dimension, such as the population, land, or economy, which hardly reflects the complexity and diversity of urbanization development in a comprehensive manner [56]. In fact, the new urbanization under the "people-oriented" concept not only pursues a high growth rate but also focuses on high-quality development. The level of urban public services, the degree of social civilization, and the quality of life of residents have been important measures of urbanization development, and scientific and technological innovation has become the core endogenous driving force for the transformation and upgrading of urban development patterns in the middle and late stages of urbanization [57]. Therefore, in addition to considering the three dimensions of population, land, and economy, efforts should be increased to evaluate social urbanization [25] and innovative urbanization [58], which would help to explain the interactions between urbanization and ecosystem services from a more integrated perspective and in a comprehensive manner. Current studies mainly deal with the overall relationship between the two systems, urbanization and ecosystem services [59], with little attention to the differences in the impact of different ULs by subsystem on ESV [60]. Therefore, this study provides a comprehensive evaluation of the UL according to demographic, landscape, economic, social, and innovative subsystems, further enriching the understanding of the connotations of new urbanization and thus providing a more comprehensive understanding of the mechanisms by which urbanization subsystems affect ESV.

Although there are studies that have addressed the spatial correlation between ESV and the UL [61–63], only a few have considered the uneven spatio-temporal distribution of the two [64]. ESV and the UL are found to be significantly spatially heterogeneous as a result of different natural environmental conditions and socio-economic development levels in different regions [65,66]. Moreover, the impacts of urbanization on ecosystem services are also time-variant, usually moving from negative effects of disturbance in the primary stage to positive effects of support in the advanced stage [67,68]. Regarding the spatio-temporal heterogeneity characteristics of urbanization and ESV, the geographically weighted regression (GWR) model can deal with spatial correlation and reflect the spatial heterogeneity and influence direction of different geographical locations via regression coefficients, overcoming the shortcomings of traditional regression models that lead to biased regression results due to ignoring individual and temporal differences [69]. Therefore, this study employs a GWR model to conduct a multi-stage comparative analysis based on panel data to explore the spatio-temporal dynamics of the mechanism of urbanization's impact on ESV.

It should also be emphasized that the interaction of the two spatio-temporally heterogeneous processes adds to their complexity [70], and it is difficult to learn from each of their findings in different regions [71], necessitating spatio-temporal heterogeneity studies of ecosystem service value and urbanization in specific regions. Most of the current studies place their focus on areas with advanced urbanization, such as urban agglomerations and metropolitan areas [72–74]. In recent years, some scholars have also shifted their attention to basin-scale studies [39]. From the perspective of spatio-temporal heterogeneity, river basins are a more appropriate and more deserving research scale, since an ecosystem is a complex open system with a strong external impact, and its ecological processes, material cycles, and energy transfers are not subject to administrative boundaries. The economic and social systems are closely related to the local natural background, historical accumulation, and development mode and have unique local identities. As a multi-level network system of "nature–economy–society", river basins have significant heterogeneity and local characteristics of the ecological environment and socio-economy, and they offer a comprehensive reflection of the main qualities of these two systems. Therefore, the study of river basins is important in guiding ecological protection and high-quality development in basin areas.

As a typical region where rapid urbanization and integrated ecosystem management are taking place simultaneously, the YRB in China plays a major role in national ecological security and the new urbanization strategy [75]. With the successive implementation of ecological restoration projects such as the Sanjiangyuan Ecological Protection Project, the Sanbei Protection Forest Project, and the Return of Cropland to Forests and Grasses Project in the past three decades, the YRB has achieved breakthroughs in ecosystem restoration and environmental pollution management [76]. Nevertheless, the YRB still faces serious problems, such as a fragile ecological environment, a degraded ecosystem, regional economic incoherence, and prominent human–land conflict. With the continued acceleration of urbanization in the YRB, the choice of its future urbanization path will be one of the determinants of the prospects for regional ecological security and sustainable growth.

The following objectives are the foci of this paper: (1) to analyze the spatio-temporal variability of ESV in the YRB from 1990 to 2020 by evaluating ESV and the UL using the equivalent factor method (EFM); (2) to construct a UL evaluation indicator system based on demographic, landscape, economic, social, and innovative subsystems to analyze the spatio-temporal change characteristics of the UL from 2000 to 2020; (3) to verify the spatial interaction of ESV with the UL using the bivariate SPAC model; (4) to explore the mechanisms and spatio-temporal dynamics of urbanization subsystems' impacts on ESV and the UL using a GWR model. This study addresses the shortcomings of current studies, such as the lack of diversity in the evaluation dimensions of ULs, the neglect of different impacts of urbanization subsystems on ecosystem service value, and insufficient consideration of the spatio-temporal heterogeneity characteristics of urbanization and ESV. From the perspective of heterogeneity, it reveals the spatio-temporal variation patterns of the impacts of urbanization on ESV, proposes targeted policy recommendations for the coordinated development of ecological environmental protection and urbanization in the YRB, and provides a reference for the sustainable development of similar basin areas.

2. Materials and Methods

2.1. Research Area: Yellow River Basin, China

The Yellow River Basin (32°10′ N~41°50′ N, 95°53′ E~119°05′ E) lies in Northern China, connected with the four major geomorphic units of the Qinghai–Tibet Plateau, Inner Mongolia Plateau, Loess Plateau, and Huanghuaihai Plain, and is one of the most important ecological barriers [77] and providers of ecosystem services [78] in China, covering a total area of approximately 1.7×10^6 ha, with 82 national nature reserves (Figure 1). The land in the basin is dominated by grassland, farmland, and forest, accounting for 51% (4.8×10^7 ha), 22% (2.1×10^7 ha), and 12% (1.2×10^7 ha) of the total, respectively. The terrain in the basin is high in the west and low in the east, with a total drop of 4448 m and significant differences in climatic conditions, average annual rainfall of 300–600 mm, an average annual temperature of −4–14 °C, and average annual runoff of 5.8×10^{10} m^3. The basin is rich in biological resources, with over 4000 plant species, over 400 bird species, and over 150 fish species. The vegetation is affected by the horizontal zonality and monsoon, and from east to west, it consists of crops, broad-leaved forests, coniferous forests, grassland, and sparse shrub–steppe. The YRB is the core economic zone and key urbanization area in China, spanning three economic zones in the east, middle, and west of China and containing four national urban agglomerations (Guanzhong Plain, Zhongyuan, Hohhot–Baotou–Ordos–Yulin, and Lanzhou–Xining). From 2000 to 2020, the YRB enjoyed rapid socio-economic development, with the GDP growing from 7.94×10^{11} CNY to 7.65×10^{12} CNY, the urban population growing from 3.75×10^7 to 7.86×10^7, and the urban population rising from 33% to 62% in proportion.

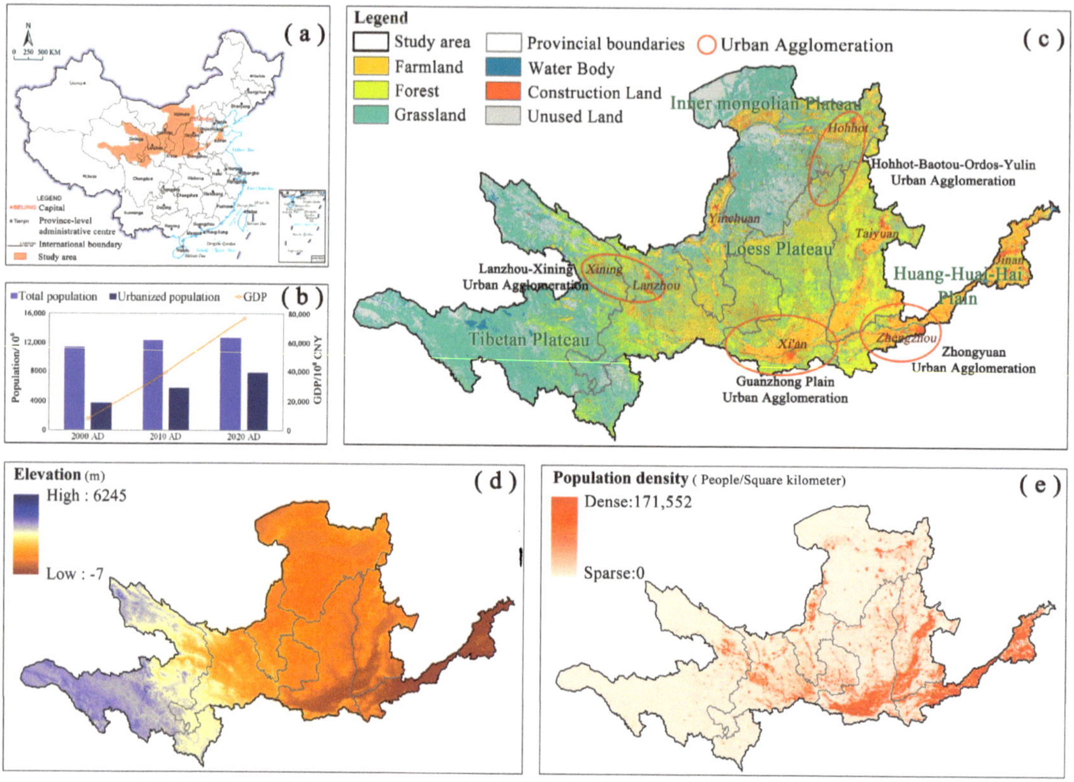

Figure 1. Study area. (**a**) Location of YRB in China; (**b**) changes in total population, urban population, and GDP in the YRB from 2000 to 2020; (**c**) land use types in 2020; (**d**) elevation; (**e**) population density in 2020.

The rapid socioeconomic development and urbanization have placed tremendous pressure on the ecological environment. Under the influence of long-term high-intensity human activities and natural disasters, the ecological environment of the YRB has become sensitive and fragile, leading to the overall and systematic degradation of the ecosystem [76], such as the decline of the water conservation function in the upper reaches and the degradation of natural grassland on a large scale of up to 60–90%; serious soil erosion and desertification in the middle reaches [9], with soil erosion on the Loess Plateau reaching 2.08×10^7 ha; and siltation, the widening of river channels, and the elevation of riverbeds in the lower reaches, with shrinkage of 52.8% in the natural wetlands in the delta of the estuary into the sea. In the past three decades, the LUCC in the basin has changed significantly, with an increase of 1.3×10^6 ha in construction land, up by 72.84%, including 1.1×10^6 ha coming from farmland, indicating that construction land continues to encroach on farmland. To address the severe challenges of ecological protection, China has elevated the ecological protection and high-quality development of the YRB to a major national development strategy and has implemented ecological restoration projects such as natural forest protection, the construction of the Sanbei protection forests, and the return of cropland to forests and grassland. It has converted 2.7×10^6 ha of unused land to grassland, 1.6×10^6 ha of cropland to grassland, and 1.5×10^6 ha of grassland to cropland within thirty years (Figure 2), which has slowed down the ecosystem's degradation in the YRB to some extent. This study is based on county-level administrative divisions as research units, and the total study area is determined to be 9.45×10^5 km^2, involving a total of 361 county-level administrative units in nine provinces.

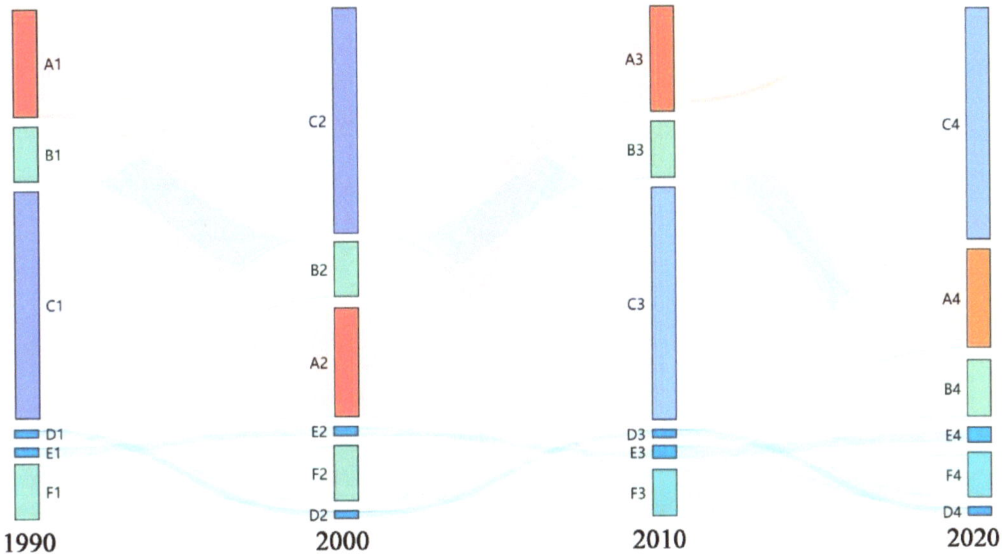

Figure 2. Transfer flows of different land use types in the YRB from 1990 to 2020 (A: Farmland; B: Forest; C: Grassland; D: Water Body; E: Construction Land; F: Unused Land).

2.2. Research Methods

2.2.1. Evaluation of Ecosystem Service Value

In this study, we introduce the Equivalent Factor Method (EFM) proposed by Costanza et al. [1] to evaluate the ESV in the YRB. According to the "Table of Ecological Service Value Equivalence per Unit Area of Chinese Ecosystems", revised by Xie et al. [79], and in view of the heterogeneity, complexity, and dynamics of ecosystem service value [80], as well as the actual grain production capacity, the ESV coefficient per unit area is corrected by the grain yield correction method (Table 1), with wheat, cotton, and rapeseed as the main grain crop species. The equation is as follows:

$$VC_{kf} = \frac{1}{7} EC_{kf} \sum_{i=1}^{n} \frac{m_i p_i q_i}{M} \quad (1)$$

where VC_{kf} is the corrected ESV coefficient corresponding to the f-th ecosystem service of land use type k in the YRB—no construction land is included; EC_{kf} is the value equivalent corresponding to the f-th ecosystem service of the land use type k in the "Table of Ecological Service Value Equivalence per Unit Area of Chinese Ecosystems", revised by Xie et al.; m_i is the total sown area of Class i grain crops; p_i is the average price of Class i grain crops; q_i is the average yield per unit area of Class i grain crops; M is the total sown area of all grain crops; n is the total number of the main grain crop series; 1/7 refers to a one-seventh share of the economic value of an ecosystem service value equivalent factor in the average market value of food production [81].

Table 1. ESV per unit area of different land use types in the YRB (CNY·ha·a^{-1}).

	Farmland	Forest	Grassland	Water Body	Unused Land
Provision service value (PSV)	2895.4	6894.8	1645.59	1541.43	124.98
Regulation service value (RSV)	8019.62	29,578.9	12,289.82	83,310.42	1083.17
Support service value (SSV)	5186.72	17,768.16	8561.22	9915.17	1187.32
Cultural service value (CSV)	354.11	4332.68	1812.23	9508.99	499.93
Ecosystem service Value (ESV)	16,455.85	58,574.54	24,308.86	104,276.01	2895.4

The value of the ecosystem services for each county in the YRB is calculated using the following equation:

$$ESV = \sum_{f=1}^{v} A_k VC_{kf} \quad (2)$$

where ESV is the value of ecosystem services; A_k is the area of land use type k in the county; VC_{kf} is the ESV coefficient corresponding to the f-th ecosystem service of land use type k in the YRB after correction; v is the number of types of ecosystem services.

2.2.2. Assessment of Urbanization Level

Urbanization is an interrelated and dynamic process of demographic, landscape, economic, and social subsystems [82,83], and deviation from any one subsystem will reduce the comprehensive UL [84]. Science and technology innovation has provided technical support and guidance for urbanization development in recent years, and it has come to be a major driver of high-quality urbanization [85]. This study further expands the connotations of urbanization according to the characteristics of county urbanization from the dimensions of the five subsystems, and we select eight indicators to comprehensively measure the UL in a scientific, objective, and comprehensive manner with consideration of data availability [86] (Table 2). Demographic growth and agglomeration are the core elements of urbanization, and the demographic urbanization level (DUL) is measured by the total population (TP) and the proportion of the urban population in the total population (UPP) [24]; the expansion of construction land is the spatial expression of urbanization, and the landscape urbanization level (LUL) is measured by the proportion of construction land in the total land area (CLP) [87]; economic development is the driving engine of urbanization, and the economic urbanization level (EUL) is measured by the GDP per capita (PGDP) and the proportion of secondary and tertiary industries in the GDP (STIP); the improvement of people's living standards is the ultimate goal of urbanization, and the social urbanization level (SUL) is measured by the number of people with a high school education and above (HSE) and the per capita living space (LS); scientific and technological innovation is the endogenous driving force of urbanization, and the innovative urbanization level (IUL) is measured by the number of domestic patent applications authorized (DPP) [88].

In this study, the indices of the urbanization level (UL) and urbanization subsystem level (DUL, LUL, EUL, SUL, IUL) are calculated for each county in each year in the YRB by the entropy method [25,89].

Table 2. Indicator system of comprehensive urbanization level.

System	Indicator	Indicator Meaning
Demographic urbanization level (DUL)	Total population (TP) (person)	Reflecting the total demographic size of the region, as the population basis of urbanization.
	Proportion of urban population in total population (UPP) (%)	Reflecting the degree of demographic agglomeration in urban areas, as a key measure of the urbanization process.
Landscape urbanization level (LUL)	Proportion of construction land in total land area (CLP) (%)	Reflecting the expansion of urban land, as a direct spatial expression of urbanization.

Table 2. *Cont.*

System	Indicator	Indicator Meaning
Economic urbanization level (EUL)	GDP per capita (PGDP) (CNY)	Reflecting the level of regional economic development, as the economic basis of urbanization.
	Proportion of secondary and tertiary industries in GDP (STIP) (%)	Reflecting the structure of the regional economy, as the driving force for urbanization in the non-agricultural sectors.
Social urbanization level (SUL)	Number of people with high school education and above (HSE)	Reflecting the level of regional education services and the quality of human resources, as a key driving force for urbanization quality.
	Per capita living space (LS) (m^2)	Reflecting the quality of life of regional residents, as a typical expression of urbanization to enhance the well-being of residents.
Innovative urbanization level (IUL)	Number of domestic patent applications authorized (DPP)	Reflecting the level of regional science and technology innovation, as the endogenous driving force of urbanization.

2.2.3. Bivariate Spatial Autocorrelation Model

This study uses the bivariate SPAC model to examine the spatial interaction between the ESV and UL in the YRB. The global bivariate Moran's I is used to examine the comprehensive association degree between the ESV and UL in the study area and its significance [90], by the following equation:

$$I = \frac{n}{\sum_{i=1}^{n}\sum_{j=1}^{n} W_{ij}} \times \frac{\sum_{i=1}^{n}\sum_{j=1}^{n} W_{ij}(x_i - \overline{x})(x_j - \overline{x})}{\sum_{i=1}^{n}(x_i - \overline{x})^2} \quad (3)$$

where x_i and x_j are observed values; w_{ij} is the spatial weight matrix between spatial units i and j.

The local bivariate Moran's I is used to identify the possible spatial association patterns at different spatial locations to acquire, by the following equation,

$$I_i = Z_i \sum_{j=1}^{n} w_{ij} Z_j \quad (4)$$

where Z_i and Z_j are the normalized values of the variance for the observed ESV and UL in spatial units i and j, respectively.

2.2.4. Geographically Weighted Regression Model (GWR)

The parameter estimation of the traditional linear regression model is performed by ordinary least squares (OLS), but it only permits global estimates of the parameters. In contrast, the theory of spatial economics assumes that almost all spatial phenomena are spatially dependent or spatially autocorrelated, and the assumption of the independence of residual terms in traditional regression models (OLS models) cannot be satisfied under this theory [65]. By incorporating the spatial location into the model and taking into account the influence of different spatial location indicators on the regression results, the geographically weighted regression (GWR) model can fully demonstrate the non-smoothness of the interaction relationship between the independent and dependent variables in different spatial geographic locations, and the results are more realistic, which can effectively address the problem that traditional regression models cannot reveal the spatial heterogeneity of regression coefficients [91]. Therefore, the GWR model [54] is used in this study to explore the spatial distribution of the impact of urbanization on the value of ecosystem services. The equation is as follows:

$$y_i = a_0(u_i, v_i) + \sum_{j=1}^{k} b_j(u_i, v_i)x_{ij} + c_i \quad (5)$$

where $b_j(u_i, v_i)$ is the variable parameter of the j-th explanatory variable x_{ij} of the i-th county.

2.3. Data Source

The data required to calculate the ESV and UL in the study include vector ranges, land use information, socioeconomic statistics, and other relevant indicators for each county in the YRB. The sources and descriptions of the data are shown in Table 3. For a small number of missing data due to incomplete statistics, the average growth rate of the last three years was used to make projections or supplementation was achieved by interpolating the plural of the indicator according to neighboring counties.

Table 3. Data description.

Data	Indicator	Source	Description
Vector ranges	UL (CLP)	Resources and Environmental Science Data Center of the Chinese Academy of Sciences (http://www.resdc.cn, accessed on 25 February 2022.)	The data of 1990, 2000, and 2010 are integrated according to the adjustment of administrative divisions, based on the county-level administrative regions of the YRB in 2020.
Land use data	ESV (PSV, RSV, SSV, CSV)	Resources and Environmental Science Data Center of the Chinese Academy of Sciences (http://www.resdc.cn, accessed on 25 February 2022.)	The spatial resolution is 30 m, and the reclassification is made according to 6 primary land classes of cultivated land, forest, grassland, water bodies, construction land, and unused land.
Grain data	ESV (PSV, RSV, SSV, CSV)	Provincial statistical yearbooks in YRB Yearbook of China Agricultural Product Price Survey (1990~2021)	Containing data on grain production, sown area, and grain prices, used for ESV calculations.
Demographic data	UL (TP, UPP, PGDP)	China Census by County (2000, 2010, and 2020)	Both total population and urban population used in the study refer to the resident population of the county.
Patent data	UL (DPP)	Statistical Annual Report of the State Intellectual Property Office of China (2000, 2010, and 2020)	The number of patent applications accepted in China covers invention patents, utility model patents, and design patents applied for in China.
Other socioeconomic data	UL (PGDP, STIP, HSE, LS)	Provincial statistical yearbooks in YRB China County Statistical Yearbook (2001, 2011, and 2021)	Including GDP, education, housing, and other data.

3. Results

3.1. Spatio-Temporal Variation Characteristics of ESV

The ESV in the YRB enjoyed a steady rise between 1990 and 2020, with a growth rate of 2.87% to 6.7 × 10^{10} CNY (Table A1). The achievement was mainly attributed to ecological restoration projects such as grass planting to control desert and returning cultivated land to forests, which effectively increased the area of forests, grassland, and other ecological land. The proportion of each ESV type remained relatively stable, dominated by RSV (52%) and SSV (32%) and supplemented by PSV (9%) and CSV (7%). In terms of change trends, RSV and CSV showed sustained and rapid growth, increasing by 3.47% (4.2 × 10^{10} CNY) and 3.62% (5.8 × 10^9 CNY), respectively, while PSV and SSV showed fluctuations with an increase and then a decrease, increasing by 2.33% (5.1 × 10^9 CNY) and

2.02% (1.5×10^{10} CNY) from 1990 to 2010 and decreasing by 0.46% (1.0×10^{9} CNY) and 0.03‰ (2.4×10^{7} CNY) from 2010 to 2020, respectively.

The spatial distribution of the ESV in the YRB showed obvious spatial heterogeneity, but the overall spatial structure remained stable, and the ESV in the northwestern plateau was much higher than that in the southeastern plain region (Figure 3). Specifically, the high-ESV regions were mainly scattered in the plateau areas, such as the southwest of the Inner Mongolia Autonomous Region and the southeast of Qinghai Province in the upper reaches, while the low-value regions were mainly in the plain areas, such as the Guanzhong Plain, Fen River Plain, and Huanghuaihai Plain, in addition to the hilly and gully areas and windy beach areas of the Loess Plateau. RSV, SSV, and CSV for the ESV types were largely identical to the ESV in spatial distribution, whereas PSV showed a decentralized network distribution because of its role in providing materials or energy directly to humans. Changes in ESV by county showed the coexistence of improvement (162 counties) and deterioration (198 counties) (Figure A1), resulting in an overall relatively stable ESV in the YRB. The counties with larger growth and impairment rates were concentrated in the high- and low-ESV areas, respectively, showing a spatial distribution pattern of "high-value area–high growth and low-value area–high impairment", which further strengthened the spatial heterogeneity of the ESV distribution in the YRB.

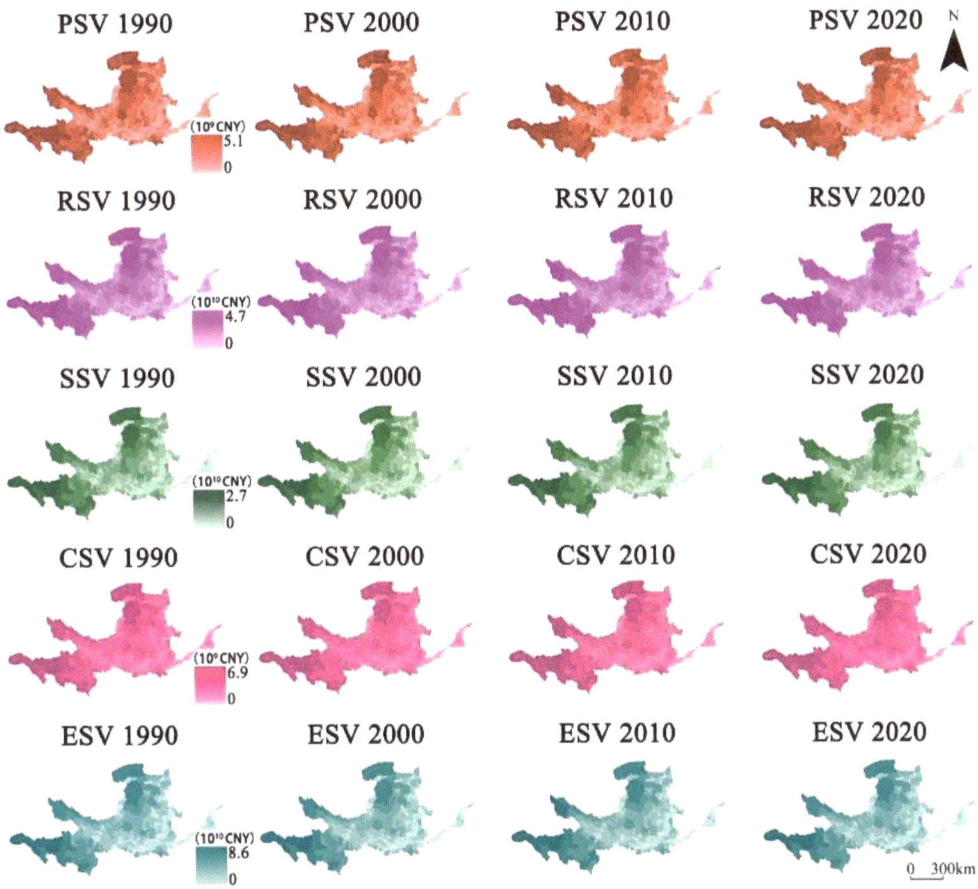

Figure 3. Spatio-temporal pattern of ESV in the YRB from 1990 to 2020.

3.2. Spatio-Temporal Variation Characteristics of UL

Urbanization has had a profound impact on China's economic development, social change, and environmental protection since 2000, when the country proposed the strategic goal of accelerating urbanization. Given the lack of reliable socioeconomic statistics before 1990, this study focuses on the evaluation of the UL in the YRB from 2000 to 2020. The result shows a significant increase in the UL in all counties, with the average index growing by 137.4% to 0.0460–0.1092 (Table A2) and clear spatial heterogeneity in the UL, with the middle and lower reaches being much more urbanized than the upper reaches (Figure 4). It also shows that regions with a high UL featured spatial agglomeration, and a "multi-center grouped" spatial pattern began to take shape around Xi'an, Zhengzhou, Jinan, Taiyuan, Hohhot, and other central cities, with the "Xi'an–Zhengzhou–Jinan" belt of contiguous urbanized areas already emerging on some scale. In general, the UL of the urban agglomeration showed the spatial pattern of a "strong center with a weak periphery" within each group.

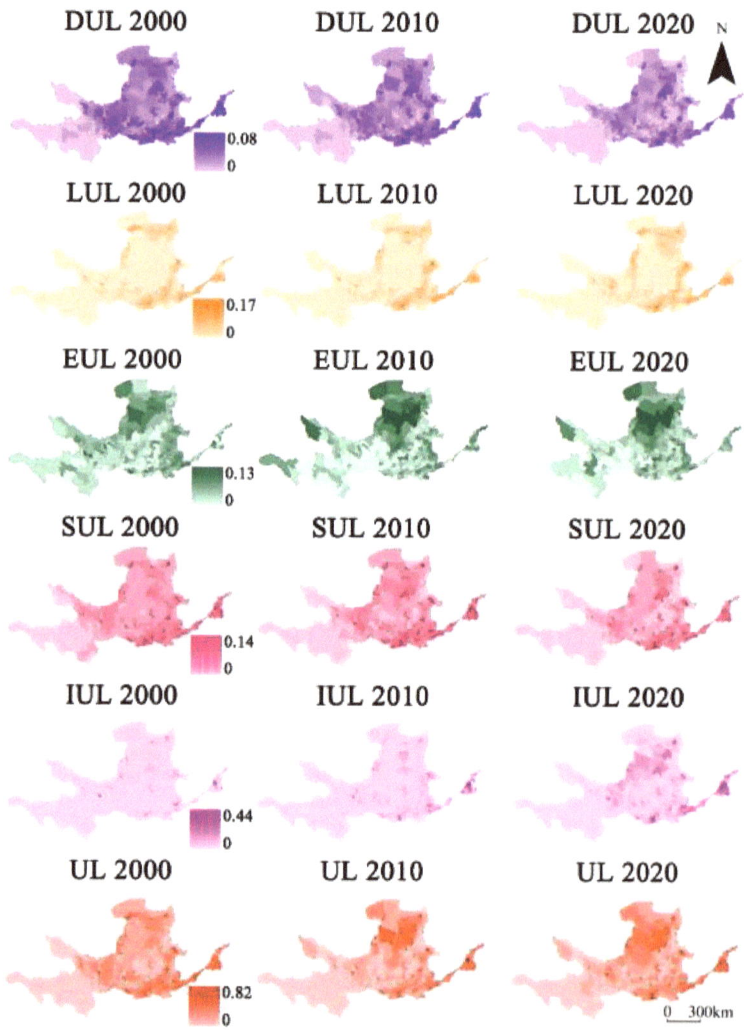

Figure 4. Spatio-temporal pattern of UL in the YRB from 2000 to 2020.

The changes in the average index of the five urbanization subsystems showed that DUL and SUL maintained a steady growth trend (Figure A2), increasing by 0.0046 and 0.0206, respectively; LUL grew overall, but at a slower rate, increasing by 0.0038 in the first decade and only 0.0021 in the second; EUL and IUL showed rapid growth, with the average EUL index of 0.0304 in 2020, 5.15 times that of 2000 (0.0059), and the IUL index rose from a low level of 0.005 to 0.0231. In terms of spatial patterns, the indices at the level of the five urbanization subsystems maintained strong consistency with those at the level of integrated urbanization, i.e., they showed both a gradient pattern of a gradual increase from west to east and a polycentric pattern around the central city. From the change in the UL of each county, the integrated urbanization and five subsystems showed growth, but there was also a phenomenon of "high level–fast growth and low level–slow growth".

3.3. The Spatial Correlation between ESV and UL

The global Moran's I of the bivariate SPAC model was −0.19, −0.17, and −0.15 in 2000, 2010, and 2020, respectively, suggesting a significant negative spatial correlation of ESV with the UL ($p = 0.001$), but with weakening strength. A bivariate, locally spatially autocorrelated LISA aggregation plot (Figure 5) based on the Z-test ($p = 0.05$) showed that the main spatial clustering patterns in the YRB were dominated by low–high (low ESV and high UL) and high–low (high ESV and low UL) types. Specifically, the low–high types were mostly scattered in densely inhabited and economically advanced areas such as the Huanghuaihai Plain and the Guanzhong Plain, where high-speed demographic concentration and rapid land expansion led to the encroachment and destruction of ecosystems. The high–low types were mainly in the middle and upper reaches, such as the Qinghai–Tibet Plateau and the Loess Plateau, where the UL was lagging behind, although they were rich in natural resources. The low–low types (low ESV and low UL) were clustered in the hilly and ravine areas, the wind–sand and grassland areas, and the Gobi Desert areas of the Loess Plateau, where there were constraints from natural conditions such as water scarcity and ecological fragility. Moreover, a few high–high types (high ESV and high UL) were found in the Yellow River Delta and the northern foot of the Qinling Mountains, showing the coordinated development of ecosystem conservation and urbanization.

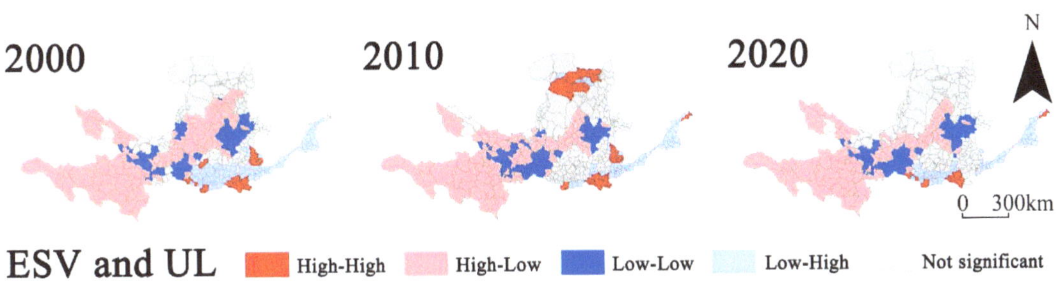

Figure 5. Bivariate LISA cluster map between ESV and UL in the YRB.

3.4. Impact of Urbanization on ESV

To objectively present the overall changes and dynamic trends of urbanization regarding the ESV and UL, and to reduce data covariance and enhance the model's robustness, two time points, 2000 and 2020, were taken for comparative analysis. The correlation of ESV with the eight explanatory variables was first examined using the OLS model (Table 4), and the regression results showed that the VIFs of all variables in 2000 and 2020 ranged from 1.09 to 13.17, with a weak effect of collinearity and a small impact on the model regression results. It was found that, in the YRB, PGDP, HSE, and DPP were positively correlated with ESV in 2000, while the other explanatory variables were negatively correlated; by 2020, the impact of UPP changed from negative to positive, while no changes were found for the other explanatory variables.

Table 4. Estimation parameters of OLS model.

Year	2000				2020			
Variable	Coefficient	p-value	t-value	VIF	Coefficient	p-value	t-value	VIF
TP	−0.17 ***	0.00	−3.41	2.27	−0.14	0.17	−1.38	8.20
UPP	−0.15 *	0.05	−1.97	1.87	0.02	0.62	0.49	2.82
CLP	−0.26 ***	0.00	−3.26	2.24	−0.28 ***	0.00	−4.98	2.56
PGDP	0.06	0.39	0.86	1.34	0.14 ***	0.00	3.23	1.54
STIP	−0.06 **	0.02	−2.30	1.09	−0.19 ***	0.00	−5.60	1.30
HSE	0.10	0.26	1.12	4.41	0.06	0.71	0.37	13.17
LS	−0.19 ***	0.00	−4.63	1.22	−0.14 ***	0.00	−3.67	1.40
DPP	0.04	0.68	0.42	1.67	0.16	0.06	1.88	3.82

Note: *** indicates significance at the 1% level, ** indicates 5%, * indicates 10%.

Although the OLS model provides a global perspective for analysis, it does not take into account the spatial variation in the impact of urbanization on ESV, so to investigate this spatial heterogeneity, we further introduced the GWR model for regression analysis. The R^2 and adjusted R^2 values of the GWR model were significantly higher than those of the OLS model, while the AICc values of the GWR model were lower than those of the OLS model (Table 5), indicating that the GWR model could better fit the true correlation between the UL and ESV. The average of the coefficients of eight explanatory variables in the regression results of the GWR model (Table 6) suggests that CLP has the strongest negative impact on ESV, followed by TP, LS, STIP, and UPP; meanwhile, HSE, PGDP, and DPP have a progressively weaker positive impact on ESV. The results imply that there are significant differences in the direction and extent of urbanization's impact via the five subsystems on ESV, and that LUL is the primary cause of ecosystem degradation.

Table 5. Model performance comparisons between GWR and OLS.

Year	2000			2020		
	AICc	R^2	R^2 Adjusted	AICc	R^2	R^2 Adjusted
OLS model	−581.47	0.25	0.23	−667.93	0.28	0.26
GWR model	−727.46	0.61	0.54	−871.88	0.66	0.61

Table 6. Summary of the estimates of GWR model.

Year		2000				2020			
Dimension	Variable	Mean	Max	Min	Median	Mean	Max	Min	Median
DUL	TP	−0.16	0.10	−1.10	−0.05	−0.21	0.17	−0.97	−0.08
	UPP	−0.05	0.22	−0.26	−0.02	−0.01	0.23	−0.08	−0.02
LUL	CLP	−0.59	−0.15	−3.10	−0.29	−0.26	−0.12	−1.06	−0.18
EUL	PGDP	0.08	0.99	−0.19	0.00	0.10	0.20	−0.02	0.10
	STIP	−0.06	0.00	−0.21	−0.03	−0.06	0.01	−0.38	−0.02
SUL	HSE	0.14	0.79	−0.08	0.06	0.21	1.19	−0.15	0.07
	LS	−0.08	0.08	−0.38	−0.06	−0.07	0.33	−0.39	−0.06
IUL	DPP	0.06	0.75	−0.34	0.09	0.08	0.36	−0.01	0.06

The results of the GWR model regression were spatially visualized and expressed, and the results showed that there were also significant spatial differences and spatio-temporal dynamics in the impact of the urbanization subsystems on ESV in the YRB (Figure 6). Overall, landscape urbanization was the primary factor resulting in the ESV decline, and it had a negative impact in all regions. The change toward a positive impact for economic urbanization and innovative urbanization reflects the positive effect of increased levels of regional economic industry and technological innovation on the ESV and UL. In contrast, demographic urbanization and social urbanization had unstable impacts, in both positive and negative directions.

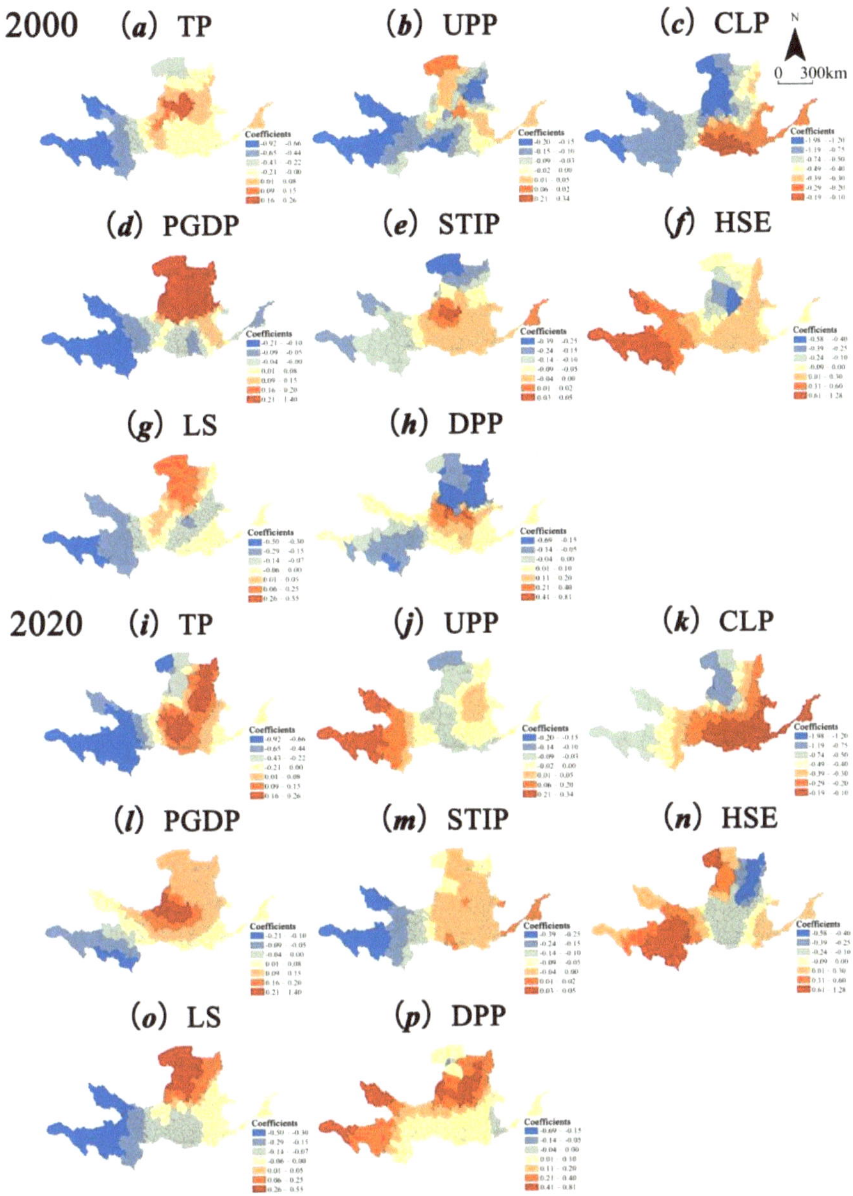

Figure 6. (**a**–**p**) Spatial distribution of regression coefficients between urbanization indicators and ESV in the YRB in 2000 and 2020.

Specifically, the direction and intensity of the impacts of the five urbanization subsystems were as follows.

(1) The impact of DUL on ESV was characterized by significant spatio-temporal dynamics. In 2000, TP had an overall negative impact on ESV, and, by 2020, the negative impact was intensified in the upper reaches and the positive impact was weakened or became negative in the lower reaches, while the negative impact was weakened or became positive in the middle reaches and the positive impact increased in intensity (Figure 6a,i).

The impact of UPP on ESV also changed, from a negative global correlation in 2000 to a weakening negative correlation or positive correlation in 2020, with a significant increase in the positive impact, especially in the upper YRB (Figure 6b,j).

(2) LUL showed a consistently negative global impact on ESV, but the intensity diminished over time. The impact of CLP on ESV was spatially distributed in a stable gradient, gradually weakening from the upper to the middle and lower reaches of the Yellow River (Figure 6c,k).

(3) EUL shifted from a negative to a positive impact on ESV, but the extent was generally weak. PGDP was mainly negatively correlated with ESV in 2000, but it had a strong positive impact in the Inner Mongolia Plateau and the north of the Loess Plateau. By 2020, the positive impact had increased in the middle and lower reaches, while the negative impact continued to weaken in the upper reaches (Figure 6d,l). STIP was negatively correlated with ESV, and its negative impact on the Tibetan Plateau continued to gain strength, while it generally had a weak impact on the middle and lower reaches (Figure 6e,m).

(4) SUL showed both positive and negative impacts on ESV, similar to the characteristics of demographic urbanization. HSE had an overall positive but unstable impact on ESV, shifting from positive to negative in the middle reaches, while a diametrically opposite trend was observed in the lower reaches (Figure 6f,n). LS had an overall negative impact on ESV, except for the positive impact on the Inner Mongolia Plateau and the north of the Loess Plateau, both of which continued to increase (Figure 6g,o).

(5) The impact of IUL on ESV was manifested as a full shift to the positive direction in 2020. The DPP was extremely low in 2000 across the counties and had almost no impact on ecosystem services, while the positive impact of DPP had increased significantly by 2020 (Figure 6h,p).

4. Discussion

4.1. The Relationship between Urbanization Transition and ESV

Urbanization in the YRB underwent a major transition in its development pattern between 2000 and 2020. The proportion of urbanization by subsystem shows a significant decrease in the share of DUL, LUL, and SUL, and a significant increase in the share of EUL and IUL. The change implies that urbanization has shifted from the rapid development driven by demographic concentration and spatial expansion to a new stage of high-quality growth led by industrial upgrading and technological innovation. This conclusion is in agreement with the previous findings of Bai et al. [53] and Liu et al. [92] and also conforms to the general trend of urbanization development in China [93]. In this process, with the rapid rise in the overall UL, ESV also took on an upward trend. From 1990 to 2020, the forest and grassland cover increased by 9.2×10^5 ha and 4.8×10^5 ha, respectively, dramatically improving the net primary productivity of the region [94]. According to Ouyang et al. [95], the trend is similar to that seen when valuing ecosystem services at the national scale in China. Tian et al. [96] also pointed out that, since 2012, China's urbanization has focused on integrated economic, ecological, and social benefits, and ecosystem services have gradually shifted to develop in synergy with urbanization.

However, ESV variation across counties also showed significant spatial heterogeneity. The counties with faster ESV growth were clustered in the Sanjiangyuan area, the Loess Plateau's hilly and ravine area, the Fen River Basin, and other regions, where the growth in ESV mainly benefited from national ecological restoration projects such as the ecological protection in the Sanjiangyuan area, the Sanbei protection forests, the return of cropland to forest and grassland, the protection of natural forests, and the restoration of mining areas [97]. The rapid expansion of construction land in the lower reaches of the YRB (an increase of 2.2×10^5 ha by 2020 from 1990) led to a general decline in ESV, but Puyang, Dongping, Dongying, and other demonstration counties for ecological civilization construction witnessed a significant increase in ESV, as they effectively offset the negative impact of urbanization on ESV by ecological restoration [25]. Bryan et al. [98], Sharma et al. [99], and Wu et al. [100] obtained similar findings in the Guangzhou–Foshan Metropolitan

Area, Delhi of India, and the Loess Plateau region, suggesting that national ecological regulation policies play a decisive role in ecosystem restoration. Additionally, it is also important to note the spatial misalignment and incompatibility between such urbanization and ecological restoration, which may further exacerbate the supply–demand imbalance of ESV and increase the risk of ecosystem security.

According to the changes in ESV by type in each county, CSV, RSV, and SSV showed essentially the same change characteristics as the ESV, but PSV decreased in 235 counties. This suggests that despite the ability to increase productivity per unit of farmland, agricultural technology improvements have not been sufficient to compensate for the loss of farmland due to the massive encroachment of construction land [101], which contributed to the significant decline in PSV in the YRB following 2010. This finding differs from the studies of Kindu et al. [71], Richards et al. [102], García-Nieto et al. [103], and Jaligot et al. [104]. For example, Kindu et al. found an increasing trend in PSV in Munessa-Shashemene; Richards et al. and García- Nieto et al. noted that urbanization led to a decrease in RSV in Singapore and Mediterranean cities; and Jaligot concluded that urbanization would lead to a decrease in CSV in Cameroon. The different findings may arise from the large differences in the national context and the stage of urbanization development in the case studies.

4.2. Spatio-Temporal Heterogeneity of Urbanization's Impact on Ecosystem Service Value

Scholars generally agree that urbanization has a predominantly negative impact on ESV and the UL. For example, Aguilera et al. [59], Eigenbrod et al. [105], and Dadashpoor [72] found that the massive replacement of natural ecosystems (farmland, grassland, forest, etc.) by artificial surfaces during urbanization has brought about a decrease in the number and quality of ecosystem service providers, which seriously affects the structure, processes, and functions of ecosystems [106,107]. Tiwari et al. [108] further argued that many urban areas in developing countries have gone beyond the permissible growth limits. However, we found that the impact of urbanization on the ESV and UL in the YRB is complex [35] (Figure 7), as it is affected by a combination of factors, such as the urbanization stage, ecological background, management policies, and regional collaboration. This conclusion is in agreement with the studies of Mitchell et al. [109] and Tian et al. [68].

Specifically, LUL was the only subsystem that showed a negative impact in all regions, indicating that LUL reduced ESV, but the negative impact diminished in intensity over time. It may be due to land expansion encroaching on natural ecosystems, thus leading to changes in land use/cover type, structure, and pattern that disrupt ecosystem functions and quality. Similar evidence was provided by Gifford et al. [110] and Mao et al. [111]. The implementation of land use policies such as permanent basic farmland and ecological red lines for the purpose of strengthening the planning and management of land use changes in China in the past 30 years has promoted the optimization of the land use structure and pattern, offsetting, to some extent, the loss of ESV caused by land expansion and maintaining the stability of ecosystem services. In their studies of the Dongting Lake Basin and Idaho, Zhao et al. [112] and Halperin et al. [113] also found that policies such as enhancing forest conservation and building high-quality agricultural land play a crucial role in improving overall ecosystem services and balancing the demand for housing, food, and other sustainable energy sources.

EUL had a predominantly negative impact on ESV in the YRB in 2000, probably due to the fact that, with the rapid growth of industrial development after the start of the development strategy in Western China, the transfer of many high-polluting and energy-intensive industries from East–Central to Western China [114] has aggravated the problems of resource depletion, ecological damage, and environmental pollution, leading to a continuous decrease in ESV. Yang et al. [115] also demonstrated that ecosystems in Western China are more sensitive to the impact of EUL. However, the middle and lower reaches saw the beginning of positive impacts of EUL in 2020, indicating that with the growth of the regional economy and social productivity, the government has the ability to invest more funds and human resources to vigorously carry out ecological restoration and protection work and strengthen the ecosystem's ability to resist the negative impacts of urbanization. However, Wang et al. [116] pointed out that economic development in neighboring cities has an overall negative effect on the local ESV, and Sannigrahi et al. [117] also suggested that the effect of economic factors on ESV is negligible, which may be due to differences in research methods and indicator selection.

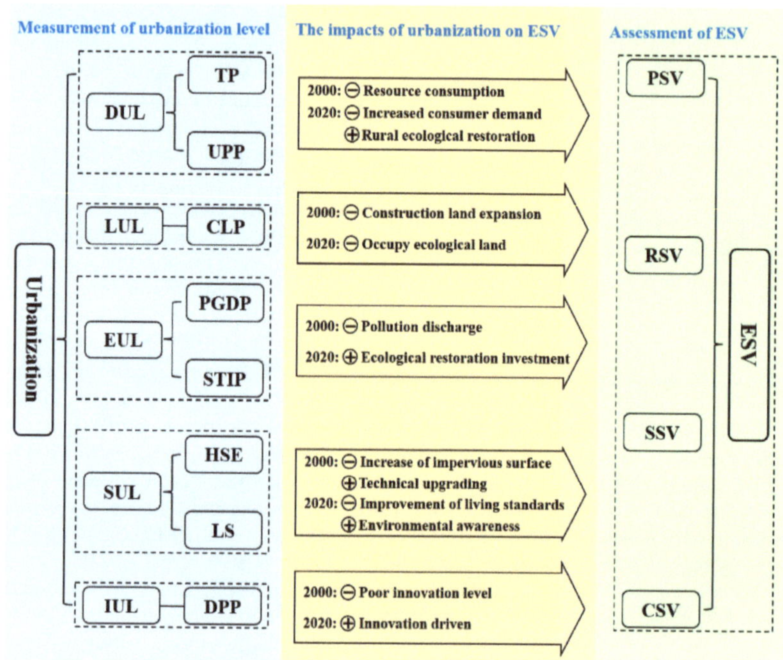

Figure 7. The impact mechanism of urbanization on ESV. "+" and "−" represent the positive and negative impacts of urbanization on ESV, with some examples.

DUL and SUL have both negative and positive impacts on ESV. Given the fragile ecological carrying capacity of the three plateau areas and the highly dense population in the Yellow and Huaihai Plains, the rapid growth of the total population and the increase in the living standards and consumption levels of urban residents drive the increasing demand for PSV and CSV [118], which leads to the overconsumption of natural resources and serious damage to ecosystems, further exacerbating the loss of ecosystem RSV and SSV. The scattered population and extremely low level of urbanization in the early years, and the continued reduction in the rural population while continuing to concentrate in towns and cities, have effectively alleviated problems such as resource consumption and environmental pollution and promoted the return of rural land to forestry and grassland, thus stabilizing the ecological environment. Moreover, with the increase in education, the regional population has a stronger desire for eco-environmental protection and a greater sense of awareness and responsibility [119], resulting in greater willingness to introduce more advanced and environmentally friendly technologies to boost the resource use efficiency and environmental management capabilities and to enhance the regional ESV. This conclusion supports the findings of Kollmuss et al. [120] and Singh et al. [121], suggesting that a higher level of education usually corresponds to greater environmental concern and that the improvement of the education level may contribute to sustainable regional development.

We also found that IUL did not begin to have a positive impact on ESV in the YRB until 2020, which may be the result of a certain "threshold" brought by scientific and technological innovation, i.e., only when innovation reaches a certain height can it effectively drive ESV to grow. In addition to the potential to improve ecosystem conservation techniques, innovation allows for the more precise assessment of the type, structure, and patterns of changes in ecosystem services through regional ecosystem monitoring and modeling, and thus the precise governance and optimization of ecosystem services. The study by Guo et al. [122] also demonstrates that innovation and technological advances will drive vertical urban sprawl and increase land use efficiency. However, in general, unlike existing studies that rarely consider the operational mechanisms by which innovative urbanization affects ESV and UL, this paper provides an in-depth discussion that contributes to the understanding of the relationship between urban technological innovation dynamics and ecosystem services, and this is consistent with the current trend of urbanization in China, where innovation-driven growth is the core of competitiveness.

Although subject to change, the correlation between UL and ESV in the YRB shows a relatively stable pattern in space. Urbanization in the Tibetan Plateau region generally has a negative impact on ESV, suggesting that the ecosystems there are more sensitive to the impact of human activities. Zhang et al. [123] also found that the ecosystem service scarcity value (ESSV) in the Tibetan Plateau region increased significantly, with the growth rate of public product-type services exceeding that of private product-type services. The main cause is that the fragile ecological environment of the Qinghai–Tibet Plateau has reduced the supply capacity of ecosystem services [70], as well as the low urbanization level, leading to the result that demographic growth, land expansion, industrial development, and other anthropogenic factors have caused serious damage to the ecosystems of alpine meadows and desert grasslands [124]. The positive impact of urbanization on ESV is more pronounced in the middle and lower reaches, mainly because China has implemented a series of ecological restoration projects and land management policies in response to eco-environmental problems such as soil erosion in the Huanghuaihai Plain and wetland degradation in the Loess Plateau. As urbanization in the region enters a new stage of high-quality development, technological innovation and environmental awareness, among others, have also contributed positively to the ecosystem. Yang et al. [125] also found evidence of this.

4.3. Policy Implications

This study finds that the quality of ecosystem services has a close connection with the complex interactions between ecosystems and human activities, and that the traditional highly energy-consuming and highly polluting urbanization is difficult to sustain [126,127]. Strassburg et al. [128] and Sirakaya et al. [129] also pointed out that changing the development pattern of urbanization and strengthening government ecological construction and management are critical paths to effectively mitigate the negative impacts of urbanization on ecosystem services. Therefore, this study recommends incorporating the protection and coordination of ecosystem services into new urbanization management policies and strengthening sustainable urban planning. It also proposes three policy recommendations: changing the urbanization development mode, carrying out ecological restoration by zone, and formulating development plans by type.

(i) Changing the urbanization development mode. The YRB, especially the middle and upper reaches, has not yet completed its urbanization process and is an important regional growth area for China's future urbanization development. To promote the transformation of urbanization to a green, low-carbon, innovation-driven, and collaborative development mode based on the characteristics of the development stage, policies should be introduced to build a pattern of coordinated development in small, medium, and large cities; strengthen the leading role of national central cities such as Xi'an and Zhengzhou; accelerate the construction of the modern Xi'an metropolitan area with national influence; systematically cultivate the Zhengzhou, Jinan, Lanzhou, Xining, Taiyuan, Yinchuan, and Hohhot metropolitan areas; develop and expand the Guanzhong Plain Urban Agglomeration and Zhongyuan Urban Agglomeration; guide the steady development of the Hohhot–Baotou–Ordos–Yulin Urban Agglomeration and Lanzhou–Xining Urban Agglomeration [28]; and simultaneously promote new urbanization with county cities as important carriers. Public service systems such as urban green infrastructure, education and healthcare, and housing security should be completed. Ecosystem services should be incorporated into the national spatial planning management system and measures should be taken to construct the ecological security pattern of the river basin by weighing the development patterns under various planning scenarios [130] and carrying out basic work such as the assessment of the current state of the YRB ecosystem, evaluation of the resource and environmental carrying capacity, and ecological risk identification. The counties involved should strictly implement a farmland protection system, strengthen the monitoring and assessment of farmland quality, and improve the efficiency of farmland use and output. An ecological economy system for the YRB should be established, with sound systems for ecological product value accounting, ecological tenure trading, ecological transfer payment, and cross-regional ecological compensation [55], as well as other systems to guide the sustainable management and equitable distribution of regional ecosystem service value [131], with focus placed on pilot demonstrations in key ecological function areas, such as the Sanjiangyuan, Qinling, and Qilian Mountains and the Yellow River Delta wetlands. A multi-level and normalized regional collaborative governance mechanism should be established to bring about the positive spillover effects of urbanization, while eco-environmental resources should be used in a complementary manner to carry out ecological governance, economic cooperation, scientific and technological innovation, and facility construction through cross-administrative coordination.

(ii) Carrying out ecological restoration by zone. Key projects of ecological protection should be implemented regionally in the upper, middle, and lower reaches of the YRB, to strengthen the protection and restoration of critical ecological functional areas. (1) The upper reaches are an important water recharge area in the YRB and also the area with the most extensive desertification (61.7%). Desertification control in key areas has achieved remarkable results in recent years, and the area of unused land in the upper reaches has been reduced by 1.9×10^6 ha in 30 years. Future efforts to protect and restore the degraded grassland ecology should be increased in key areas such as the Yellow River source area, Sanjiangyuan, and the Ruoergai grassland wetlands. In addition, stronger measures should be taken to manage the wind and sand desert in the Ordos Plateau area and continue to promote the construction of the Sanbei protection forests, returning farmland to forest, and other key projects to control the expansion of the Ulanbuh Desert and Tengger Desert. (2) The middle reaches are faced with ecological problems such as increased soil erosion, fragmentation of the landscape, and the declining production capacity of animal husbandry. In the future, there should be policies to encourage the implementation of national key projects for soil and water conservation in the hilly and ravine areas of the Loess Plateau, such as the comprehensive management of small river basins, comprehensive management of sloping land, sand interception in the concentrated source areas of coarse sediment, and land protection of the Loess Plateau. In Eastern Ningxia, Northern Shanxi, Northern Shaanxi, Central Inner Mongolia, and other coal-rich areas, key projects such as geological environment management, comprehensive land improvement, and continuous industry cultivation should be arranged as a whole. In addition, because of the loss of 7.6×10^5 ha of the grassland area in the middle reaches in the past 30 years, the implementation of policies such as grazing bans and rotational grazing should be intensified in overgrazing areas such as the riverain-irrigated regions. (3) Most of the lower reaches are in low-ESV zones and face ecological problems such as flooding in beach areas and shrinking wetlands. In the future, policies should be developed to promote the management of the Yellow River beach area; strengthen the water ecological space control in the beach area, flood control and sand sedimentation, and other functions; and build urban forest parks along the Yellow River according to the local conditions. Priority should be given to restoring the wetland ecosystem of the Yellow River Delta, systematically carrying out the work of returning farmland to water bodies and beaches, and building a green ecological corridor in the lower reaches of the YRB, integrating flood control and bank protection, water conservation, and biological habitats.

(iii) Formulating development plans by type. County ecological protection and high-quality development plans should be formulated according to the spatial clustering pattern of the ESV and UL in the YRB by type, to explore a new path of urbanization tailored to the local conditions [115]. Specifically, (1) the 77 low–low-type counties should continue the construction of the desert shelter forest system, implement critical projects such as the return of grazing land to grassland and saline land management, carry out pilot sand control based on the photovoltaic industry relying on policy support, build a long-term mechanism to guarantee funds for desertification control, improve the ecosystem service capacity, and create conditions for urbanization. (2) The 54 high–low-type counties should, following the ecology-based functional zoning, try to develop ecological tourism, special agricultural products, and other green industries and make efforts to complete the ecological product value accounting and ecological tenure trading system to achieve ecological product value [132], to transform natural ecological advantages into economic advantages and promote urbanization in harmony with ecological protection. (3) The 72 low–high-type counties should focus on building a multi-level ecological network system, slowing down urban expansion [66], promoting the ecological and green transformation of industries, boosting sustainable and healthy economic and social development centered on supply-side structural reform, and improving the quality of the urban ecological environment. (4) The six high–high-type counties should strengthen the positioning of the "ecological economic zone" [133], encourage the development of high-tech industries, eliminate polluting industries, establish a number of ecological economic demonstration bases and recycling industrial parks, and set up green development samples to lead the process of ecological civilization construction.

5. Conclusions

Urbanization is one of the most significant features of human social development, and it has a profound impact on the value of regional ecosystem services. Taking counties as the study units, this study analyzed the spatio-temporal variation characteristics and spatial interactions of the ecosystem service value and urbanization levels of 361 county units in the YRB from the perspective of spatio-temporal heterogeneity at the study scale of the basin using the GWR model. It also explored the mechanisms of influence of five urbanization subsystems—population, land, economy, society, and innovation—on the value of ecosystem services and the patterns of spatio-temporal dynamics.

The results showed that the ESV in the YRB experienced a steady rise of 2.87% from 1990 to 2020, high in the northwest plateau region and low in the southeast plain region. Regulation services and support services were the main components of the ESV, holding 84% of the total, and determined the overall trend of the ESV. The YRB enjoyed a significant increase in the UL and also underwent a major transformation in its development pattern, eventually forming a "multi-center grouped style" urbanization spatial pattern with central cities as the core. There was a gradually decreasing negative spatial correlation between the ESV and UL, with low–high and high–low types as the dominant spatial clustering patterns. The impact of urbanization subsystems on ESV showed significant spatial differences and spatio-temporal dynamics. Landscape urbanization showed a significant negative impact across the board; economic urbanization and innovative urbanization changed to have a positive impact; and demographic urbanization and social urbanization had both positive and negative impacts. On this basis, this paper proposes three policy recommendations based on the actual situation of the basin, such as changing the urbanization development mode, carrying out ecological restoration by zone, and formulating development plans by type, to provide a theoretical basis and practical experience for the high-quality development of similar basin areas.

On the basis of existing studies, this paper attempts to expand the vision and depth of urbanization and ESV research from three points. First, in the comprehensive evaluation of the UL, this paper takes into account the three traditional dimensions of population, land, and economy, and two additional important driving factors for new urbanization, namely social services and technological innovation, enriching the connotations and indicator system of new urbanization. Second, from the perspective of the spatio-temporal heterogeneity of urbanization and ESV, this paper reveals the spatio-temporal variation pattern of the urbanization subsystem's influence mechanism on ESV, which compensates for the deficiency of previous studies that ignore the uneven spatio-temporal distribution characteristics of the two and enriches the theoretical understanding of the interrelationships between natural systems and human systems. Third, this paper proposes policy recommendations to promote the coordinated development of ecological environmental protection and urbanization in the YRB, which will provide a reference for the sustainable development of similar basin areas with great practical value.

This paper advances the theory and methodology of urbanization and ESV research to a certain extent, but it still has the following shortcomings for further improvement in future research. First, this study does not delve into the mutual or synergistic effects among the five urbanization subsystems. More suitable methods, such as geographic probes, can be used in future studies to analyze the interactions between urbanization subsystems and the combined effects on ESV. Second, although the use of county-level administrative divisions as the study units in this paper facilitates county-level administrative entities to formulate regional ecological environmental protection and sustainable development policies according to local conditions in response to research findings [134], there is no comprehensive knowledge of the mechanisms by which urbanization affects the value of ecosystem services at the municipal scale, basin scale, or raster scale. The impact of urbanization on ecosystem services at different spatial scales and its variability should be explored in depth in the future. Finally, with the increasing richness of China's new urbanization [135], limitations in data sources and quality make it difficult to cover all influences on urbanization that may be relevant to the value of ecosystem services. The framework for UL evaluation can be further improved in future studies with efforts to acquire more reliable data.

Author Contributions: Conceptualization, Y.C. and P.C.; methodology, Q.K., K.Z. and Q.N.; software, Q.K., K.Z. and Q.N.; validation, Q.K., K.Z. and Q.N.; formal analysis, Y.C. and Q.K.; investigation, Q.N. and X.M.; resources, Q.K. and X.M.; data curation, Q.K. and X.M.; writing—original draft preparation, Y.C., Q.K., Q.N. and X.M.; writing—review and editing, Y.C. and Q.K.; visualization, Q.K. and Q.N.; supervision, P.C. and K.L.; project administration, K.L. and J.L.; funding acquisition, J.L. All authors have read and agreed to the published version of the manuscript.

Funding: This research was funded by the National Natural Science Foundation of China (grant number 42071211) and the National Key R&D Program of China (grant number 2022YFC3802804).

Data Availability Statement: The LUCC data and vector ranges of the Yellow River Basin were obtained from the Resources and Environmental Sciences and Data Center, Chinese Academy of Sciences (http://www.resdc.cn/, accessed on 12 January 2023). The data on the grain sown area, output, and grain price were obtained from the provincial statistical yearbook and the Yearbook of China Agricultural Product Price Survey. Data on other influencing factors were obtained from the "China Census by County", "China County Statistical Yearbook", "Statistical Annual Report of the State Intellectual Property Office of China", and the Statistical Yearbook of the Yellow River Basin.

Conflicts of Interest: The authors declare no conflict of interest.

Appendix A

Table A1. Change in total value of ESV in the YRB from 1990 to 2020 (CNY).

YEAR	PSV	RSV	SSV	CSV	ESV
1990	2.194×10^{11}	1.217×10^{12}	7.388×10^{11}	1.604×10^{11}	2.336×10^{12}
2000	2.222×10^{11}	1.226×10^{12}	7.411×10^{11}	1.615×10^{11}	2.351×10^{12}
2010	2.245×10^{11}	1.254×10^{12}	7.537×10^{11}	1.652×10^{11}	2.397×10^{12}
2020	2.235×10^{11}	1.259×10^{12}	7.537×10^{11}	1.662×10^{11}	2.403×10^{12}

Table A2. Change in mean values of UL in the YRB from 1990 to 2020.

YEAR	DUL	LUL	EUL	SUL	IUL	UL
2000	0.0139	0.0059	0.0095	0.0108	0.0005	0.0406
2010	0.0164	0.0179	0.0146	0.0146	0.0024	0.0659
2020	0.0185	0.0304	0.0206	0.0167	0.0231	0.1092

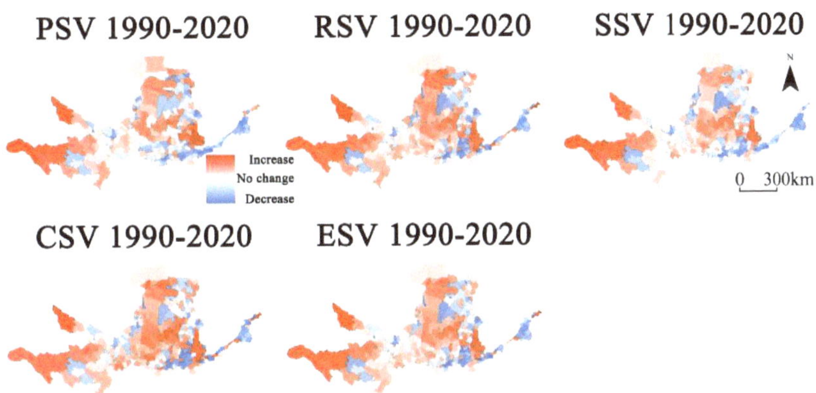

Figure A1. Spatio-temporal change in ESV in the YRB from 1990 to 2020.

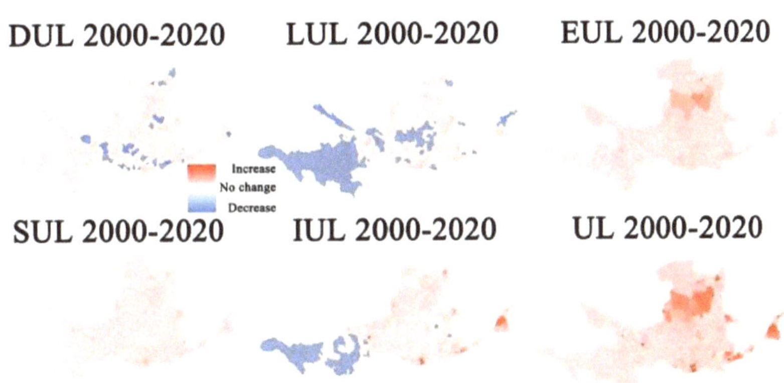

Figure A2. Spatio-temporal change in UL in the YRB from 2000 to 2020.

References

1. Costanza, R.; d'Arge, R.; de Groot, R.; Farber, S.; Grasso, M.; Hannon, B.; Limburg, K.; Naeem, S.; O'Neill, R.V.; Paruelo, J.; et al. The value of the world's ecosystem services and natural capital. *Nature* **1997**, *387*, 253–260. [CrossRef]
2. Nelson, E.; Mendoza, G.; Regetz, J.; Polasky, S.; Tallis, H.; Cameron, D.; Chan, K.M.A.; Daily, G.C.; Goldstein, J.; Kareiva, P.M.; et al. Modeling multiple ecosystem services, biodiversity conservation, commodity production, and tradeoffs at landscape scales. *Front. Ecol. Environ.* **2009**, *7*, 4–11. [CrossRef]
3. Gómez-Baggethun, E.; Barton, D.N. Classifying and valuing ecosystem services for urban planning. *Ecol. Econ.* **2013**, *86*, 235–245. [CrossRef]
4. Li, G.; Fang, C.; Wang, S. Exploring spatiotemporal changes in ecosystem-service values and hotspots in China. *Sci. Total Environ.* **2016**, *545–546*, 609–620. [CrossRef] [PubMed]
5. Tao, Y.; Wang, H.; Ou, W.; Guo, J. A land-cover-based approach to assessing ecosystem services supply and demand dynamics in the rapidly urbanizing Yangtze River Delta region. *Land Use Policy* **2018**, *72*, 250–258. [CrossRef]
6. de Groot, R.S.; Wilson, M.A.; Boumans, R.M.J. A typology for the classification, description and valuation of ecosystem functions, goods and services. *Ecol. Econ.* **2002**, *41*, 393–408. [CrossRef]
7. Grimm, N.B.; Faeth, S.H.; Golubiewski, N.E.; Redman, C.L.; Wu, J.; Bai, X.; Briggs, J.M. Global change and the ecology of cities. *Science* **2008**, *319*, 756–760. [CrossRef]
8. Pan, Z.; Wang, J. Spatially heterogeneity response of ecosystem services supply and demand to urbanization in China. *Ecol. Eng.* **2021**, *169*, 106303. [CrossRef]
9. Wang, C.; Wang, L.; Zhan, J.; Liu, W.; Teng, Y.; Chu, X.; Wang, H. Spatial heterogeneity of urbanization impacts on ecosystem services in the urban agglomerations along the Yellow River, China. *Ecol. Eng.* **2022**, *182*, 106717. [CrossRef]
10. Taylor, L.; Hahs, A.K.; Hochuli, D.F. Correction to: Wellbeing and urban living: Nurtured by nature. *Urban Ecosyst.* **2018**, *21*, 1227–1228. [CrossRef]
11. Zhang, M.; Tan, S.; Zhang, C.; Han, S.; Zou, S.; Chen, E. Assessing the impact of fractional vegetation cover on urban thermal environment: A case study of Hangzhou, China. *Sustain. Cities Soc.* **2023**, *96*, 104663. [CrossRef]
12. Yang, K.; Hou, H.; Li, Y.; Chen, Y.; Wang, L.; Wang, P.; Hu, T. Future urban waterlogging simulation based on LULC forecast model: A case study in Haining City, China. *Sustain. Cities Soc.* **2022**, *87*, 104167. [CrossRef]
13. Liu, X.; Pei, F.; Wen, Y.; Li, X.; Wang, S.; Wu, C.; Cai, Y.; Wu, J.; Chen, J.; Feng, K.; et al. Global urban expansion offsets climate-driven increases in terrestrial net primary productivity. *Nat. Commun.* **2019**, *10*, 5558. [CrossRef]
14. Grêt-Regamey, A.; Galleguillos-Torres, M.; Dissegna, A.; Weibel, B. How urban densification influences ecosystem services—A comparison between a temperate and a tropical city. *Environ. Res. Lett.* **2020**, *15*, 075001. [CrossRef]
15. Wang, S.; Song, S.; Zhang, J.; Wu, X.; Fu, B. Achieving a fit between social and ecological systems in drylands for sustainability. *Curr. Opin. Environ. Sustai.* **2021**, *48*, 53–58. [CrossRef]
16. Deng, C.; Liu, J.; Nie, X.; Li, Z.; Liu, Y.; Xiao, H.; Hu, X.; Wang, L.; Zhang, Y.; Zhang, G.; et al. How trade-offs between ecological construction and urbanization expansion affect ecosystem services. *Ecol. Indic.* **2021**, *122*, 107253. [CrossRef]
17. Peng, J.; Shen, H.; Wu, W.; Liu, Y.; Wang, Y. Net primary productivity (NPP) dynamics and associated urbanization driving forces in metropolitan areas: A case study in Beijing City, China. *Landsc. Ecol.* **2016**, *31*, 1077–1092. [CrossRef]
18. Zhang, D.; Huang, Q.; He, C.; Wu, J. Impacts of urban expansion on ecosystem services in the Beijing-Tianjin-Hebei urban agglomeration, China: A scenario analysis based on the Shared Socioeconomic Pathways. *Resour. Conserv. Recy.* **2017**, *125*, 115–130. [CrossRef]
19. Dang, K.B.; Pham, H.H.; Nguyen, T.N.; Giang, T.L.; Pham, T.P.N.; Nghiem, V.S.; Nguyen, D.H.; Vu, K.C.; Bui, Q.D.; Pham, H.N.; et al. Monitoring the effects of urbanization and flood hazards on sandy ecosystem services. *Sci. Total Environ.* **2023**, *880*, 163271. [CrossRef]
20. Zhu, C.; Zhang, X.; Zhou, M.; He, S.; Gan, M.; Yang, L.; Wang, K. Impacts of urbanization and landscape pattern on habitat quality using OLS and GWR models in Hangzhou, China. *Ecol. Indic.* **2020**, *117*, 106654. [CrossRef]
21. McFarlane, R.A.; Horwitz, P.; Arabena, K.; Capon, A.; Jenkins, A.; Jupiter, S.; Negin, J.; Parkes, M.W.; Saketa, S. Ecosystem services for human health in Oceania. *Ecosyst. Serv.* **2019**, *39*, 100976. [CrossRef]
22. Song, W.; Deng, X. Land-use/land-cover change and ecosystem service provision in China. *Sci. Total Environ.* **2017**, *576*, 705–719. [CrossRef] [PubMed]
23. Tian, Y.; Jiang, G.; Zhou, D.; Li, G. Systematically addressing the heterogeneity in the response of ecosystem services to agricultural modernization, industrialization and urbanization in the Qinghai-Tibetan Plateau from 2000 to 2018. *J. Clean. Prod.* **2021**, *285*, 125323. [CrossRef]
24. Zhang, Y.; Liu, Y.; Zhang, Y.; Liu, Y.; Zhang, G.; Chen, Y. On the spatial relationship between ecosystem services and urbanization: A case study in Wuhan, China. *Sci. Total Environ.* **2018**, *637–638*, 780–790. [CrossRef]
25. Xiao, R.; Lin, M.; Fei, X.; Li, Y.; Zhang, Z.; Meng, Q. Exploring the interactive coercing relationship between urbanization and ecosystem service value in the Shanghai–Hangzhou Bay Metropolitan Region. *J. Clean. Prod.* **2020**, *253*, 119803. [CrossRef]
26. Li, W.; An, M.; Wu, H.; An, H.; Huang, J.; Khanal, R. The local coupling and telecoupling of urbanization and ecological environment quality based on multisource remote sensing data. *J. Environ. Manag.* **2023**, *327*, 116921. [CrossRef]
27. Peng, J.; Wang, X.; Liu, Y.; Zhao, Y.; Xu, Z.; Zhao, M.; Qiu, S.; Wu, J. Urbanization impact on the supply-demand budget of ecosystem services: Decoupling analysis. *Ecosyst. Serv.* **2020**, *44*, 101139. [CrossRef]

28. Ouyang, X.; Tang, L.; Wei, X.; Li, Y. Spatial interaction between urbanization and ecosystem services in Chinese urban agglomerations. *Land Use Policy* **2021**, *109*, 105587. [CrossRef]
29. Taylor, L.; Hochuli, D.F. Creating better cities: How biodiversity and ecosystem functioning enhance urban residents' wellbeing. *Urban Ecosyst.* **2014**, *18*, 747–762. [CrossRef]
30. Wan, L.; Ye, X.; Lee, J.; Lu, X.; Zheng, L.; Wu, K. Effects of urbanization on ecosystem service values in a mineral resource-based city. *Habitat Int.* **2015**, *46*, 54–63. [CrossRef]
31. Zhou, D.; Tian, Y.; Jiang, G. Spatio-temporal investigation of the interactive relationship between urbanization and ecosystem services: Case study of the Jingjinji urban agglomeration, China. *Ecol. Indic.* **2018**, *95*, 152–164. [CrossRef]
32. Yao, X.; Kou, D.; Shao, S.; Li, X.; Wang, W.; Zhang, C. Can urbanization process and carbon emission abatement be harmonious? New evidence from China. *Environ. Impact Assess. Rev.* **2018**, *71*, 70–83. [CrossRef]
33. Liu, Y.; Liu, S.; Sun, Y.; Sun, J.; Wang, F.; Li, M. Effect of grazing exclusion on ecosystem services dynamics, trade-offs and synergies in Northern Tibet. *Ecol. Eng.* **2022**, *179*, 106638. [CrossRef]
34. Wang, Z.; Liang, L.; Sun, Z.; Wang, X. Spatiotemporal differentiation and the factors influencing urbanization and ecological environment synergistic effects within the Beijing-Tianjin-Hebei urban agglomeration. *J. Environ. Manag.* **2019**, *243*, 227–239. [CrossRef]
35. Peng, J.; Tian, L.; Liu, Y.; Zhao, M.; Hu, Y.; Wu, J. Ecosystem services response to urbanization in metropolitan areas: Thresholds identification. *Sci. Total Environ.* **2017**, *607–608*, 706–714. [CrossRef] [PubMed]
36. Wang, J.; Zhou, W.; Pickett, S.T.A.; Yu, W.; Li, W. A multiscale analysis of urbanization effects on ecosystem services supply in an urban megaregion. *Sci. Total Environ.* **2019**, *662*, 824–833. [CrossRef]
37. Faulkner, S. Urbanization impacts on the structure and function of forested wetlands. *Urban Ecosyst.* **2004**, *7*, 89–106. [CrossRef]
38. Wang, S.; Ma, H.; Zhao, Y. Exploring the relationship between urbanization and the eco-environment—A case study of Beijing–Tianjin–Hebei region. *Ecol. Indic.* **2014**, *45*, 171–183. [CrossRef]
39. Zhou, S.; Huang, Y.; Yu, B.; Wang, G. Effects of human activities on the eco-environment in the middle Heihe River Basin based on an extended environmental Kuznets curve model. *Ecol. Eng.* **2015**, *76*, 14–26. [CrossRef]
40. Saboori, B.; Sulaiman, J. Environmental degradation, economic growth and energy consumption: Evidence of the environmental Kuznets curve in Malaysia. *Energy Policy* **2013**, *60*, 892–905. [CrossRef]
41. Chen, W.; Zeng, J.; Li, N. Change in land-use structure due to urbanisation in China. *J. Clean. Prod.* **2021**, *321*, 128986. [CrossRef]
42. Chen, W.; Chi, G.; Li, J. The spatial aspect of ecosystem services balance and its determinants. *Land Use Policy* **2020**, *90*, 104263. [CrossRef]
43. Peng, J.; Liu, Y.; Corstanje, R.; Meersmans, J. Promoting sustainable landscape pattern for landscape sustainability. *Landsc. Ecol.* **2021**, *36*, 1839–1844. [CrossRef]
44. Sallustio, L.; De Toni, A.; Strollo, A.; Di Febbraro, M.; Gissi, E.; Casella, L.; Geneletti, D.; Munafo, M.; Vizzarri, M.; Marchetti, M. Assessing habitat quality in relation to the spatial distribution of protected areas in Italy. *J. Environ. Manag.* **2017**, *201*, 129–137. [CrossRef]
45. Claris Fisher, J.; Emmerson Bicknell, J.; Nesbitt Irvine, K.; Fernandes, D.; Mistry, J.; Georgina Davies, Z. Exploring how urban nature is associated with human wellbeing in a neotropical city. *Landsc. Urban Plann.* **2021**, *212*, 104119. [CrossRef]
46. Wang, Q.; Lan, Z. Park green spaces, public health and social inequalities: Understanding the interrelationships for policy implications. *Land Use Policy* **2019**, *83*, 66–74. [CrossRef]
47. Chen, W.; Chi, G.; Li, J. The spatial association of ecosystem services with land use and land cover change at the county level in China, 1995–2015. *Sci. Total Environ.* **2019**, *669*, 459–470. [CrossRef]
48. Ahmed, Z.; Asghar, M.M.; Malik, M.N.; Nawaz, K. Moving towards a sustainable environment: The dynamic linkage between natural resources, human capital, urbanization, economic growth, and ecological footprint in China. *Resour. Policy* **2020**, *67*, 101677. [CrossRef]
49. Zank, B.; Bagstad, K.J.; Voigt, B.; Villa, F. Modeling the effects of urban expansion on natural capital stocks and ecosystem service flows: A case study in the Puget Sound, Washington, USA. *Landsc. Urban Plann.* **2016**, *149*, 31–42. [CrossRef]
50. Wang, L.; Li, Q.; Bi, H.; Mao, X.Z. Human impacts and changes in the coastal waters of south China. *Sci. Total Environ.* **2016**, *562*, 108–114. [CrossRef]
51. Luo, Q.; Zhou, J.; Zhang, Y.; Yu, B.; Zhu, Z. What is the spatiotemporal relationship between urbanization and ecosystem services? A case from 110 cities in the Yangtze River Economic Belt, China. *J. Environ. Manag.* **2022**, *321*, 115709. [CrossRef] [PubMed]
52. Shi, Y.; Feng, C.-C.; Yu, Q.; Guo, L. Integrating supply and demand factors for estimating ecosystem services scarcity value and its response to urbanization in typical mountainous and hilly regions of south China. *Sci. Total Environ.* **2021**, *796*, 149032. [CrossRef] [PubMed]
53. Bai, Y.; Deng, X.; Jiang, S.; Zhang, Q.; Wang, Z. Exploring the relationship between urbanization and urban eco-efficiency: Evidence from prefecture-level cities in China. *J. Clean. Prod.* **2018**, *195*, 1487–1496. [CrossRef]
54. Cao, Y.; Kong, L.; Zhang, L.; Ouyang, Z. The balance between economic development and ecosystem service value in the process of land urbanization: A case study of China's land urbanization from 2000 to 2015. *Land Use Policy* **2021**, *108*, 105536. [CrossRef]
55. Yu, Q.; Feng, C.-C.; Shi, Y.; Guo, L. Spatiotemporal interaction between ecosystem services and urbanization in China: Incorporating the scarcity effects. *J. Clean. Prod.* **2021**, *317*, 128392. [CrossRef]

56. Deng, C.; Liu, J.; Liu, Y.; Li, Z.; Nie, X.; Hu, X.; Wang, L.; Zhang, Y.; Zhang, G.; Zhu, D.; et al. Spatiotemporal dislocation of urbanization and ecological construction increased the ecosystem service supply and demand imbalance. *J. Environ. Manag.* **2021**, *288*, 112478. [CrossRef]
57. Villamagna, A.M.; Angermeier, P.L.; Bennett, E.M. Capacity, pressure, demand, and flow: A conceptual framework for analyzing ecosystem service provision and delivery. *Ecol. Complex.* **2013**, *15*, 114–121. [CrossRef]
58. Xu, Z.; Peng, J.; Qiu, S.; Liu, Y.; Dong, J.; Zhang, H. Responses of spatial relationships between ecosystem services and the Sustainable Development Goals to urbanization. *Sci. Total Environ.* **2022**, *850*, 157868. [CrossRef]
59. Aguilera, M.A.; Tapia, J.; Gallardo, C.; Nunez, P.; Varas-Belemmi, K. Loss of coastal ecosystem spatial connectivity and services by urbanization: Natural-to-urban integration for bay management. *J. Environ. Manag.* **2020**, *276*, 111297. [CrossRef]
60. Dong, F.; Pan, Y.; Li, Y.; Zhang, S. How public and government matter in industrial pollution mitigation performance: Evidence from China. *J. Clean. Prod.* **2021**, *306*, 127099. [CrossRef]
61. Li, B.; Chen, D.; Wu, S.; Zhou, S.; Wang, T.; Chen, H. Spatio-temporal assessment of urbanization impacts on ecosystem services: Case study of Nanjing City, China. *Ecol. Indic.* **2016**, *71*, 416–427. [CrossRef]
62. Luo, Q.; Zhou, J.; Li, Z.; Yu, B. Spatial differences of ecosystem services and their driving factors: A comparation analysis among three urban agglomerations in China's Yangtze River Economic Belt. *Sci. Total Environ.* **2020**, *725*, 138452. [CrossRef]
63. Yuan, Y.; Wu, S.; Yu, Y.; Tong, G.; Mo, L.; Yan, D.; Li, F. Spatiotemporal interaction between ecosystem services and urbanization: Case study of Nanjing City, China. *Ecol. Indic.* **2018**, *95*, 917–929. [CrossRef]
64. Li, R.; Shi, Y.; Feng, C.-C.; Guo, L. The spatial relationship between ecosystem service scarcity value and urbanization from the perspective of heterogeneity in typical arid and semiarid regions of China. *Ecol. Indic.* **2021**, *132*, 108299. [CrossRef]
65. Su, S.; Li, D.; Hu, Y.N.; Xiao, R.; Zhang, Y. Spatially non-stationary response of ecosystem service value changes to urbanization in Shanghai, China. *Ecol. Indic.* **2014**, *45*, 332–339. [CrossRef]
66. Yu, P.; Zhang, S.; Yung, E.H.K.; Chan, E.H.W.; Luan, B.; Chen, Y. On the urban compactness to ecosystem services in a rapidly urbanising metropolitan area: Highlighting scale effects and spatial non–stationary. *Environ. Impact Assess. Rev.* **2023**, *98*, 106975. [CrossRef]
67. Guo, X.M.; Fang, C.L.; Mu, X.F.; Chen, D. Coupling and coordination analysis of urbanization and ecosystem service value in Beijing-Tianjin-Hebei urban agglomeration. *Ecol. Indic.* **2022**, *137*, 108782. [CrossRef]
68. Tian, Y.; Zhou, D.; Jiang, G. Conflict or Coordination? Multiscale assessment of the spatio-temporal coupling relationship between urbanization and ecosystem services: The case of the Jingjinji Region, China. *Ecol. Indic.* **2020**, *117*, 106543. [CrossRef]
69. Xing, L.; Zhu, Y.; Wang, J. Spatial spillover effects of urbanization on ecosystem services value in Chinese cities. *Ecol. Indic.* **2021**, *121*, 107028. [CrossRef]
70. Wang, J.; Zhai, T.; Lin, Y.; Kong, X.; He, T. Spatial imbalance and changes in supply and demand of ecosystem services in China. *Sci. Total Environ.* **2019**, *657*, 781–791. [CrossRef]
71. Kindu, M.; Schneider, T.; Teketay, D.; Knoke, T. Changes of ecosystem service values in response to land use/land cover dynamics in Munessa-Shashemene landscape of the Ethiopian highlands. *Sci. Total Environ.* **2016**, *547*, 137–147. [CrossRef] [PubMed]
72. Dadashpoor, H.; Azizi, P.; Moghadasi, M. Land use change, urbanization, and change in landscape pattern in a metropolitan area. *Sci. Total Environ.* **2019**, *655*, 707–719. [CrossRef] [PubMed]
73. Narducci, J.; Quintas-Soriano, C.; Castro, A.; Som-Castellano, R.; Brandt, J.S. Implications of urban growth and farmland loss for ecosystem services in the western United States. *Land Use Policy* **2019**, *86*, 1–11. [CrossRef]
74. Zhang, H.; Deng, W.; Zhang, S.; Peng, L.; Liu, Y. Impacts of urbanization on ecosystem services in the Chengdu-Chongqing Urban Agglomeration: Changes and trade-offs. *Ecol. Indic.* **2022**, *139*, 108920. [CrossRef]
75. Wang, S.; Fu, B.; Piao, S.; Lü, Y.; Ciais, P.; Feng, X.; Wang, Y. Reduced sediment transport in the Yellow River due to anthropogenic changes. *Nat. Geosci.* **2015**, *9*, 38–41. [CrossRef]
76. Li, Y.; Xie, Z.; Qin, Y.; Zheng, Z. Responses of the Yellow River basin vegetation: Climate change. *Int. J. Clim. Chang. Str.* **2019**, *11*, 483–498. [CrossRef]
77. Wang, S.; Fu, B.; Liang, W.; Liu, Y.; Wang, Y. Driving forces of changes in the water and sediment relationship in the Yellow River. *Sci. Total Environ.* **2017**, *576*, 453–461. [CrossRef]
78. Wang, S.; Fu, B.; Liang, W. Developing policy for the Yellow River sediment sustainable control. *Natl. Sci. Rev.* **2016**, *3*, 162–164. [CrossRef]
79. Xie, G.D.; Zhang, C.X.; Zhang, L.M.; Chen, W.H.; Li, S.M. Improvement of the evaluation method for ecosystem service value based on per unit area. *J. Nat. Resour.* **2015**, *30*, 1243–1254.
80. Xing, L.; Hu, M.; Wang, Y. Integrating ecosystem services value and uncertainty into regional ecological risk assessment: A case study of Hubei Province, Central China. *Sci. Total Environ.* **2020**, *740*, 140126. [CrossRef]
81. Zhang, B.; Wang, Y.; Li, J.; Zheng, L. Degradation or Restoration? The temporal-spatial evolution of ecosystem services and its determinants in the Yellow River Basin, China. *Land* **2022**, *11*, 863. [CrossRef]
82. Ahern, J.; Cilliers, S.; Niemelä, J. The concept of ecosystem services in adaptive urban planning and design: A framework for supporting innovation. *Landsc. Urban Plann.* **2014**, *125*, 254–259. [CrossRef]
83. Ma, L.; Cheng, W.; Qi, J. Coordinated evaluation and development model of oasis urbanization from the perspective of new urbanization: A case study in Shandan County of Hexi Corridor, China. *Sustain. Cities Soc.* **2018**, *39*, 78–92. [CrossRef]
84. Friedmann, J. Four theses in the study of China's urbanization. *Int. J. Urban Regional* **2006**, *30*, 440–451. [CrossRef]

85. Ahmad, M.; Jiang, P.; Murshed, M.; Shehzad, K.; Akram, R.; Cui, L.; Khan, Z. Modelling the dynamic linkages between eco-innovation, urbanization, economic growth and ecological footprints for G7 countries: Does financial globalization matter? *Sustain. Cities Soc.* **2021**, *70*, 102881. [CrossRef]
86. Ahani, S.; Dadashpoor, H. A review of domains, approaches, methods and indicators in peri-urbanization literature. *Habitat Int.* **2021**, *114*, 102387. [CrossRef]
87. Zhou, Y.; Chen, M.; Tang, Z.; Mei, Z. Urbanization, land use change, and carbon emissions: Quantitative assessments for city-level carbon emissions in Beijing-Tianjin-Hebei region. *Sustain. Cities Soc.* **2021**, *66*, 102701. [CrossRef]
88. Zhang, L.; Huang, S. Social capital and regional innovation efficiency: The moderating effect of governance quality. *Struct. Chang. Econ. Dynam.* **2022**, *62*, 343–359. [CrossRef]
89. Li, Y.; Li, Y.; Zhou, Y.; Shi, Y.; Zhu, X. Investigation of a coupling model of coordination between urbanization and the environment. *J. Environ. Manag.* **2012**, *98*, 127–133. [CrossRef]
90. Chen, T.-L.; Lin, Z.-H. Impact of land use types on the spatial heterogeneity of extreme heat environments in a metropolitan area. *Sustain. Cities Soc.* **2021**, *72*, 103005. [CrossRef]
91. Zhang, M.; Tan, S.; Zhang, X. How do varying socio-economic factors affect the scale of land transfer? Evidence from 287 cities in China. *Environ. Sci. Pollut. Res.* **2022**, *29*, 40865–40877. [CrossRef]
92. Liu, W.; Zhan, J.; Zhao, F.; Wei, X.; Zhang, F. Exploring the coupling relationship between urbanization and energy eco-efficiency: A case study of 281 prefecture-level cities in China. *Sustain. Cities Soc.* **2021**, *64*, 102563. [CrossRef]
93. Bai, X.; Shi, P.; Liu, Y. Society: Realizing China's urban dream. *Nature* **2014**, *509*, 158–160. [CrossRef]
94. Mu, W.; Zhu, X.; Ma, W.; Han, Y.; Huang, H.; Huang, X. Impact assessment of urbanization on vegetation net primary productivity: A case study of the core development area in central plains urban agglomeration, China. *Environ. Res.* **2023**, *229*, 115995. [CrossRef]
95. Ouyang, Z.; Zheng, H.; Xiao, Y.; Polasky, S.; Liu, J.; Xu, W.; Wang, Q.; Zhang, L.; Xiao, Y.; Rao, E.; et al. Improvements in ecosystem services from investments in natural capital. *Science* **2016**, *352*, 1455–1459. [CrossRef]
96. Tian, P.; Liu, Y.; Li, J.; Pu, R.; Cao, L.; Zhang, H. Spatiotemporal patterns of urban expansion and trade-offs and synergies among ecosystem services in urban agglomerations of China. *Ecol. Indic.* **2023**, *148*, 110057. [CrossRef]
97. Fu, B.; Wang, S.; Liu, Y.; Liu, J.; Liang, W.; Miao, C. Hydrogeomorphic Ecosystem Responses to Natural and Anthropogenic Changes in the Loess Plateau of China. *Annu. Rev. Earth Planet. Sci.* **2017**, *45*, 223–243. [CrossRef]
98. Bryan, B.A.; Ye, Y.; Zhang, J.E.; Connor, J.D. Land-use change impacts on ecosystem services value: Incorporating the scarcity effects of supply and demand dynamics. *Ecosyst. Serv.* **2018**, *32*, 144–157. [CrossRef]
99. Sharma, S.; Nahid, S.; Sharma, M.; Sannigrahi, S.; Anees, M.M.; Sharma, R.; Shekhar, R.; Basu, A.S.; Pilla, F.; Basu, B.; et al. A long-term and comprehensive assessment of urbanization-induced impacts on ecosystem services in the capital city of India. *City Environ. Interact.* **2020**, *7*, 100047. [CrossRef]
100. Wu, X.; Wang, S.; Fu, B.; Feng, X.; Chen, Y. Socio-ecological changes on the loess plateau of China after Grain to Green Program. *Sci. Total Environ.* **2019**, *678*, 565–573. [CrossRef]
101. Wang, S.; Hu, M.; Wang, Y.; Xia, B. Dynamics of ecosystem services in response to urbanization across temporal and spatial scales in a mega metropolitan area. *Sustain. Cities Soc.* **2022**, *77*, 103561. [CrossRef]
102. Richards, D.R.; Law, A.; Tan, C.S.Y.; Shaikh, S.F.E.A.; Carrasco, L.R.; Jaung, W.; Oh, R.R.Y. Rapid urbanisation in Singapore causes a shift from local provisioning and regulating to cultural ecosystem services use. *Ecosyst. Serv.* **2020**, *46*, 101193. [CrossRef]
103. García-Nieto, A.P.; Geijzendorffer, I.R.; Baró, F.; Roche, P.K.; Bondeau, A.; Cramer, W. Impacts of urbanization around Mediterranean cities: Changes in ecosystem service supply. *Ecol. Indic.* **2018**, *91*, 589–606. [CrossRef]
104. Jaligot, R.; Kemajou, A.; Chenal, J. Cultural ecosystem services provision in response to urbanization in Cameroon. *Land Use Policy* **2018**, *79*, 641–649. [CrossRef]
105. Eigenbrod, F.; Bell, V.A.; Davies, H.N.; Heinemeyer, A.; Armsworth, P.R.; Gaston, K.J. The impact of projected increases in urbanization on ecosystem services. *Proc. R. Soc. B-Biol. Sci.* **2011**, *278*, 3201–3208. [CrossRef]
106. Seto, K.C.; Guneralp, B.; Hutyra, L.R. Global forecasts of urban expansion to 2030 and direct impacts on biodiversity and carbon pools. *Proc. Natl. Acad. Sci. USA.* **2012**, *109*, 16083–16088. [CrossRef]
107. Estoque, R.C.; Murayama, Y. Landscape pattern and ecosystem service value changes: Implications for environmental sustainability planning for the rapidly urbanizing summer capital of the Philippines. *Landsc. Urban Plan.* **2013**, *116*, 60–72. [CrossRef]
108. Tiwari, A.; Tyagi, D.; Sharma, S.K.; Suresh, M.; Jain, K. Multi-criteria decision analysis for identifying potential sites for future urban development in Haridwar, India. In Proceedings of the International Conference on Communications and Cyber Physical Engineering 2018, Hyderabad, India, 24–25 January 2019; Springer: Singapore, 2019; Volume 500, pp. 761–777. [CrossRef]
109. Mitchell, M.G.E.; Schuster, R.; Jacob, A.L.; Hanna, D.E.L.; Dallaire, C.O.; Raudsepp-Hearne, C.; Bennett, E.M.; Lehner, B.; Chan, K.M.A. Identifying key ecosystem service providing areas to inform national-scale conservation planning. *Environ. Res. Lett.* **2021**, *16*, 014038. [CrossRef]
110. Gifford, R.; Nilsson, A. Personal and social factors that influence pro-environmental concern and behaviour: A review. *Int. J. Psychol.* **2014**, *49*, 141–157. [CrossRef]
111. Mao, D.; He, X.; Wang, Z.; Tian, Y.; Xiang, H.; Yu, H.; Man, W.; Jia, M.; Ren, C.; Zheng, H. Diverse policies leading to contrasting impacts on land cover and ecosystem services in Northeast China. *J. Clean. Prod.* **2019**, *240*, 117961. [CrossRef]
112. Zhao, Y.; Wang, M.; Lan, T.; Xu, Z.; Wu, J.; Liu, Q.; Peng, J. Distinguishing the effects of land use policies on ecosystem services and their trade-offs based on multi-scenario simulations. *Appl. Geogr.* **2023**, *151*, 102864. [CrossRef]

113. Halperin, S.; Castro, A.J.; Quintas-Soriano, C.; Brandt, J.S. Assessing high quality agricultural lands through the ecosystem services lens: Insights from a rapidly urbanizing agricultural region in the western United States. *Agr. Ecosyst. Environ.* **2023**, *349*, 108435. [CrossRef]
114. Liao, S.; Wu, Y.; Wong, S.W.; Shen, L. Provincial perspective analysis on the coordination between urbanization growth and resource environment carrying capacity (RECC) in China. *Sci. Total Environ.* **2020**, *730*, 138964. [CrossRef] [PubMed]
115. Yang, M.; Gao, X.; Siddique, K.H.M.; Wu, P.; Zhao, X. Spatiotemporal exploration of ecosystem service, urbanization, and their interactive coercing relationship in the Yellow River Basin over the past 40 years. *Sci. Total Environ.* **2023**, *858*, 159757. [CrossRef] [PubMed]
116. Wang, J.; Gao, D.; Shi, W.; Du, J.; Huang, Z.; Liu, B. Spatio-temporal changes in ecosystem service value: Evidence from the economic development of urbanised regions. *Technol. Forecast. Soc.* **2023**, *193*, 122626. [CrossRef]
117. Sannigrahi, S.; Zhang, Q.; Pilla, F.; Joshi, P.K.; Basu, B.; Keesstra, S.; Roy, P.S.; Wang, Y.; Sutton, P.C.; Chakraborti, S.; et al. Responses of ecosystem services to natural and anthropogenic forcings: A spatial regression based assessment in the world's largest mangrove ecosystem. *Sci. Total Environ.* **2020**, *715*, 137004. [CrossRef]
118. Andersson, E.; Tengö, M.; McPhearson, T.; Kremer, P. Cultural ecosystem services as a gateway for improving urban sustainability. *Ecosyst. Serv.* **2015**, *12*, 165–168. [CrossRef]
119. Yu, X. Is environment 'a city thing' in China? Rural–urban differences in environmental attitudes. *J. Environ. Psychol.* **2014**, *38*, 39–48. [CrossRef]
120. Kollmuss, A.; Agyeman, J. Mind the Gap: Why do people act environmentally and what are the barriers to pro-environmental behavior? *Environ. Educ. Res.* **2002**, *8*, 239–260. [CrossRef]
121. Singh, P. Environmental education: Enhancing learning and awareness through assessment. *Syst. Pract. Act. Res.* **2012**, *26*, 299–314. [CrossRef]
122. Guo, J.; Zhang, M. Exploring the patterns and drivers of urban expansion in the Texas triangle megaregion. *Land* **2021**, *10*, 1244. [CrossRef]
123. Zhang, S.; Wu, T.; Guo, L.; Zou, H.; Shi, Y. Integrating ecosystem services supply and demand on the Qinghai-Tibetan Plateau using scarcity value assessment. *Ecol. Indic.* **2023**, *147*, 109969. [CrossRef]
124. Tian, G.; Qiao, Z. Assessing the impact of the urbanization process on net primary productivity in China in 1989–2000. *Environ. Pollut.* **2014**, *184*, 320–326. [CrossRef] [PubMed]
125. Yang, J.; Xie, B.; Zhang, D. Spatial-temporal evolution of ESV and its response to land use change in the Yellow River Basin, China. *Sci. Rep.* **2022**, *12*, 13103. [CrossRef]
126. Pan, Y.; Dong, F. Design of energy use rights trading policy from the perspective of energy vulnerability. *Energy Policy* **2022**, *160*, 112668. [CrossRef]
127. Zhang, P.; Yuan, H.; Tian, X. Sustainable development in China: Trends, patterns, and determinants of the "Five Modernizations" in Chinese cities. *J. Clean. Prod.* **2019**, *214*, 685–695. [CrossRef]
128. Strassburg, B.B.N.; Iribarrem, A.; Beyer, H.L.; Cordeiro, C.L.; Crouzeilles, R.; Jakovac, C.C.; Braga Junqueira, A.; Lacerda, E.; Latawiec, A.E.; Balmford, A.; et al. Author Correction: Global priority areas for ecosystem restoration. *Nature* **2022**, *609*, E7. [CrossRef]
129. Sirakaya, A.; Cliquet, A.; Harris, J. Ecosystem services in cities: Towards the international legal protection of ecosystem services in urban environments. *Ecosyst. Serv.* **2018**, *29*, 205–212. [CrossRef]
130. Pham, K.T.; Lin, T.-H. Effects of urbanisation on ecosystem service values: A case study of Nha Trang, Vietnam. *Land Use Policy* **2023**, *128*, 106599. [CrossRef]
131. Wellmann, T.; Lausch, A.; Andersson, E.; Knapp, S.; Cortinovis, C.; Jache, J.; Scheuer, S.; Kremer, P.; Mascarenhas, A.; Kraemer, R.; et al. Remote sensing in urban planning: Contributions towards ecologically sound policies? *Landsc. Urban Plann.* **2020**, *204*, 103921. [CrossRef]
132. Zhang, Z.; Gao, J.; Fan, X.; Lan, Y.; Zhao, M. Response of ecosystem services to socioeconomic development in the Yangtze River Basin, China. *Ecol. Indic.* **2017**, *72*, 481–493. [CrossRef]
133. Zhao, Y.; Shi, Y.; Feng, C.-C.; Guo, L. Exploring coordinated development between urbanization and ecosystem services value of sustainable demonstration area in China- take Guizhou Province as an example. *Ecol. Indic.* **2022**, *144*, 109444. [CrossRef]
134. Wang, H.; Liu, L.; Yin, L.; Shen, J.; Li, S. Exploring the complex relationships and drivers of ecosystem services across different geomorphological types in the Beijing-Tianjin-Hebei region, China (2000–2018). *Ecol. Indic.* **2021**, *121*, 107116. [CrossRef]
135. Chen, M.; Liu, W.; Lu, D.; Chen, H.; Ye, C. Progress of China's new-type urbanization construction since 2014: A preliminary assessment. *Cities* **2018**, *78*, 180–193. [CrossRef]

MDPI AG
Grosspeteranlage 5
4052 Basel
Switzerland
Tel.: +41 61 683 77 34

Land Editorial Office
E-mail: land@mdpi.com
www.mdpi.com/journal/land

www.ingramcontent.com/pod-product-compliance
Lightning Source LLC
Chambersburg PA
CBHW041414210926
51850CB00013B/3266

* 9 7 8 3 7 2 5 8 4 1 2 7 1 *